A Tree Trimmer's Guide to the Universe

Forest Primeval

"Smoke and mirrors"!
With 'lotsa' Love,
Ari

by

John Ari Stepp

First Edition

ISBN: 1484152921

ISBN 13: 9781484152928

Forest Primeval is also available at Amazon.com

Made in California, USA

Introduction

The Adventure That Awaits Us Within (Forest Primeval), besides presenting the latest advances in understanding our universe, is primarily a modern "updated" exploration into the latest, more human side of science and technology, as well as into what it really means to be "human".

To be human is to be aware that we exist, and thus can make independent observations and draw conclusions from our experiences. Who we think we are, what we believe the universe is, and our apparent place within it, are all shaped by our experiences.

Yet in spite of our seeming independence, life inevitably presents us with unacceptable situations that are, disturbingly at times, quite beyond our control! Then our experience of existing becomes difficult to bear. Moreover, being cognizant of our own mortality, we naturally want our lives to have some sort of an enduring meaning.

But an honest appraisal of our true-life situation reveals many contradictions that challenge our search for meaning. Unsettling quandaries soon appear. Therefore it seems important first to understand what is actually going on in our consciousness that might be contributing to them. Only when this is done does it make sense to try to understand whatever follows.

Eventually, and without bias, we need to ask, "What exactly brings on our various experiences of life, be they what we feel to be good or bad, happy or sad, meaningful, or meaningless?"

This is a deceptively simple question and some of the "answers" might just surprise you! Because for us human beings, life is certainly much more than a mere struggle for survival. Forest Primeval explores why this is so. However, this is *not a book for children*, or for those seeking the sort of easy answers that might feel good or seem meaningful at first, until you examine them more closely!

But I must warn you to be prepared for a frank examination of some of the most puzzling, and occasionally perhaps even a *bissel* "unsettling", mysteries of our life and consciousness.

Because this is not just another "feel good" book written to uplift and inspire hope for its own sake, if you are not prepared to examine the deeper nature of our universe, the mysterious appearance of life within it, and the astounding phenomenon of our human existence directly, then I suggest that this may not be the book for you.

If you are, it is an exploration into life's infinite possibilities and delightful probabilities, and it also offers some engaging new perspectives on our human condition.

When we enter our adult years, most of us are soon so caught up in the details that life demands of us in order to earn a living, let alone to attend to our necessary interpersonal relationships, that we have almost no free time left over to grasp the "big picture" in our fast paced and increasingly complex modern world. (Where it seems as if we are constantly spending ever greater amounts of our ever shrinking "free time", just trying to keep up with the ever-more-rapid changes, and to maintain our exponentially growing "world-wide-web" of social interconnectivity!).

Indeed, most of us have (prompted primarily by necessity) become so involved in our busy modern world that we now have precious little time left to think about *anything* that's not directly related to "earning-a-living"; in today's, ever more technical, "pressure cooker" world!

Therefore, as I on occasion try to present the "Big Picture", I have also endeavored to keep foremost in mind that just because an articulate presentation is made, even if it is compellingly presented, that does not necessarily mean that it is correct! Hence, I have also endeavored to stick to the verifiable facts at hand, especially when we venture down a new trail.

But in the end, no matter what is said, we all must live our own lives, learn our own lessons and draw our own conclusions. Because it will never do to accept blindly the suppositions of others where such important things are concerned, regardless of whatever authority or evidence that they might cite in forming their opinions.

Indeed, where such important things are concerned, *nullius in verba* "take no one's word for it!" So please consider this forestland tale to be a discussion, rather than a lecture or an exposition of unremitting absolutes and final truths. Because knowledge in these intimate matters soon becomes "worse-than-useless", if it does not lead to the awakening of a deep understanding that is truly yours alone.

I have spent a good deal of my working life happily trimming amongst the boughs of green trees, while working "shoulder to shoulder" with the strong and brave hearted, so-called "blue collar," tree people* who care for them (id est, the people*, who care for the trees).

Consequently, over the course of a lifetime, I have discovered, just as the labor honoring American born poet Walt Whitman and composer Aaron Copeland evidently did, that a higher education is not necessarily an indication of a person's character, nor of their ability to grasp even the deepest of complex concepts, when they are presented in a manner that the average person can more easily understand.

Accordingly, while I have gone to great lengths to simplify and demystify the more complex and inscrutable concepts that are at times discussed in Forest Primeval, I also do not avoid even the most arcane of modern scientific speculations, mystifying metaphysics, or perspicacious philosophies! Nonetheless, any *responsible* metaphysics is bounded by the same test of "fact based" reasonableness (awaiting empirical scientific verification), as the theoretical physics that it is quite often based upon.

Moreover, contrary to the often prevalent pessimism of many older men, despite life's inevitable panoply of unhappy occurances, I have developed a deep and abiding faith in the uncommon ability of the decent and hard-working "common man", to improve both himself and the world. (Please note that "man" is being used here in a literary fashion to indicate humankind, and most certainly this includes the female gender as well!).

Because, time and again, armed with nothing more than humility, kindness, and an unquenchable thirst to understand life's deeper mysteries, they somehow accomplish the formerly "impossible!" As the wise captain Piccard, of Gene Roddenberry's visionary "Star Trek: The Next Generation", so aptly put it, they quite often, even against the greatest of odds, do indeed, "Make it so!"

I

Traditional Zen Buddhism, Taoism, and even the exacting teachings of Jñana Yoga all deal with the matter of our existential suffering. However, beyond their very "person oriented" esoteric teachings, might something quite unanticipated be currently going on in the world around us that they are even more directly related to? (I.e. Since everything that we experience is contained within the consciousness that they are attempting to explain.)

But with all of the "bad news" in the present, and on the near horizon, all of this esoteric stuff seems just a bit too frivolous. When it seems not unreasonable to wonder if our entire living world, like a runaway train, is beginning to lose its grip on the environmental "rails" that ensure its more vital and life-allowing functions!

Or is something more subtle going on that is going to change our world in ways that are so radical that even our near future is fast becoming like a real world of science fiction? In Forest Primeval we explore these "ancient-but-new" astonishing possibilities, and even a few unexpected *novus* probabilities!

The beloved American author Samuel Clemons (Mark Twain) once humorously observed, "First get your facts straight, and then you can distort them at your leisure!" I have endeavored to observe the former, while avoiding the latter, but in order to tell a story of interest a storyteller also needs to use imagination and creativity.

Therefore, I must beg my more seasoned and erudite reader's indulgence in the occasional use of whimsy, somewhat fanciful "higher" mathematics, and even a *bissel* "little bit" of artistic license, in the telling of a tale whose true aim is to reveal a deeper truth.

But, because of the sometimes obscure, yet nonetheless compelling nature of the topics under discussion, a sincere author on these matters should also include a cautionary warning to their more novice, and therefore sometimes more gullible "freshman" readers.

As a rather poignant example I offer a cautionary tale. Not only to you, but to my fellow writers as well! Once upon a time, the (apparently financially frustrated) science fiction author L. Ron Hubbard quizzically advised a science fiction fan who wanted to become a writer, "You don't get rich writing science fiction. If you want to get rich, you start a religion!"

Which is exactly what the savvy Mr. Hubbard (evidently) then did, although we must decide for ourselves about his claim of its legitimacy! Perhaps he was referring to the "tax exempt" status that many rather questionable "religions" enjoy in our wonderfully liberal land of unfettered religious freedoms?

But whatever the unspoken intent of his original statement was, the "better angels" of his conscience were (also evidently) an insufficient deterrent to the deeper dictates of his own integrity. And despite being quite fantastical, glaringly illogical, and even thoroughly impossible to substantiate, Hubbard's now infamous nouveau "Sci-Fi Religion" has nonetheless attracted some rather famous Hollywood stars! And (one can only speculate) it has no doubt also netted its (evident) Author a rather handsome "tax free" profit!

Nonetheless, while this is perhaps a good financial strategy, from a business standpoint, it is (comparatively) a rather inequitable situation if you happen to own a financially challenging business, like a Silicon Valley "mom and pop" Tree Service, or a small startup Software Firm comprised of an entrepreneurial band of goodhearted and like-minded friends. (Or if you happen to work for our fiscally struggling government, and are trying to collect fair business taxes, so we can fix our countries failing infrastructure!)

Yet this inequitable situation, no-doubt seems "fair enough" if you are only interested in making money or in being entertained, rather than in learning whatever truths might be revealed to an honest and inquiring mind!

But I am neither interested in starting a new religion, or in using what many believe should be a "sacred" subject like a common garden spade to turn a profit. Hence, in respect to the aforementioned cautionary "business ethics" tale, I have tried to limit my inquiries to reasonable logic, verifiable facts, and the better opinions of noted experts on the subjects at hand, during the long and careful writing of Forest Primeval.

And whenever the subject that we are discussing still remains uncertain, please notice that I use modifiers like "evidently" and "apparently" to alert you to the fact, so you can take the time to "vet" the facts that it is based upon, and leisurely ponder its deeper meaning for yourself. Nonetheless, it is always good to keep in mind that, as British Physicist Brian Cox once rather trenchantly observed, "Certainty is the enemy of science!"

Moreover, while it is clearly necessary to examine and reject the superstitions and ignorance of the past, it seems unwise to do so with that which may be difficult to define simply because it is couched in archaic terminology, or has a more ephemeral and less logical sort of style. As the Jewish Talmud and many other wisdom traditions teach, "You must first understand from whence you have come, before you can know to where it is that you are going!"

Thus, besides giving us an important, if not essential, connection to our ancestors, the wisdom teachings of the past are likely as valuable today as they ever were, because humanity (within its individual and collective psyches) remains essentially, almost unchanged. And as we move ever-faster into an exciting and uncertain future, not losing our humanity along the way may well become one of our greatest challenges!

II

The main body of Forest Primeval was written during a series of seven forays into the rugged, yet meteorologically benign, redwood rainforests of Northern California's enchanting Santa Cruz Mountains. Subsequently, new ideas, experiences and discoveries were added to its literary "bones" to "flesh out" the interior portions of its wilderness themed back stories.

Each of Forest Primeval's seven chapters begins and ends with the ongoing story of a shared wilderness adventure. But like a storybook "sandwich", the essential "meat" of its meaning is found in the pages that lie inbetween the book's sandwiching camping stories, and intervening wilderness adventures "of mind". (That seldom exercised freedom of rational imagination that Einstein once praised, and which Stephen Hawking so eloquently exemplifies in our present time!)

Each chapter is also titled with the scientific names of some of my favorite trees, not all of which are true natives of the Northwestern Santa Cruz Mountains. Cedar for instance is more properly called *Cedrus*, but since I am really referring to *Thuja plicata* "Western Red Cedar", the shorter term "Cedar" is used to preserve the book's continuity of Chapter Titles.

And while imitation may well be the "highest form of flattery", artistic plagiarism is actually a subtle form of theft that is harmful to the artist. Therefore I feel compelled to explain that while most all of the accompanying artwork is by my hand, since I am not a consummate artist, in addition to the usual artistic chicaneries of implied perspective, shading, and form, I did at times use photographs, tracery and even Adobe Photoshop® to achieve the desired end result.

Moreover, when I did on occasion incorporate other artists work into my own creation, I have, to the best of my ability, tried to use only the older art and seasoned photographs that lie in the democratic provenance of the "public domain", and to give artistic credit where it is due.

Contrary to the general rule, as you read on, you will also soon discover that some familiar terms like "spacetimelessness" have been combined in a similar fashion to the (sometimes rather longish) Modern German Language *begriff* "concepts", in order to tell the story better.

Like the whimsical word "acquisitionalism", for example, which is meant to convey out of control capitalism, thereby yielding a new conglomerate concept. Others have on occasion been joined in a similar fashion to Modern Hebrew Language *smichut* "contractions", such as "Toranic", which refers to a biblical concordance of meaning.

Also, please note that when the term "body-mind" is used that our body portion inevitably includes the brain by which it is known. And while the hyphenated term "subject-object" refers to our dualistic consciousness, when you see "subject/object" (with a forward slash), it always signifies the (veridical) object that masquerades as our personal ego.

Additionally, whenever you see Consciousness, with a capital "C", it always denotes impersonal Universal-Consciousness, and whenever you see Self, with a capital "S", it always indicates our *Atman* or "True-Self".

Accordingly, I encourage you to make liberal use of the diligently researched Glossary and Appendix at the end of this book, which precisely define the special terms that are occasionally used, in a crystal clear "user friendly" conversational manner, rather than consulting an un-vetted "on line dictionary", or "Wikipedia".

You will also find the Index useful for a deeper understanding of specific words and concepts as they are used in various ways. And while it is clearly impossible for any author to anticipate in advance the extent of every readers vocabulary, the first time that words and terms that may be new to you are used, an explanation (in parenthesis) most often immediately follows.

Indeed, in order to fully understand some of the concepts that may be new or unfamiliar to you, and those that are presented in lyrical or "storybook" form, it is also a good idea to follow the occasional informational page references (that also appear in parenthesis) when you are first reading about it, and then read it again, while the new concept is still fresh in your mind.

I have endeavored throughout Forest Primeval to make the recondite more easily understood. Thus, on the rare occasions that arcane scientific, metaphysical, or philosophical terms, mysterious mathematical concepts, or unfamiliar foreign words are encountered that might make the book more difficult to read or to understand, you should not be overly concerned! Because they have been rendered into an easy to understand format (generally into "concept paradigms" that are similar in structure to Einstein's creative "thought experiments").

And, please note, that despite the recent trend of choosing only "fresh" references to lend the appearance of "keeping pace with the times", the references used herein were chosen solely based upon their timeless veracity, and clarity of elucidation.

However, just trying to keep pace with the ever-more-rapid advance of science and technology that have occurred in Silicon Valley (and elsewhere) during the writing of this book has been a major challenge! Therefore, at the end of the Bibliography section, you will find a short list of visionary scientists/authors with recent books on the various scientific topics and technologies that are discussed herein.

As you read Forest Primeval you will also occasionally see some small numbers in brackets – (1,2,3) – at the end of a word or sentence that refer to a specific information source in the End Notes section at the back of the book. Additionally, a comprehensive Bibliography is included that contains books and articles by other authors on the newer subjects that are covered in this book.

The "o" has also been removed from the word G-d as a formal reminder that any sort of truly transcendent "Supreme" is always indescribable. While the word "god", with a small "g", is used to describe the "god of our own [mind's] creation"...; and [brackets], like the former, indicate any clarifications that may be necessary to preserve a quotes continuity of purpose.

Also, please note that the Acknowledgements section, with the names of people who either assisted, inspired or encouraged me in Forest Primeval's eventual preparation, is found on page 493. And, contrary to the general rule, I encourage you to read it first, so you will become aware of the many people (living and "otherwise") who, in a fashion, will also be accompanying us during our shared wilderness adventure.

Finally, I suggest that Forest Primeval is probably best read in small installments, and savored as if you were sipping a tasty cup of hot, aromatic tea, rather than attempting to take it in all at once.

III

I do not pretend to be the master of all, or indeed of any of the science, philosophy or esoteric traditions that are being discussed herein, even if it were possible! But I will admit to seeing our astonishing universe with, as Gallagher the perspicacious (watermelon smashing) comic would often say, "fresh eyes", as a result of the study, practices and thoughts that have finally culminated in the book that you now hold in your hands (or view on the screen before you).

Nonetheless, I also must confess that, despite my best efforts to the contrary, in ever so many ways, like the unfolding phenomenon of our lives, this book has, in a fanciful sense, "written itself".

But in our busy, often information dominated lives, unlike the difficult but nonetheless rewarding "Tree Service" business, where it's easy to spot acts of courage and valor when they are taking place in the canopy of a tall tree, it is easy to forget that we Americans "stand on the shoulders of Titans"!

Therefore I have also endeavored to give credit to the great minds and heroic hearts of those who, often through great personal sacrifice, have ensured our modern freedoms by blazing the military, political, business, scientific, philosophical, artistic, teaching and technological trails that we are presently pursuing.

But, beyond what's been bequeathed to us, and even beyond all of our own prior achievements, something entirely new is currently going on that most people don't yet well understand. As America continues to advance ever-more-rapidly into an exhilarating, but nonetheless disturbingly unpredictable, "Singularity of Consciousness". Because the sudden unfolding "synthetic" consciousness is beginning to touch almost every aspect of our lives!

Also, in addition to the occasional quotation of some famous scientist's topically relevant statement, on occasion you may as well encounter a stanza of familiar lyrics from a popular musician's song, or perhaps even some well-known lines from a famous poet's musings.

But please note, that in addition to their intrinsic entertainment value, that they have also been chosen for the puissant reason that understanding ourselves, and the universe in which we currently find ourselves, is currently about as much of an ephemeral *art,* as it is an exact science!

Forest Primeval is primarily a science and metaphysically themed book, but you will also discover as you read on (chiefly in Chapters 4 and 5) that I have applied its primary observations to an examination of the difficult contemporary problems that both America and the world at large are facing today. I did so reluctantly, and thus have endeavored to do so both subtly, and succinctly.

But nonetheless I felt compelled to do so, because of the clearly serious nature of our immediate problems, and those soon to come. (I.e. That our talented "Millennial" youth of today will soon be compelled to resolve!).

And whether you politically lean, to the "left" or to the "right" of our great country's persistently "moderate" core, because the world still relies upon politics and economics, as well as the much more fascinating topics of science and technology, in order to get anything done, we will be *briefly* discussing the former as well!

After a lifetime of searching for reliable answers, exhausted, and in the early stages of a deadly cancer, I finally returned to the "enchanted" redwood rainforests of my youth to rediscover what I was looking for. (And with all due respect to our state's thirty third (past) Governor, as a native Californian that grew up amongst the subtle "magic" of the massive ancient Sequoias, I can assure you that just because you may have "…seen one redwood tree", that in no way means that you have "…seen them all!").

So please join me as we take a closer look at the green "miracle" of trees, our astounding universe, the latest in "bioinvolved" science and technology, our complex hominid species, and our own mysterious "metaphysical" Selves. I hope that you will enjoy sharing this (perhaps at times, unabashedly patriotic) American forestland journey with me, as we explore some new ideas, with some venerable old roots!

"Nothing is too wonderful to be true, if it be consistent with the laws of nature".

Michael Faraday

Table of Contents

Chapter 2

Chapter 3

Chapter 4

Chapter 5

Chapter 6

Printed with light on 100% recycled stardust!

List of Illustrations

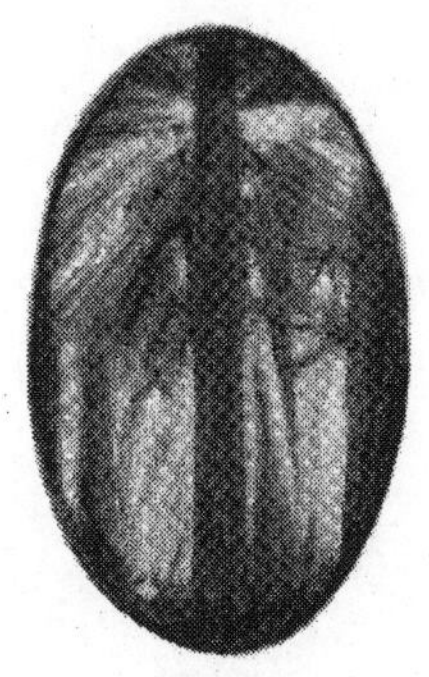

114 - Cedar Cone
Mountain View, California, USA
Pen and ink; number 2 graphite pencil

Chapter 3

117 - Live Oak Acorn
Palo Alto, California, USA
Pen and ink; number 2 graphite pencil

171 - Blue Oak Acorn
Woodside, California, USA
Pen and ink; number 2 graphite pencil

171 - Live Oak Branch Tip
Santa Cruz Mountains, California, USA
Pen and ink; number 2 graphite pencil

Chapter 4

172 - Madrone Flowers
Santa Cruz Mountains, California, USA
Pen and ink; number 2 graphite pencil

186 - Orb Spider
From my garden, Mountain View, California, USA
Pen and ink

234 - Millennium Moon
San Jose, California, USA
From a photograph by Ken Brown
Number 2 graphite pencil

Chapter 5

Chapter 6

Chapter 7

416 - Redwood Clone Ring
Sketch from USDF 1900's photograph
Number 2 graphite pencil

476 - Sempervirens Leaf
Santa Cruz Mountains, California, USA
Pen and ink

Appendix

553 - Möbius "String" Paradigm
Pen and ink; Inspiration, Allan Watts

An Adventure Amongst the Redwood Trees

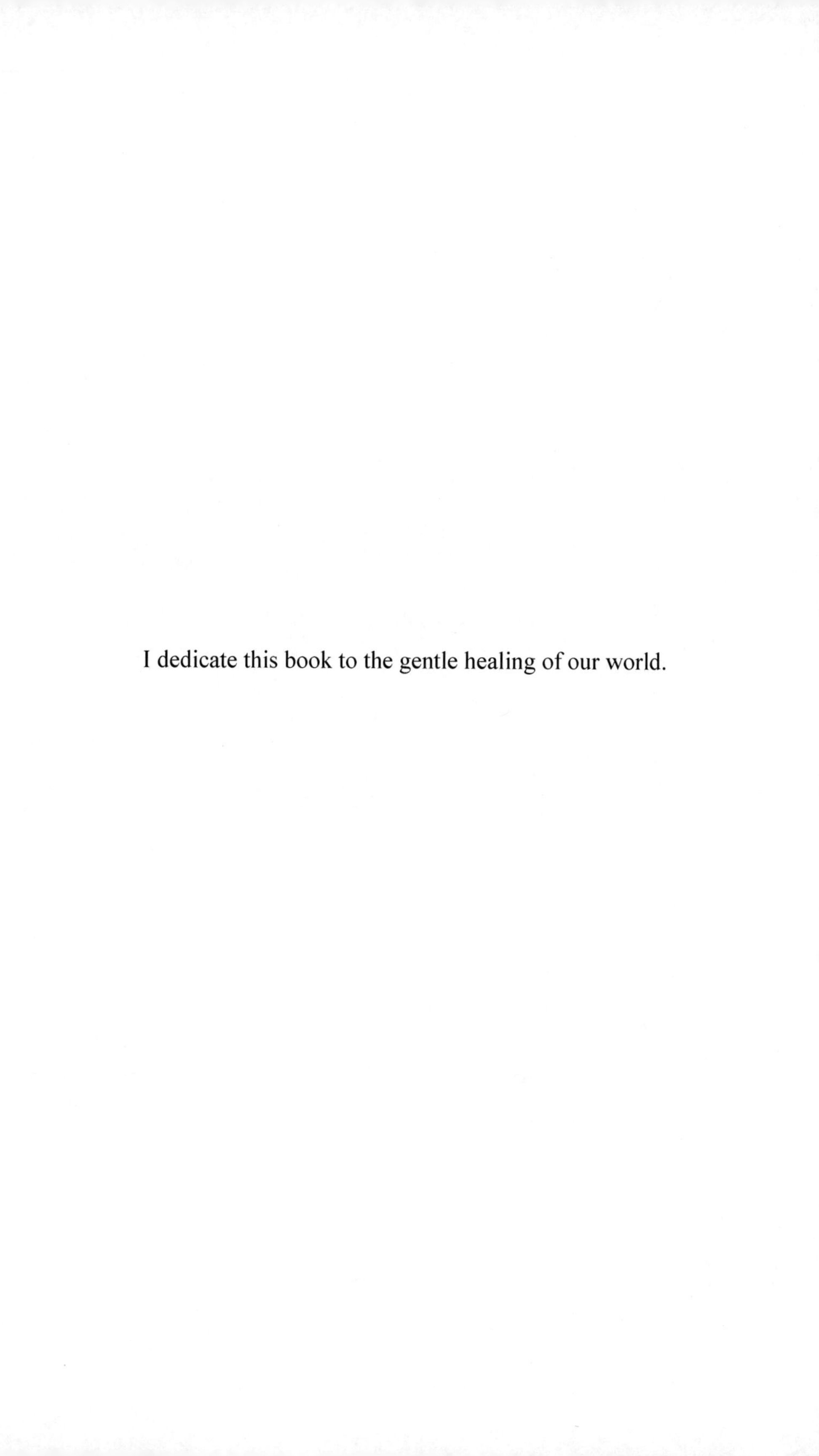

I dedicate this book to the gentle healing of our world.

Prologue

On This Sunlit Northern California Day, we are taking a little "space-time" trip together, in Argo, my (seldom-driven) midnight blue, 1965 Ford Mustang. Accompanied by the deep driving bass of Argo's rumbling "muscle car" exhaust, we soon leave the bizarre, utterly quiet, self-driving "Google Cars" of sunny suburban Mountain View, far behind us!

Continuing our joint adventure, we cautiously "snake up" the sinuous mountain curves of the upper Sand Hill Road (above Leland Stanford University), until we at last enter the golden dry grassland beauty of the oak wooded Los Altos Hills.

As Argo throttles back, we carefully negotiate the weathered old asphalt road's dangerous sharp curves and testy switchbacks, until we finally reach the summit at Skyline Boulevard. We are poised now, on the crest of the "continental-drift" formed Santa Cruz Mountain Range.

Directly across the highway's hot black tarmac Alpine Road beckons us to descend precipitously toward the distant chilly coolness of the dusky-blue Pacific Ocean.

To the East, in the big green valley that stretches out below us like a colorful Muslim "prayer rug", the great grey-blue waters of the salty San Francisco Bay reach out so far and wide that they nearly fill the entire basin of the great San Andreas "fault valley" that (for now) "sleeps" peacefully below.

At the North end of the bay, the "Forty-Niner" Gold-Rush City of San Francisco gleams white in the bright sunlight, like a carefully arranged pile of old bleached whale bones.

Next to the City's bright reflected light, the constituent ships of a colorful, but nonetheless dissimilar looking, Lilliputian "nautical fleet" (actually comprised of busy gargantuan cargo ships and massive oil tankers), pushes past the overarching Golden Gate Bridge. Which "welcomes" them into our briny San Francisco Bay, with its "wide open" steel cabled arms, as they ply their way to and from the bustling markets of the Far East.

And in the roiling San Francisco Bay waters, that are just visible beyond the sharp tip of the distant Transamerica Pyramid Building, the triangular white sails of a few small sailboats sail slowly past the infamous, but now derelict and slowly rusting, prison that stands forlornly (still) on San Quinton Island.

The big tankers are delivering their crude oil to the refineries that are at the bay's far Northern end. Where the bay is being steadily filled by the great Sacramento River. (With crystal-clear waters that are being constantly refreshed, even during California's dry summer months, by the gravity fed remnants of the winter alpine snowmelt.). Which originates from the high Sierra Nevada Mountains that lie on the far Eastern side of California's long Central Valley.

This is the great agricultural "rift valley" that stretches almost the full length of our "Golden State". From the redwood forests, and profitable clandestine (billion dollar) "gorilla" marijuana farms of Humboldt County in the North, to the dry Mojave Desert that lies over the mountains, just inland from busy metropolitan Los Angeles, in the South.

Consequently, although most people don't know it, because they generally think of "Silicon Valley" and "Hollywood" when they imagine California, we are also a busy Western "Agricultural State". But despite the massive monocrop corporate farming concerns that today dominate most of the valley, we still have our fair share of independent hardworking cowboys and ranchers, and farmers and farmhands, of all kinds. Including at least one native Californian who, in addition to William Shatner, lives with the challenging and self-contradictory conundrum of being a real Jewish American Cowboy!

Indeed, my first "bicycle" was a big chestnut-brown Quarter Horse, and I learned to drive by plowing the "endless" green corn fields with my

Uncle Sherman's little red Ford Tractor. I also learned to shoot, by protecting the crops that we grew. From a distance with a twenty two rifle that was fitted with a 12x riflescope, and "up close" with the fearsome blast of a twelve gauge shotgun (varmints beware!)

But also "on-the-farm", at an early age, I learned about the terrible price that is actually being paid for the bread and meat that graces our American tables. Watching a steer being slaughtered, that you kept alive as a newly weened calf with a bucket of fresh warm cow's milk, is a sobering experience that stays with you for a lifetime. (I cried myself to sleep that night!). And ironically, it often teaches country boys and girls to be kinder to the animals that we share our lives with. Because we learn that they suffer in life, and in death, just as we "human" animals do!

And moreover, you learn that our food does not just magically appear on some grocery store shelf, all wrapped up in shiny plastic, and clean of any trace of telltale blood and dirt. Just like us human creatures, the food that we eat was also once a living thing. And moreover, it has quite recently given its all, so that we may continue to live! But such is the difficult, and self-contradictory nature of life struggling to survive.

Nonetheless, once I came to terms with the inevitable suffering and death that are found on a ranch, it seemed like an idyllic childhood for an outdoorsy sort of boy. Indeed, I would not exchange my country childhood memories for "all the gold in the world!"

But, returning to our adventure at hand, the distant cargo ships are delivering their goods to our busy Oakland docks. Whose waiting fleets of trucks supply the genteel people that live around the San Francisco Bay with all of the necessary food, fuel, and accoutrements of a modern society. And directly below us, back along the way that we just came, lies the hub of our emerging new world, "Silicon Valley".

Directly to our West, at the faraway coastal subduction zone "toe" of the Santa Cruz Mountains, lie the sandy golden beaches, cold restless waters, and gigantic "Maverick" surfing waves, of the great blue-gray Pacific Ocean.

Below us, the lower mountains and slopes are completely submerged beneath a gently rolling blue-green carpet of giant redwood trees. And on the distant horizon line of the great Pacific Ocean, a second dividing line of cold gray fog stretches out, to the North and South, for about as far as the unaided human eye can see!

But our camping goal lies far below us, in the midst of the remaining "old growth" redwood trees. (In the redwood rainforest, about a dozen miles to the north of the bright and busy little coastal (cotton candy) "Boardwalk Town" of Santa Cruz).

As we descend we drive slowly past the brightly glowing wild oat grass that covers Russian Ridge like a golden yarmulke. And then, we descend even more slowly, as if we were "slipping" between the towering blue-green walls of the Douglas Fir trees, which grow on either side of the narrow road, not too far below Russian Ridge.

Cautiously we emerge, and begin to carefully negotiate the sinuous road's "hairpin" switchbacks, and dramatically exposed, acrophobia inspiring curves! And then, as we drive carefully down, and yet even further down, we come at last within reach of the welcoming outstretched limbs of the awaiting "primeval" redwood tree wilderness. (Which grows quite verdantly on the mountain slopes' maritime influenced rainy Western side!).

And so begins our slow, final stepwise descent, down into the massive "Jurassic Era" redwood trees that adorn the slopes and peaks of the rugged mountains that stretch all the way to the coastline in the distance before us.

As we carefully descend into the antediluvian redwood rainforest, it seems as if we are going back in time, to a quieter, "simpler time". To a time without noise and cell phones and high flying drones!

After what seems like a good deal of cautious "mountain driving", being careful to turn on our headlights, and to give a little *honk!* to warn any cars, motorcycles, bicyclists, or hikers that might be on the other side of a "blind turn" in the narrow roadway that lies ahead, we finally arrive safely at our destination in the "primeval" redwood forest.

Soon, with very little fuss, tent poles slip easily into their sockets, and with the sudden "*pop*" of nylon fabric snapping taught, our grey domed two person tent is quite suddenly up! Next, we stash our cache of tasty camping food away, to protect it from any wilderness varmints that might be looking for an easy nighttime snack.

Then we split and stack our firewood, and build a fire-ring of sound, heat shatter proof stones, and inside it we erect our American Indian "tee-pee" style campfire. (To keep us warm during the darkness of the now fast approaching night.)

The forest here is so very peaceful, that it beckons us to rest and sleep. But just as the American born poet Robert Frost so poignantly observed, in his lovely style of lyrical prose, "...we have miles to go, and promises to keep!"

Indeed, we too have come here with a great resolve, to gain a greater understanding of our universe, and of our truer place within it. As it is being almost daily impacted by the rapid, and unpredictable changes of our modern "technologically intelligent" world.

So, as the sun begins to slowly set in the West, and the "robin's-egg-blue" sky above us, which is now dotted with fluffy pink, "cotton-candy-like" clouds, begins to take on its customary rosy evening glow (and before the wet fog descends upon us), we might as well get started on our journey of discovery, into the mystery of the universe that surrounds us. And concomitantly, our "inward" journey, into the rather poorly understood universe, of our own humanoid minds!

Chapter 1
Sequoia "Redwood"

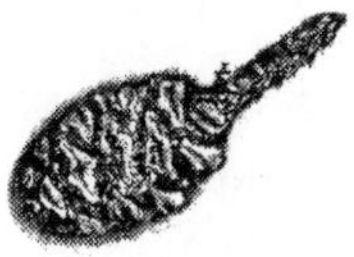

The Guide

A Spirit Guide appears when a pilgrim in the universe requires guidance; sometimes, even when we at first believe it's unnecessary and unwanted. But the Guide must go, even when wanted, when they are no longer needed. (Such is the mysterious way of guidance in our universe.)

A wise traveler is ever vigilant, for guidance may come at any moment and can take many forms. But in the end, the pilgrim learns that they and the Guide are ever one, and that they are also never truly alone![1, 2]

A Breeze Among the Trees

Should We Not Define reality as the experience of what is real, and not something that may exist, but is not verifiable? But does not even this definition beg a more precise one? It also seems reasonable, that if in this definition the components of reality are themselves found to be insubstantial, that the reality which they are supposed to compose, must surely be lacking in substance as well.

It is our common experience that the objects that we perceive seem to possess a certain substantiality. But, upon reflection, what we experience appears to be a composite of both subject and object, and thus it only seems to possess an independent existence!

Moreover, whatever we experience appears to be separated, in an independent existence, as objects that are suspended both in space, and in time. But perhaps the truth of things is so close to us that it easily evades our attention?

Upon closer "introspection"; that is, inner inspection of the witnessing of the events that we perceive as person and universe; something very different may be observed to occur. What appears to have substantiality, as separate objects within the seriality (one event after another) of space-time, is seen to be merely the projection of our own consciousness.(3)

We are both the subject and the object that appear before the Witness of our own "subject/object" selves, within the backdrop of our own consciousness. Which is the apparent space-time in which the objects of our consciousness make their appearances.

But what we actually experience is the *um velt* of our own experiencing Self. (I.e. *Um velt,* is the world as it is experienced by a particular organism.) Hence, everything that we experience has the authenticity of our own presumed reality of being. Nonetheless it is only our reflection in the biological machinery of our body-minds that generates our familiar experiences of "person and universe". (I.e. This experience is synonymous with the old German philosophical "um velt"). Therefore what we see, is who we are!

Yet the true nature of our every experience stubbornly eludes our understanding, until it is finally completed by the inclusion of our essential Witness, as both the source and sum of our every experience.

The truth of who we are is literally right before our eyes! And so, there is never a need to search any further for the "ultimate truth", than our own experiential front door. All that is required is to see rightly what is always right before us! If you fully grasp this, then there may be no need to read any further.

However, "fully grasping" means that you not only intellectually understand, but that you also apperceive its meaning (see "Apperception", pgs. 504, and 517).

This is only possible if there has been a *metanoesis*, a "shifting of the locus" of our identity, away from our pretend self and back to the independent awareness of our veridical (truer) witnessing Self. If this "grace" has not yet fully come, you may find something of use as you read on. (See "Grace", pgs. 507, and 539).

> *"When a superior man hears of the Tao, he immediately begins to embody it. When an average man hears of the Tao, he half believes it, half doubts it. When a foolish man hears of the Tao, he laughs out loud. If he didn't laugh, it wouldn't be the Tao!"*

~ Tao Te Ching by Lao-tzu ~
(Original text in Appendix, on pg. 599)

The Calling

MOST PERSONS ARE BORN into this world unknowingly. They slowly grow and develop according to nature's quiet plan, quite content to pursue their lives as the persons that they believe themselves to be. But, in the rare individual, a discontent begins to stir.

Often some event in life causes them to begin to question what they had always simply accepted as factual. However, they usually soon dismiss their feelings of discontent; "After all, who am I to stand against the powerful opinions of the majority?"

Yet, there is always the exception, the iconoclast (nonconformist) who simply cannot still the quiet voice, which tells him that something is amiss.

A society generally tolerates the talented iconoclast because it recognizes excellence and is also hungry for some relief from the tedium of a predictable life and, perhaps in some, even the gnawing fear of the unknown.

But if insufficiently talented, in a communally valued way, the artistic iconoclast is soon dismissed by the "public court", often to suffer a miserable existence on the outer fringes of society.

In other words, the talented discontent may indeed find a channel of expression in the "Arts", if they are lucky! This is because human society often takes great pride (and even pleasure!) in the inspired work that results from the labors of a sometimes tortured artist, who attempts to reveal their inspiration to them in a creative fashion. But artists, be they performing artists, authors, or seekers of truth in some other medium apparently outside of themselves, are often then trapped between two worlds.[4]

They long to stay in touch with the world that their talent and dedication opens to them, but invariably, like Daedalus, they fall again, painfully to earth (see "Daedalus", pgs. 506 and 524)!

I

Oh how lonely at times is the world of the dedicated artist, and those privileged to see and feel what most around them do not allow themselves to directly see! Nonetheless, for a brief period in everyone's youth, we are all "called". This uninvited calling is, no doubt, often the root cause of suicide in our often discontent and disillusioned teenage years (see "Van Gough", pg. 594).

The angst is usually more keenly felt by those sensitive and intelligent enough to feel its powerful pull. An attraction so often truncated by the circumstances of our young lives. The easy answers offered by society are often seen through by the intelligent young, and they soon come to suspect that either no one understands them, or that there really are no substantive answers at all!

The common root of our youthful dilemma is generally clothed in the complex garb of our developing egos, but is nonetheless a leading cause of death in people from the ages of 15 to 24.[5] The unpleasant thought of a lifetime of quiet struggle, in a world seemingly bounded by hopelessness and meaningless suffering, is simply too much for some to bear.

However, most people safely transit these uneasy early years, and by conforming to the paradigm (model) of the dominant group consciousness, they eventually take their place, playing an acceptable role within the society that they are born into.

II

But there is always the very rare individual, who for some reason enters the world with the understanding that they are neither the body nor the mind (see לו *Lamed Vav*, pg. 509). They may take on a complex persona and develop a functional ego, but they also never quite lose touch with the fact that they are not whatever appears before them.

People like these cannot be moved by any amount of persuasion to accept that they are anything but whatever is observing that which goes on within their own minds. They simply understand that all of their experiences are present within their consciousness, and that since they are the observer of their own consciousness, that they cannot in essence be their consciousness either![(6)]

Intuitively, they understand that they exist, even prior to consciousness. Such people are not only called by דבה טבא *Deva Teva* "Mother Nature" to more fully awaken, but when they feel the calling "in their bones", they cannot dismiss the call, no matter how hard that they might try!

They likely represent the next step in Mother Nature's creative evolution, which is evidently the evolution of human consciousness.[(7)] If our human species is to survive, and not destroy itself with the newly acquired gifts of emerging consciousness within matter that is the hallmark advantage of our species, it seems imperative that we somehow learn to recognize what Mother Nature is apparently attempting to accomplish!

Simply put, "She" is evidently trying to reconnect our (only illusory) individual human consciousness, with the overarching impersonal Universal-Consciousness from which we have all somehow emerged.

And although our present world can still be a rather violent place, at least we are no longer burning non-conformists alive! (As the Catholic church did to poor mystically inclined Frater Giordano Bruno in the fifteen-hundreds!). We are also (thankfully), not banishing, or otherwise castigating, those among us who exhibit the unique talents that often accompany the awakening of a higher awareness; and historically speaking this is, to say the least, encouraging!

As a result, science and academia are today closer to the reality of man's existence and truer relationship with nature than the imaginary mental constructions of our past. The fantastic supernatural elaborations that, all too soon, became powerfully problematic and dogmatic religious/political institutions suborning and prolonging elitist superstitious practices within almost every sort of human society on our planet.

And although science and academia are evidently not immune to such elementary nonsense, their emphasis on reasonable "proofs" has, so far at least, proven to be reasonably effective.

But Mother Nature's rather elegant solution to this recalcitrant dilemma is self-evident, because our collective human consciousness is clearly still evolving! The past ten thousand years of recorded history present ample evidence of the torturous struggle of our humanities emerging awareness, to the observant eye.

Moreover, nature's poor bumbling *mentschnechite* "humanity" is clearly still attempting to climb out of the insentient "clay" of our own mysterious creation! And we are doing so with an ungainly "two steps forward and one step back" sort of evolutionary motion, through the history of our own species' experiential space-time.

And as far as we currently know this is still nature's best method of progress, at least on our planet! But society is also not yet fully cognizant of what this awakening portends, for the world at large.

III

But it also makes no sense to suppose that only one race or group within humanity is meant to carry this progression of consciousness out, since nature has already used "innumerable" forms to reach her present human expression.

Moreover, exclusionary intellectual notions, like that of Plato's "Philosopher Kings", only point out the curious admixture of animal and higher awareness that is the difficult and peculiar legacy of our evolving human race.

In other words, what we so often think of as our very own "will to power" is simply a slightly more evolved expression of nature's more primitive urge for the Life-Force within us to survive, dominate and flourish in the animal world.

Rather than leading humanity to the utopian "Kallipolis" that Plato had envisioned, the more deeply disturbing realities of recent history, such as the brutal dystopia of the genocidal Nazi "Aryan", and sterile "Ayn Rand" (pg. 517) type of elitist philosophies, demonstrates the self-defeating nature of these tragically narrow sorts of erroneous and destructive exclusionary notions![(8)] Indeed, greater awareness seems to quite randomly express itself amongst humanity as a whole, in much the same way that genius does.

Simply put, aside from the more obvious moral implications, denying any segment of non-felonious humanity the right to life and social participation diminishes the chances that the whole of humanity has of flourishing and surviving! (See "Dystopia vs Utopia", pg. 525).

Hence, this "calling" has nothing to do with the superstitious traditions and musings of the past that have been responsible for the problematic development of many of the world's religions.

It also has little to do with our modern psychiatry or *psychology* "study of the mind" that are far too often based upon the shifting, shallow premises of the dominant social paradigm, but nonetheless whose laudatory aim is to achieve a so-called "healthy" and functional ego.

Instead, this calling is, in incipient fact, about the relentless and remarkable, for want of a better word, "Will" of Universal-Consciousness to more wholly emerge within what, at least on its surface, appears to be wholly physical energy and matter. But it takes courage, determination, and even a measure of faith for the individual to answer this calling.

And it takes perhaps even more than that for a consciously evolving human society to be willing to let go of the familiar superstitions and comfortable past paradigms that will likely destroy the habitatibility of our entire planet, if we continue blindly to cling to them![9]

We must first acknowledge, and then encourage, the transformation of awareness that is apparently ever endeavoring to take place within humanity. There is nothing unusual or alarming about this event. It is simply the inevitable success of the timeless, and infinitely patient will of "Mother Nature" to preserve her most precious form, that of Life itself!

But in Life-forms that have been laboriously brought forth, most likely "time and again", in different cosmic places, and eventually even upon this most special of planets, our very own planet Earth! A unique and rare "life-friendly" planet, which floats alone, like a miraculous blue-green diadem, in the unimaginable vastness of our ever-expanding universe (see "The Seeds of Life", pg. 586). (Piae Memoria, the promising young artists of the tragic "Ghost Ship" fire in Oakland California.)

Space, Time & Eternity

"There was a young man named Bright who travelled much faster than light. He left one day in a relative way, and returned home the previous night!"

~ From the laboratory notes of a perplexed physicist ~
(See "Frank Tippler", pg. 590).

Several Years Ago I decided to discover whatever it was that Albert Einstein had figured out in 1916 that so abruptly turned classical Newtonian physics upon its head! I talked to my local librarian, checked out several volumes about Einstein's life, his beliefs and, of course, his renowned dual Theory of Relativity. (One of my favorite television shows at that time was "Quantum Leap", because it dabbled in the subject of time and the possibility of return; a continuation of sorts beyond our physical deaths).

I was fortunate to have a good neighbor who, as luck would have it, is also a nuclear physicist. If I hit a rough patch, where I was *stymied* "stumped" despite my best efforts, he would help me to understand the elusive point with the elegant simplicity of someone who really understands particle physics!

After I had finally digested enough of Albert Einstein's theories to suit my purposes, my kind neighbor unexpectedly rewarded me one day with an unforgettable tour of the Stanford Linear Accelerator "SLAC". Dave was nearing retirement after a long and rewarding career with SLAC from its inception.

A brilliant and kind scientist, Dave is the rare kind of "stand-up guy", who was so devoted to his partner that he could never bring himself to remarry. Because in his heart he knew that no other woman could ever take the place of the "love-of-his-life" that he had so tragically lost in an automobile accident. And so he never remarried.

But even more importantly, he still did a good job of raising his daughter all on his own after his young wife's tragic death. I like Dave, but it's also not good to be alone, and I'm happy to say that before the writing of this book was through, another decent woman (herself a widow) discovered the treasure of a good man, and neither one of them is either lonely, or without love, anymore!

Dave also regaled me with some very interesting, but nonetheless disturbing, stories about testing nuclear bombs in the arid American wilderness of the Nevada desert. And because of my past experience with Nike Hercules missiles in the American Army, he certainly had a rapt audience!

As good fortune would have it, we arrived on a day when the SLAC facility was actually being used. And despite our being precluded from venturing down into the bowels of the accelerator, actually seeing it in action more than made up for it!

At the Northeast end of the mile-long linear accelerator a "cyclotron" nestles snugly beneath a huge bunker of grass covered earth. Dave explained that the cyclotron adds tremendously to the power of the linear accelerator.

To my amazement, I discovered that subatomic particles of matter and antimatter are accelerated in the cyclotron in opposite directions. A strong magnetic containment field makes this possible, but the substantial electromagnets that are required to accomplish the job are about the size of a small pickup truck!

When the particles are accelerated to near light speed, around 299,792,458 meters per second, they are released in opposite directions, where they shoot down and back the mile long tunnel. (A meter is around 3.28 feet long). They collide inside a chamber where sensors record and digitize the resulting "shower" of subatomic particles.

The energy released when the particles finally collide is astounding! Indeed, it is said to duplicate the theorized conditions in the first few moments after the "Big Bang" began our universe. But fortunately, the explosion is proportional in size to the tiny particles that it produces!

The subatomic particles that are generated by the near-light-speed collision of matter and antimatter are digitally recorded by a bank of high speed computers located in a control room that is directly above the cyclotron. I soon discovered that physics has entered a "Brave New World", where high energy subatomic particles behave in rather astonishing ways!

Some of the researchers that I talked with believe that we are approaching the end of the "subatomic trail", while others think that we will continue to find new particles, "all the way down." Nonetheless, eventually we will reach the point of our inevitable limitation to produce ever higher energy collisions, and our ability to detect ever smaller subatomic particles.

But of course humankind continues to push the ever expanding envelope of our understanding ever deeper, by building ever bigger, faster and more powerful colliders, such as the Large Hadron Collider "LHC" facility located near Geneva Switzerland; also known as *CERN* Conseil Européen pour la Recherche Nucléaire (see "Higgs Field", pg. 543).

I

However, it is precisely at this ever diminishing "breaking point" that things begin to get quite interesting. Because more puissant (formidable) questions are being generated than credible answers can be readily provided! Indeed, today's theoretical physicists are even beginning to look at our universe in an entirely new way that postulates the existence of several dimensions, which may be involved in the "evolving" processes of a basic universal substance.

No one is in complete agreement, because not enough is yet known. But of course any such principal substance, or energy, probably contains an element of awareness, in order for its existence to be present in any meaningful sense. This fact, of an independent Awareness that is evidently essential to any sort of cognition, certainly casts a rather serious doubt upon the reality of any sort of independent and entirely physical existence.

In other words, if a physicist is not present to perform his exacting observations, the subatomic universe that underlies the universe that we can see may or may not even exist! Indeed, in the absence of an essential experiencer, whatever is experienced, including the entire universe, cannot truly even be said to "exist" at all![(10)]

Consciousness is thus obviously involved in any meaningful process of phenomenal manifestation, which invites the obvious question; "Just how much might consciousness be involved in the existence of energy and matter?"

Some theoretical physicists envision energy "strings" which could provide some answers, and have even generated rigorous mathematical proofs for their visionary hypotheses, which will remain unfortunately unsubstantiated until we are able to actually detect anything on such a minuscule scale.[(11)]

Therefore it seems wise to keep in mind that despite our incredible recent advances in physics, there yet exist a multitude of unsolved mysteries in our universe. For instance, Einstein postulated that if it were possible to pass the speed of light, you would likely reverse the direction of time; at least until you slowed back down to light speed![(12)]

Interestingly, when we approach the quantum level of "strings" we are apparently on such a small scale that by just correctly winding around a doughnut shaped "closed string" it is theorized that we might even begin to slip back and forth in the fourth dimension of space-time![13]

Any such short-range time travel, however, would involve a distance commensurate (matching) with our minuscule scale of measure; in other words, not very far! And because it apparently moves quite well in either direction, evidently even for theoretical physicists, "time" remains a rather slippery subject! (See "String Theory" pg. 581, and "Time Travel" pg. 589).

II

But time travel is evidently not the fundamental issue on the quantum scale. Because as we approach a more basic level in our universe, space-time, infinity, and eternity begin to remerge. And even apparently physical energy begins to display recognizable signs of consciousness, as spatio-temporal events start to become increasingly irrelevant!

When the universe enters into our consciousness, it is no longer merely a "physical" thing, because it is now a manifestation within the consciousness that contains it.

Moreover, this is evidently true of our universe's most minuscule part, as well as its phenomenal whole, wherein eternity is endless space, and infinity is timelessness.

Indeed, when we attempt to factor consciousness into the existence of our universe, we soon enter into the "Twilight-Zone©"-like metaphysical domain of the sages of antiquity.

Remarkably, the wondrous time may not be too distant, where metaphysics, physics, science, and (rational) science-based science fiction will all finally meet!

And when they do, it will most likely be upon our Primeval Singularity's very first "crossroads", in the explosive simultaneous emergence of space-time and energy from within a sub-quantum, spacetimeless Universal-Consciousness. (But a Consciousness that is most likely much different than anything that we can currently even imagine!)

But by now, with all of my musings, 'though at first unnoticed, evening has at last arrived! And it's now time to stoke up our poor flagging campfire, and to put our trusty little antique fire-blackened hand-crafted Japanese tea-kettle back on the fire grate to bubble and warm.

I can feel the damp coolness of the fog as it starts its nightly inland journey, and I must now don my warmest jacket, and turn its down collar up, against tonight's invasive evening chill!

Tonight's gray misty fog is accompanied by a fresh cool sea breeze. Which trembles the rainforest's innumerable leaves, as it meanders through the branches of the towering coastal redwood trees, above our humble little "backwoods" camp.

In its gentle passing, innumerable moist droplets begin to slowly gather on the rapidly cooling surfaces of the rainforest's vast panoply of constituent needles and concealing leaves.

In the beginning, they are absorbed directly by the enveloping redwood trees' thirsty canopies. But very soon, under their own steadily growing weight, these brimming little water droplets will begin to slip off the leaves and fall, to re-wet and re-new the thirsty forest floor that lies in wait below.

Thus, our lovely coastal redwood rainforest even has its own effective watering system! And no electricity, timers, or irrigation nozzles are required to make it function! Because, like most everything else on our remarkable planet earth, it is powered by solar energy. The wind, cycle of rain, indeed, the entire climate of our planet, are all secretly "solar powered"!

As the slowly cooling "fogdrops" finally begin gently to fall, our steadfast little fire-blackened teakettle happily begins to sing its steamy little song. Announcing to the cold indifferent darkness, the promise of a comforting warm cup of tea.

But now I hear a strange stirring sound, and a ghostly figure slowly emerges, appearing to somehow float out from within the surrounding dark mist, and into the glowing circle of tonight's golden campfire light.

Campfire Tales

Ari: "Master Wu, I am honored that you have decided to visit our humble camp. I wonder if you might have some words that you would like to share? (And, just in case you don't already know – but you probably do – this "Ari" character is me, and the other camper that's listening here, of course, is you!).

Wu Roshi: "Yes, my dear friend, and I am happy to warm my old bones by your campfire! The forest is most beautiful in the moonlit fog this night, but also quite cool. If I could have a cup of hot tea, I think I might recall a couple of tales".

Ari: "Yes Sensei, here is your tea, just as you like it. I am honored to hear you speak once more. The tape recorder is on, and you may begin whenever you are ready".

WR: "I will begin with a venerable old story, which sheds some light on the true nature of `the universe that we are presently travelling in, I call it:

The Parable of the Spider

"The Universe Exists within the experience of the person. Experience exists within the mind. The mind exists within Universal-Consciousness, and Universal-Consciousness exists within the Eternal Witness. Each is an ephemeral 'sheath' of Consciousness/Life-Force. Around the שהורש *Shoresh* 'root source' of the Life-Force, is a living gossamer web, woven of concept and thought.

Ever at the heart of this phantasmagorical web, lies the Eternal Witness. Like a spider that has woven a web from within its own body, only to become entangled in the conceptual intricacies of the web's subjective 'warp', and objective 'woof'.

In a panic, foolishly fearing for its own survival, the Eternal Witness has forgotten its true nature as the Creator of the Web of Illusion by which it fancies that it is now somehow caught and is being held fast!"

WR: "That is the first tale, and it prepares the way for another to reach its completion, as is the way of the great Tao".

Ari: "What a wonderful Buddhist story! It even has some Western Biblical elements! Please tell another tale whenever you are ready Sensei".

WR: "Of course; the poor spider would eventually start to wake up, and the story of how he became a beautiful lotus blossom explains how this came about, it is called:

The Parable of the Lotus Flower

"One Day, Believing itself to be a vulnerable and fragile creature, forever trapped within a web of space and time, the foolish conceptual spider feared that it might die away, and so, 'in time' it did! But the Eternal Witness, outside of time and space, of course, survived.

Yet the Witness knew not of its own existence, and so came to 'Be' in time again, and was instantly drawn back into its own web, and stuck once more in space as well! Endlessly this cycle of being and nonbeing went on and on, until one day something quite small and insignificant happened. A tiny 'wrinkle' in the fabric of space and time occurred.

But the cycle continued to move blindly, mechanically on, giving no notice at all to any change. With each cycle, the tiny wrinkle warped the fabric of time, just a bit more, until time's 'space' finally wrapped right back onto itself! The loop of time now stood outside of the subjective warp and objective woof of the conceptual fabric of the spider's web.

The loop had become a kind of spatial 'seed', the seed of change, of something different, something new. Now, each time that the cycle moved mechanically on, weaving and reweaving the web of space-time that seemed to hold the Cosmos and Eternal Witness fast within its grasp, the seed began to sprout.

As a flower truly grows, not unfolding, but by growing new and lovely parts, in new and different places, so too the seed of emerging Awareness gently grew. In a timeless loop of self-reflection, the Eternal Witness slowly began to recognize its own true nature!

Once the process of reawakening had begun, with that first tiny wrinkle in time, it could not be stopped, until one day a golden lotus flower of incomparable beauty grew, where the web once held the tiny, frightened little spider.

Within its wondrous Divine beauty, the lotus now contained the images of all that had gone before. The heavy rot of suffering and injustice became the 'loam of life' that anchored and fed its roots.

And the good and sublime higher truths of life were like life-giving waters that nourished the lotus as it grew. The love to 'Be', which flowed around it like warm sunshine, gave the golden lotus all of the strength that it finally needed to unfurl its Divine beauty within the world.

The homely little spider now realized that it had become a magnificent golden lotus flower! And the lotus contained within its being all the secret meanings of the past, present and future, which had always been there, but were temporarily hidden away within the timeless structure of the great and mysterious Cosmic Web."

Ari: "Thank you Master Wu! There is much wisdom in these parables, could you tell a Zen tale?"

WR: "Certainly, but only after I drink my tea, which I am afraid is now quite cool."

Ari: "Please pass me your cup and I will refill it with some more hot tea".

WR: "Thank you Ari" (Master Wu quietly sips the steamy tea). "I will now tell you the story of how a once illiterate tree trimmer became the esteemed reverend Master of the renowned Ch'an Jade Mountain Temple in the rugged Northern Mountains of old China".

Ari: "I am honored". (Bows).

WR: (Smiles and returns bow) "I call this tale:

The Parable of Behind the Beyond

"BEFORE HIS PASSING THE MASTER SAID; 'The world is but a symbol of the Self. At every moment the opportunity exists for the symbol to open its meaning to you. But if you seek it, you will not find it!'

The Master gave a clue to unlock the illusive symbol, and then he died; 'What is it that reflects everything, yet retains nothing, and has no perceptible existence in itself? When you understand its true nature, by doing nothing at all, the symbol of the world will open like a flower in the sunlight on a warm spring day!'

Then the Master sighed, closed his eyes, stopped his breath, and died. Silence descended heavily upon the quiet room.

The assembled monks looked at one another in bewilderment, and then stared forlornly at the floor. The meaning of the Master's final teaching remained a dark matter, it had, sadly, eluded them all!

After a time, the old tree trimmer stood up from where he had been sitting quietly near the doorway at the very back of the meditation hall. At the sound of his rising, all of the monks turned toward the master's steadfast old helper.

Everyone knew him, he was the one who cooked everyone's meals and unselfishly cared for their aging Master. When he was not helping with the monks' laundry, he quietly tended the temple garden, trimming the trees when they needed it and doing other menial chores as required.

In a quiet, but firm voice, the old tree trimmer said; 'Perhaps our venerable Master was saying that the behind of beyond is in front of what's ahead, and that each is behind and in front of both itself and the other'. He then bowed to the astonished monks, turned and walked out into the temple's garden, and softly sighed the words again, into the crisp, cool evening air of a passing mountain breeze.

The apperception of his meaningful understanding suddenly leaped like wildfire around the meditation hall. In an instant, all of the assembled monks were fully awakened!

The faithful old servant lived out the rest of his enlightened life as the much beloved and esteemed Master of the 'Open Secret', Jade Mountain Ch'an (Taoist) Temple. It is said that his immortal words can still be heard, by the pure of heart, in the occasionally 'sighing' coolness of an evening mountain breeze.

Some who hear them will immediately understand, when the apperception of their meaning suddenly leaps up, enlightening them from within, like a wildfire – Wu Roshi's tale ends, and the forest quiet again descends (see "Roshi", pg. 512).

Ari: "Thank you Roshi Wu" (Ari bows) – "more tea?"

WR: (Smiles and returns bow) "You are most welcome Ari! Ah yes, let us sit together quietly and sip tea, while we listen to the sounds of the night floating around on the cool evening breeze.

After a time, Wu Roshi bows, bids *sayonara* (goodbye) and then disappears back into the dark night.

Truth's Measure

The "Heart" Of Truth Is So Simple that even a child can understand it, for it is best seen by the pure of heart. The rest will dismiss it as the foolishness of children, or perhaps as a mere child's tale, but the pure of heart will not.[14]

It is for these few, first and last, that these campfire tales are recorded, with much love, for you are the veritable "seeds" of Master Wu's "Golden Lotus"! Perhaps unknowingly, you bless our troubled world with the peace and beauty that are the benediction of the pure of heart. Indeed, what else could this figurative "heart" be, if not the truer heart our own authentic Self?

So, pray, let us keep the connection to our own innocent hearts pure, and be ever ready to help heal the suffering hearts of others. So that they too may someday hear the profound and simple wisdom, that is ever waiting to be heard, in the simple sighing of a cool mountain breeze.

(In loving memory of my late uncle, USMC Major Wayne Houghton, and of our country's brave World War II Navajo "Wind Talkers"; Semper Fidelis, my "Westerly" ᎠᎾᏝᏅᏟ *Digineli!* See pg. 506).

Sensing Space-Time

As A Teenager, Albert Einstein dreamt of what it would be like to, "...ride on a beam of light." His waking formulae later substantiated what he had intuitively sensed in his teenage daydream. He called his vision of universal law the "Theory of Relativity". Einstein postulated that, "Time is the fourth dimension of Space", and that, "Our universe is a space-time continuum".

His observations explained certain anomalies, which were either previously missed or dismissed, such as the phenomenon of light traveling at the same speed relative to an observer, whether the observer is moving towards or away from the source of the light.

But how is this possible? And just what was Einstein trying to explain? To better understand, we need to allow ourselves to take a closer, freer, and more informal "heuristic" (common sense) look at consciousness, time and eternity.

The eighteenth century German philosopher Immanuel Kant reasoned, "We create time ourselves, as a function of our receptive apparatus". Yet we can only sense time's relatively constant cognitive flow in segments, for as Einstein's elegant "Theory of Relativity" implies, time's corollary space is a cognitively demonstrable dimension that extends throughout our universe.

But it extends in sensorially imperceptible "right angles" to us, in a "fourth dimension" that we inescapably perceive as the automatic succession of time. Because, being perceptually limited beings, we simply cannot perceive the fourth dimension, except through our sensory apparatus, which interprets it as a succession of three-dimensional phenomena.

We experience this in nature, which displays signs of the fourth dimension all around us. Consider then, that what we are seeing is actually the third dimension, as it is being organizationally created within our own minds. And remarkably, our minds do this on a regular basis, at near light speed, before we can even notice it!

Our everyday experience of "reality" is thus created automatically by a dynamic process of emerging pattern recognition, retention and repetition, within the absolutely necessary fourth dimension of our own universal experience.

Growth, duplication, the arrangement of crystals, frost patterns, snowflakes, the symmetry of trees, motion, light, sound, radiation, electromagnetism and the motion of waves of all kinds; all of these things and more, are manifestations of, and within, an ostensible "fourth dimension".[(15)]

But unless some portion of us also stands outside of our experience, of an ever-changing time, perhaps even somewhere within an unchanging eternity, we would not be able to cognize either stillness or motion, or any sort of an apparently unchanging space, or an ever-changing time.

And remember, time is the serial measurement of motion in space, and as we commonly experience it "time" is an artifact of our sensory apparatus and brain. This is because both expansion and contraction, even in consciousness, are still motions, even if they are apparently only conceptual in nature!

I

Imagine that you have just climbed almost to the top of a two-hundred-foot tall redwood tree, growing in a park campground, somewhere in Northern California. You decide to stop climbing, to rest your tired arms and catch your breath, and are now poised at a motionless point, one of an almost infinite number of possible points, along the tree's massive trunk. The ground far below you contains people, cars, and other objects that are in constant motion.

But then, as you continue to look down, you begin to notice that all of the objects seem to move at about the same speed, relative to your point of observation, as their subject, poised at the top of our immense imaginary parkland tree.

Whether they approach or move away from the base of the tree that is in alignment with your reference point, which it supports high above, the speed of the objects below always seems to be about the same!

Notwithstanding the "redshift" phenomenon of the "Doppler Effect", this is the way that light appears to travel, at approximately the same speed relative to an observer, whether they are moving towards or away from the source of the light, or it away from them. It appears that light may actually be moving in a dimension that is at "right angles" to its observer! (See "Doppler Effect", pg. 525).

But this bold assertion certainly begs a further explanation! My good neighbor Dave, the particle physicist, once shared with me his belief that light, which is either a lepton "wave" or a photon "particle", probably moves as a lepton undulation in the fourth dimension, where we can't see it, and manifests as a particle shower of photons in the third dimension, where it can be seen as a photon wave, or a lepton particle!

In any case, photons are almost massless and exist in our three dimensional perceptions of four dimensional space-time primarily as intersecting waves (i.e. vibrations) of electricity and magnetism! (See "Standard Model, lepton", pg. 580).

It seems important, therefore, to keep in mind what the fourth dimension actually represents. By following Dr. Einstein's renowned example of "thought experiments", let us use some mental imagery to clarify our understanding of what's likely happening.

Imagine for instance, that you now descend from the top of our towering parkland tree, using a "taughtline hitch" on your climbing line. The appearance of your rate of descent will remain relatively unchanged, like that of light, to observers on the ground. It will not seem to speed up or slow down, whether they are moving towards the redwood tree that you are descending from, or away from it!

In other words, if the two-dimensional movement on the parking lot tarmac represents passing time, and eternity, immutability, or duration, is represented by the redwood tree, then any three-dimensional or "vertical movement" in the tree, ascending or descending, being in another dimension, is at "right angles" to all of the other movements on the two-dimensional plane of the parking lot's surface.

Hence, any "three-dimensional" or vertical movement will appear to be fairly constant, relative to the two-dimensional motions on the parking lot tarmac. And of course, every dimension can only be seen in its entirety from a "superior" dimension! (With gratitude to the late Wei Wu Wei, for this important understanding!).

II

It seems as if our experience of space-time, which is necessary to all of our common perceptions, might in reality be based upon a complex perceptive and conceptual *illusion*. Indeed, everything that we perceive or experience is in actuality only a mental construct, an ideation, or a concept, all of which are merely thoughts!

Our conceptions and experiences of birth, life and death are all based upon the perceptive illusions of "space-time". It therefore seems not unreasonable to ask; "How then can we possibly survive death by 'reincarnation', or any other notion that conceives of time as something which exists outside of ourselves, and that goes on in a strictly 'external' spatial dimension, whether we are here or not?"[(16)]

Evidently the notion is absurd, yet it also seems quite likely, if not obvious, that the only "real" portion of us is probably a witnessing Presence. Which forever stands outside of both space and time. However, this occult (hidden) portion of us most certainly also has no independently substantive means to support the phenomenon of our individual consciousness. Yet, having never actually been born, it could also never truly die!

Apparently this is our real portion, an immutable, ineffable (indescribable) essence of pure Awareness. But it is then a pure space-timeless Presence, existing even prior to the consciousness by which we know that we and our universe exist.

It is a principal Awareness in whose absence our sentience, and perhaps even the existence of the universe itself, seems quite impossible! Therefore any existence, conceptual or otherwise, is naturally fraught with self-contradiction (see "Reincarnation", pg. 571).

But is space-time then "all in our head?" Perhaps the best answer is, yes, *and* no! Evidently it must exist within a broader, simultaneously "local/non-local", Universal Consciousness. Which thus allows its existence both *in*side and *out*side of our heads! (See "Einstein's Original 'Speed of Light' Thought Experiment", pg. 528).

The Primal Oneness

LIFE THEN, it seems, is not unlike the creek that I presently hear gurgling happily about in the darkness, down below our mountainside campsite. Its waters begin and end in the mighty primal oneness of our planet's deep, blue, planet-wide sea. Which we ineffectively try to contain in ideationally separated "oceans", by simply calling it by different names, in different places, around our world!

Comprised of moving waters that constantly change form, the sea is sometimes liquid, and at other times vapor, or even clouds and rain. Which become in turn, rivulets, streams, ponds and lakes, and sometimes even mighty roaring rivers!

Water, for a brief while, even comprises the bulk of most living things. But water always returns back again, to its primary source in the deep, blue, planet-wide sea.

This sort of theme, the manifestation of recurring cycles, so common and essential to nature, is probably what the Buddha, Masters and Sages meant when they spoke of "reincarnation". Simply because, no independent entity ever really exists at all!

Our ego, or "person"-ality, is therefore only a concept. Like everything else, it seems to exist, yet only for a little space-time while, within our utterly space-time involved individual consciousnesses.

But inevitably, even our essential egos and all of the other phenomenal thought concepts, which actually comprise our every experience of life, will dissipate entirely with the dissolution of our body-mind instruments of three dimensional perception!

Eventually everything disperses back into the primal substrate from which it originally manifested, like water steadily returning to a sub-quantum sea. Yet any recurring "cycle of return" is also only a repetition of the segments or frames of our experience, that when spliced together by our cognition, like a movie film, represents the real extent of our lives.

In other words, since there is always a time factor involved in every space-bounded experience, there must also be an immutable "receptive apparatus", such as the 18th century philosopher Immanuel Kant proposed.

This is to say that when we observe the universe with our psychosomatic instrument (i.e. our so-called "body-mind"), we utilize sense perceptions and a brain apparatus that creates the powerful illusion of sequentially passing through a convincingly "real" space-time.

But, whether real or not, space-time is both dialectically and biologically necessary to the very existence of our purportedly "dualistic" subject-object consciousnesses! This is because every apparent "object" of our experience is not only comprised of consciousness but must also be expanded in sufficient dimension, and over enough duration, for it to be experienced by us.

The sea, for example, is always present in each and every molecule of the water that comprises it. Water is thus always "anchored" in the sea from which it emerges. It journeys for a while and then returns back to the sea, in a constantly recurring cycle.

Thus the sea source is ever present, even when its attired in different arrangements of molecular "clothing", in the endless dynamic play of water. In like fashion, we have materialized tridimensionally and therefore must also exist, perhaps eternally, in the fourth dimension, or beyond.

And although it may at first sound rather strange, we necessarily must exist in a dimension that is at "right angles" to every moment that we are apparently materialized!

This is what allows us to view the tridimensional objects of our consciousness within a four dimensional continuum into which they seem to expand, jostle about, and then like springtime ice, slowly melt away from our momentary cognition, to join our mind's hidden cache of slowly dissipating ephemeral memories.

However, it also then follows that because our "receptive apparatus" either exists "immutably", or perhaps even operates within eternity, that the illusion of a consecutive life may be eternal as well!

Thus the curvature of time makes every life a self-created experiential circle, or sphere, perhaps continuing indefinitely in a "lateral" fashion, in endlessly changing iterations of parallel dimensions.

Ever-extending, two dimensionally, our daily cycle of life repeats itself as an apparently separated aspect of eternity, which is automatically expanded within our seemingly "tridimensional" experience.

As a result, we automatically experience such motion in three apparent dimensions, as it manifests subliminally (i.e. below the level of our present noticing) from within the experience generating biological machinery of our body-minds.

It is allowed to do so as we view it from a four-dimensional perspective, which is also the grand view from the top of our two-hundred-foot tall parkland redwood tree! In other words, our everyday recurring cycle of life, by extending in three dimensions, allows for the generally unexamined, but infinite conceptual variations, which we necessarily experience from within the quadridimensional space-time aspect of our own observing consciousness.

Evidently the universe "at large", and its manifestation within our consciousness, is primarily a matter of energy and consciousness. Or perhaps more accurately, it is a "self-creative" energy, which is most likely comprised of a Universal-Consciousness. But perhaps this can be conveyed more clearly, if not more succinctly in an "algebraic" sort of paradigm?

A Cycle of Recurring Manifestation

"When Time intersects Eternity the present Moment appears."

This statement is nominally represented by the following non-rigorous "equations":

$$T = M\,x\,\Sigma, \quad \Sigma = T\,x\,M, \quad M = \Sigma\,x\,T$$

T = Time, Σ= Eternity, and M = the present Moment

(Note: *x* = Intersects; as a matter of "perspective")

Thus it appears that an eternally recurring "Cycle of Manifestation" is not only possible, and perhaps even probable, but indeed, it does seem to be quite as inevitable as it is apparently necessary! Therefore, when the Mystics, Masters and Sages say that, "There is no death", they obviously have apperceived the inescapable truth of the matter![17]

And please note that many brilliant scientists (such as Sir Isaac Newton) were also pejoratively labelled as "mystics". Nonetheless, in all fairness, poor misunderstood mysticism is also evidently only a scientific "problem" when it is used as an excuse to stop pursuing a more accurate and complete understanding of our persons and universe!

Surfaces

TODAY, WHILE WALKING through the forest, touching trees, I am moved to marvel at their bark, while noting the wide variation in its appearance, despite the almost identical biological role that it plays for all trees. This theme is repeated time and again in nature.

As I look around at the complex redwood forest biome that surrounds us, for example, it is difficult not to notice that everything has one thing in common, a surface! Indeed, our ability to perceive seems to be based upon the surfaces that we seem to sense all around us. But only because they appear (to be) within our range of sensorial perception!

Yet, it is clearly futile to wonder about which is principal, perception, or surfaces, as each is clearly necessary to the existence of the other. (Since our individual consciousness, by which the existence of every probable "surface" is known, is also (most likely) the cognitively reflective surface of some-sort-of primeval and universal Consciousness (with a capital "C").

In fact, this interdependence seems to be a fundamental property of all of the surfaces that comprise our quantumly bizarre "outside/inside" universe!

Surfaces must be able to reflect our consciousness in order to be perceived as veritable "objects", with dimension and an implied volume, and also be permeable to the consciousness of which they are comprised, in order for their volume to be perceived.

Further, surfaces must also be malleable (moldable), since they are expanded and shaped by our cognition into the objects of experience that comprise the manifestation of the universe within our consciousness.

This only sounds strange because we commonly think of the universe as being external to us, and apparently comprised of uncompromisingly solid, liquid, or gaseous "material" substances.

Indeed, the very idea that the energy and space of which the universe is comprised, may actually be a matter of perspective within, and essentially consisting of, some sort of impersonal "Energy" that is comprised of "Universal-Consciousness", at first seems quite fantastic, and even difficult to grasp!

Nonetheless, as Einstein has already proven, matter is actually just the remains of vibratory energy that was initially released by the "Big Bang" that is being held together by various forces of nature.

Yet we commonly believe (and perhaps presumptuously) that the universe came into being at some particular "time". But as the philosopher Kant also wisely observed; "We create time ourselves, as a function of our receptive apparatus".

"Well and good", you might say; "but I most certainly know when something happens!" Without a doubt, but just exactly what does happen, either inside, or outside of our minds?

I

Einstein's Theory of Relativity tells us that, "Time is the fourth dimension of space". "Time", therefore, must be our incomplete sense of four-dimensional space, and because our experience of space-time is merely representational, it is also only "hearsay evidence" for whatever reality that it implies!

For example, we perceive only the tridimensional surfaces of any four-dimensional spatial volume, which is actually the *in*-side of all that we can sense. Without this "inside" spatial expansion, of the fourth dimension, which "pushes" the tridimensional surfaces (that we do see) into existence within an enveloping, but invisible to us, four-dimensional "outside" space, nothing could be perceived.

This is because we live within the fourth dimension without being able sensorially to perceive it. In spite of the fact that it is "space" that connects all of the universe's tridimensional surfaces together, which are then expanded into an existence within time, which allows them to be sensed!

Thus, the fourth dimension, although invisible to us, is interpreted by inference, as a measurement of the distance between the ideational three-dimensional surfaces that we are able to sense.

Yet space seems to be an invisible and mysteriously empty "somethingness". Which, like some sort of invisible gelatinous substance, appears to hold things together and apart. And as if this were not strange enough in itself, space can be effortlessly traversed, because it has no apparent beginning, definable surface, or even any discernable ending!

Indeed, the prosaic, evidently "physical" space that seems to always be around, within, and upon us, with a more diligent inspection, is revealed to be a somewhat baffling mix of consciousness and substance! (But then, what sort of "substance" could we possibly be addressing in our inquiry?)

Thus it may also be said that the tridimensional surfaces of things are themselves the reflective surfaces of an invisible fourth dimension, which is in turn inferred by their very presence! Therefore, as a consequence of our mode of perception, what we interpret as the "distance" between things, we also experience, whenever we shift our point of perspective, as tridimensional "motion".

But in order to accomplish this ideational inference, we must "slice" our implied four-dimensional space into perceptible three-dimensional segments. In other words, these inferential segments have no real independent existence; they are purely conceptual. But they do allow us to inferentially measure four-dimensional space within our consciousness. It is this measurement that we interpret as "time."[(18)]

Ouspensky tells us that the universe is multidimensional, and that time, as discussed, exists spatially (see "Ouspensky", pg. 562). So perhaps a satisfactory answer to the stubbornly trenchant question; "What sort of 'magic' or 'miracle' could possibly occur in our minds that allows this to happen?" Is that there most likely exists a multidimensional universe whose existence depends on the presence of an intrinsically Aware noumenal substrate or "membrane".

Moreover, it is also (most likely then) a self-sentient Awareness, which exists in a manner that is also independent of our individual consciousnesses and the dimensions that we perceive in a fully phenomenal fashion within them (see "Noumenality", pg. 510).

Therefore, within a noumenally inclusive universe, such as ours, time exists spatially, and temporal (time bound) events, like "Creation" or "Manifestation", don't "happen", but nonetheless do "exist"!

This is to say that effects always coexist with their causes, and that all moments exist simultaneously, and contiguously (are separate but touching). Hence, as we commonly think of them, there is neither "matter" nor "movement".

And what appears to us tridimensionally as points far apart, noumenally touch one another, by the replacement of conceptual proximity and separation, with actual affinity and repulsion, or "sympathy and antipathy".

Boundaries

In A Park, Unlike In The Open Woods of nature, boundaries are soon encountered that seem to speak mutely of the limitations imposed by the consciousness of man.

Just the sort of rather insubstantial "boundaries" that also seem to beg us to ask the question; "Are the boundaries that are created by the surfaces of the thoughts that we see really *limitations*, or are they simply subtle symbolic (or 'ideational'), aspects of our own sentience imbued consciousness?" And because boundaries are inherently conceptual, they are also generally an ephemerality of limitations that are meant to be overcome.

Hence, a legitimate complaint (if there ever truly is one of any real substance in our 'bio-algorithmic' life), may be the matter of our purported "limitations". That is, are the questions, inquiries, experiences and transitory answers that we pursue in search of understanding, truly meaningful outside of the interior realm of the sensorially constrained human consciousness in which they transpire?

If not, then our purported "reality", like almost everything else within our fleeting conceptual-consciousness, is equally meaningless, and it is therefore rendered effectively nonexistent or un-"real"! The problematic natures of time and space, for instance, can be shown to be an illusion imposed by our consciousness, and bounded by its limitations. They do not exist in nature as we perceive them, any more than perfect geometric shapes do.

But like mathematical concepts its approximation exists as a useful, if not necessary, cognitive tool within our individual consciousness. Is it not then at least absurd, to assume that the phenomenal manifestation of our universe can be understood without reference to the noumenal by which it is known!

For even if we set these rather exotic appearing conclusions aside for a moment, certainly even the most ardent "Positivist" or "Objectivist" must admit to the inevitable limitations that our dimensionally challenged dualistic subject-object consciousness imposes upon us.

Indeed, in the final analysis, it is the inevitable limitations of our "binary" (i.e. both as "subject-object", and in dualistic cognitive function) human consciousness that will likely halt even the determined and laudable advances of pure science!

Consequently, our limitations may eventually render science sadly impotent in being able to generate the complete understanding of the universe, which we so ardently seek (see "Phenomenality" pg. 510).

It can only be hoped, that when our species scientific "end game" is finally played, that our determined scientists will be able to resist the urge to fabricate a "final truth". Ironically, this may not be as difficult as it at first sounds, given the arcane (mysterious and elite) language and complicated instruments that have already emerged, as we probe ever deeper into the enigmatic workings and secretive structure of our compelling cosmos.

As science's advance is slowly truncated, the illusion of having reached the end of the "research trail", rather than of our own abilities, may even convince us that we have finally understood everything!

The final "fatal flaw" will most likely not be the empirical scientific method, but the limitations of the human mind that is using it. Simply put, to dig deeper into any ultimate understanding, the limitations of our present human consciousness must be somehow transcended! This forces an earnest inquirer to probe ever deeper into a pure Awareness, which is likely present even prior to our consciousness.

For those who really want to know, this boundary must be fearlessly crossed. For what does Mother Nature know, after all, of the arbitrary subjective lines and objective boundaries that we attempt to place upon "Her" innate boundlessness?

Perhaps the illusory line, or "surface", that we have both consciously drawn, and subliminally inserted between our existence and that of our boundless universe, should cause us to pause and reconsider what our existence truly is.

Existence

Our Tridimensional Body-Mind is the phenomenal vehicle by which we experience our existence. The boundaries of our existence are set by the limitations of this vehicle. Which is (evidently) in "Essence", G-d's instrument of worldly experience.

But because of this inherently self-contradictory arrangement, our tendency is to erroneously believe that our personal realm of experience defines a veritable corporeal reality! And that we are also nothing, more or less, than the amazing (body-mind) instruments that automatically manifest our everyday experiences of life!

However, many of the woes within our mental manifestation of the universe result from the powerful illusion of our separate individual existences. So self-evident is this, that it bears almost no further consideration!

Except to say that peace and harmony will likely not fully manifest within our phenomenal group manifestation, of such an individually fragmented reality, until our species persistent but subtle illusion of strict individuality is finally revealed, and eventually outgrown.

Indeed, the subtle insight that our perceived universe(s) and persons might actually possess only a conceptual reality, which is generated by the confluence of a body, mind, and independent Awareness, almost never occurs to us. If it did, we might begin to wonder just "who" and "what" we truly are!

Some claim that the universe, our experience of its existence, and the vehicle of our experience are all objectively real. While others insist that they are only the nullity of a pure illusion!

But is it not more likely that they (and "we") are only a small portion of a greater (but largely unseen) and perhaps even "deeper" reality? Which we are precluded from directly experiencing by the very presence and limitations of our everyday body-mind consciousness?

Therefore, a good place to begin any deeper inquiry into "reality", is with the consideration that we cannot, either in essence *or* Essence, be any-"thing" that appears to be within the field of our common experience.

The *Stellers Jay* (i.e. California "Mountain Mocking Jay") for instance, presently perched upon a branch high above my head, somehow understands that he is not a portion of the branch that he is sitting on.

In other words, how could he possibly be aware that he is having any experience at all, if his experience, or even the vehicle of his experience, is the sum of whom and what he is?

So, we might also legitimately ask, "What is the nature of 'That', by which we know that we exist?" The most credible explanation, as to why we do know that we exist, takes us outside of our busy body-minds (i.e. bodies + minds). This is to say that we "exist" because our body-mind unerringly generates our perception of existence, but not our *awareness* of existing. Therefore, the seat of our Awareness must lie elsewhere.

If such is the case, as it most certainly seems to be, then how do we reconcile the appearance of existence and nonexistence, except in that same self-same "Elsewhere"? (Which must also transcend our illusory personal identity).

Hence, in our cursory examination of the mysterious appearance of "our" sentience within a space-time bounded dualistic consciousness, "Elsewhere" trumps not only any*time* and any*where*, but any*thing* as well!

This is why Sri Krishna says to Arjuna, in the ancient East Indian Bhagavad-Gita; "That which is nonexistent can never come into being, and that which is, can never cease to be". He continues, "There never was a time when I did not exist, nor you… nor is there any future in which we shall cease to be."

Or, if you prefer closely to observe jovial Jays, as they play freely about in the branches of ancient American redwood trees, it may serve your understanding just as well! (See "Bhagavad Gita" pg. 505, and Universal-Consciousness pg. 593, and 594).

Misery

THE SOURCE OF ALL MISERY, which is itself an unnatural state, is the belief that we are the phantom of our body-minds. So powerful is this ignorance, in producing the experience of suffering, that just the attainment of a thorough understanding that it is demonstrably *not true*, may be enlightenment enough to deliver us from our existential misery!

Which seems a sufficient enough reason to further enquire into our fundamental existential conundrum. The odd "trapped" sort of situation that we seem to currently find ourselves in. That lies between our supposed births and deaths. Which is brought on by the simple fact of our existence. And so we are compelled to directly ask, "Who or what then am "I"?

In the forest, there is clearly purposeful death and meaningful transitory suffering. But misery seems to be a malady that haunts Nature's more intelligent sentient creatures. For there does seem to be a difference between "intelligence" and "awareness".

Awareness seems to be more the natural state of affairs. For awareness is only partially dependent upon intelligence, which is apparently only one aspect of its naturally more comprehensive view.

Awareness evidently incorporates many levels, from the subatomic and the cellular, to the intellectual, the affective (emotional) and the intuitive.

And while everything may not be "intelligent" in exactly the same way that we are, every phenomenon in the Universal Manifestation, from stones to human beings, evidently shares in some level of the same pre-primeval singularity Awareness (i.e. "Elsewhere") that our universe, and eventually even our dualistic intelligence, emerges from!

The reason for this apparent quandary is very simple, everything in the cosmos is the manifestation of an impersonal Universal-Consciousness that we commonly experience in tridimensional interior space-time places, as the apparently external phenomenon of Nature.

Misery arises when we forget that we are the Awareness/Observer of the ubiquitous (present everywhere) Universal-Consciousness. Which underlies, supports, and connects all things on a יסוד *Yesōdic* "foundational", subquantum level. Therefore, a tree, a stream, the deep blue sea, or even a dark and starry night sky, may serve as a "fast track" back to our own vital primeval Awareness!

Three Things

"THREE THINGS CANNOT BE LONG HIDDEN, the sun the moon and the truth", said Siddhartha Gautama Buddha! Evidently the Buddha understood the essence of our modern scientific quest for understanding; to learn the whole truth about whatever the universe is, which of course includes "us" as well.

His teaching clearly contains the idea that the ultimate truth of things can be discovered, if not rediscovered. Since we can never really lose the truth of being "That" by which we know, through our bodies and minds, that we and the universe exist.

But he is also saying that, in a very real sense, "Sentience" is not only everywhere, it is also everything! Which is really just another way of saying that our Universe is experiencing "*it*Self" (i.e. its "Self"), through us all.

Epistemology

ONCE UPON A TIME, a psychiatrist asked a rather pointed question of a retired arborist. (Who may have gone a little "daft"; no doubt, from spending, "far too much time alone"… in the tranquil company of trees.)

Psychiatrist: "So, you say you that you are 'That', by which you know that you are! But how do you know that *'That'* is not just a figment of your imagination?"

Arborist: "In the same way that you know that *'I'* am not just a figment of yours!" (See "Epistemology", pg. 507).

Singularities

A SINGULARITY is a grand event that has no precedent. And because it has never happened before, no one can say what its full implications might really be! The so-called "Big Bang" is a monumental primeval singularity, the full implications of which are still unfolding.

But there are other singularities which continue to show up as our universe "unfolds", the emergence of life from within insentient matter, for example. Apparently each singularity event eventually gives birth to another.[19]

Billions of years ago, before life appeared on our planet, there were only around 250 minerals in the known universe. But when life appeared on our planet, around four billion years ago, it began to "terraform" our world into a more habitable domain that favored the development of increasingly complex forms of life.

For example, early microbes created CO^2, which then combined with hydrogen and formed water (H^2O). Which then began to release oxygen into our atmosphere, long before photosynthetic cyanobacteria were even present in our seas!

As considerable time slowly passed, water began to "rust" the iron that was within our seas. Which then formed into the iron ore that eventually became the oxygen carrying hemoglobin in our blood stream.

Incredibly, today, because of life's constant "geoengineering", we have over 5,000 minerals on planet earth that allow the habitability of the world as we presently know it!

With the planetary stage thus prepared, eventually conscious mind emerged from within life. Then, in short order, in geological terms, sentience began to unfold from within primitive mind; both are "singularities".

Evidently, as a singularity unfolds, it eventually reaches a plateau. In order to proceed, the plateau must somehow be overcome. The breakthrough event, which marks Nature's success, is another singularity.

The ability to create and use tools, the discovery of agriculture, and the emergence of language are all examples of singularities that are still continuing to unfold! Like steps on an evolutionary ladder, the true origin of words, for example, like everything else in our universe, lies herein.[20]

The next singularity will likely be occasioned by a "metacognitive processor", of some fashion, which will hasten the arrival of a new form of sentience that most computer scientists now call an "A.I." (Artificial Intelligence).

But because the first Singularity arose from a precondition of pure spacetimeless Self-Awareness, the physicist's "Elsewhere", all subsequent events are in essence the result of its appearance within our individual consciousness. Thus, the dual aspects of space-time, and energy and matter, are (evidently) polar expressions of the same universal Consciousness-Energy!

Therefore our entire universe, including any sort of subsequent "Singularity", is essentially an expression of an impersonal Consciousness that continues to unfold the cosmos, as it will.

Language, for instance, which comprises much of our internal as well as our communicative worlds, arises from within so-called "precognition". In other words, every thought that becomes a word, just like everything else within our super-experiential universe, evidently emerges from within a Universal-Consciousness that somehow emerges from within a spacetimeless, pure Awareness. And then initiates the appearance of our whole universe, and all of its permutations, including us as well!

But the primary problem that many people have, in trying to imagine what the nature of such a primary Awareness (and its extended Consciousness) might be, is why it is even necessary!

Where our mind is concerned, because our mind is our "everywhere and everything", sentient awareness is necessary because it is not enough simply to have the tools of language, they must first be known within our consciousness, in order to be understood, and only thereby do they become creatively useful.

Hence, words and concepts are first and foremost just thoughts! Currently, this is the major difference between computers and us. They simply compute without giving a "thought" to what they are doing. In other words, there is no internal witness of the events that transpire.

Nonetheless, when the level of complexity and subtlety of our human mind is finally reached, a humanly recognizable degree of sentience will undoubtedly emerge within computers. (But this does not preclude the existence of a humanly *un*recognizable sort of "sentience" either, which some computer experts say already subtly exists, within the electronic provenance of our world-wide "Internet"!)

Then computers will no longer simply compute like a complex calculator, because they will now be *aware*! If this should somehow happen, and many A.I. researchers believe that it eventually will, it will be nothing more, or less, than Nature's next singularity "domino" to fall, with predictably, unpredictable results!

But, as the Nature-wise American Indian Shamans once taught, perhaps our ancient friends, the steadfast "old-growth" trees, may better understand what the primeval Life-Force within them more directly sees?

Yet when we ask them, they simply cannot say in words, what we ourselves can neither directly see, adequately say, nor even completely understand! (Dedicated to the brave environmental warriors of Standing Rock!).

Tranquility

Trees May Not Speak Directly To Us, but when we quietly listen, we may soon notice that they are tellingly tranquil, and that groves of them radiate a peacefulness that is truly pervasive.

Possessing neither fears nor desires, they are free to be themselves, and when we allow ourselves to follow their tranquil example, so are we. Indeed, some of my greatest teachers have been strong and quiet trees!

(See "Wanderung: Aufzeichnungen", pg. 603).

The Lesson of the Leaf

Is It Not Likely that within the primary mechanism, which determines the appearance of the tridimensional surfaces that we see, also lies that which creates a certain range of sensorial foci?

To achieve this, the all-inclusive potential of our universe's foundational Noumenon is viewed with an artful combination of subject-object "split-mind" perspectives.

By focusing upon implied differences in position and in scale, our binary subject-object minds thus bring particular aspects of our consciousness into our cognition of them, while excluding a near infinitude of others. In this fashion, the phenomenal universe, which we perceive, is manifested within our consciousness.

But because the universe of our experience is also manifested primarily by a subliminal process of informational exclusion from our awareness, it is also likely incorrect to say that whatever we experience, is merely an illusion!

It is much more likely that the aspects of our experience that are illusions result from a misinterpretation, and subsequent misapplication of this imprecision of mind, upon what we normally experience within our consciousness.

This is likely so, since our subliminally edited reality is only a minuscule part of a far greater noumenal infinitude or *superuniversal* Totality.

In other words, if enough parts of a picture are missing, or if it is simply just too big to ever be seen in its entirety, it becomes easy to make mistakes about what the overall picture might actually be!

Moreover, what we experience as motion and "change", also an aspect of motion, in our own fantastic, stereophonic, tactual, and polychromatic representations of the universe, is made possible more by what we don't even know that we are ignoring, than by what our mind automatically brings into our cognitive focus.

I

When we look at a leaf, for example, we see that the beauty and wonder of its form and function, so basic to life as we know it, is allowed more by what is not present, than by that which is. However, within the full reality of the Noumenon ("Elsewhere"), of which the leaf is only a partial representation within our consciousness, nothing of the sort is even possible!

For the noumenal contains the phenomenal universe of our experience, in addition to every other possibility. It is therefore an absolute *plenitude* (i.e. as used in Forest Primeval, a "plenitude" represents a repository of endless potential).

Within the Noumenon, for instance, what we experience as "cause and effect" are actually linked together as one whole. And because every apparent effect in our universe is actually the obverse of endless causes, then noumenally, no effect has a single cause within space-time. It follows therefore, that everything is finally, and fundamentally, inter-linked.

And thus, we could even say that our entire universe is quantumly "folded" together, much like an interdimensionally pleated accordion might be! This is because, as inferential space-time closes up or disappears, all possible points touch one another. Thereby rendering motion, change, and individuality literally impossible, although not virtually so, as we measure it into "existence" within our dualistic subject-object consciousness.

Our entire familiar universe, and the space-time that allows its experience, is therefore essentially only a manifestation within an intimately recurring illusion. Which then masquerades as our discretely individual consciousness.

There is no doubt "something" outside of our individual consciousness. But it is nothing at all like what our constrained consciousness informs us. The beauty of color, fragrance of smell, sweetness of taste and warmth of the sun are entirely lacking.

Evidently what does "exist", is an infinite and spacetimeless universe comprised of Consciousness-Force. This not only contains every constantly expanding possibility of our experiential universe, but it also far exceeds our ability to either sense fully, or to completely understand!

This noumenal Essence is actually an "eternal infinitude". Which we might best imagine as being one solid "block" that is substantially comprised of its own Self-aware, but nonetheless insubstantial, "Substantiality"!

But then it must also be a boundless block that is entirely composed of its own endless possibilities. Therefore it is also not only a remarkable conundrum, but it is also an unimaginably rich cornucopian Plenitude!

Moreover, if our universe is "created", it is done so not by creating something out of nothing, and then using it to build up the cosmos as a mason does with bricks. Nor, conversely, is our universe created by "carving it down", like a sculptor does with marble.

Instead, our common experience of a "personal" universe most likely occurs through our binary minds "precipitation", "permeation", and subsequent ideational manipulation, of its own cerebral data. Which it accomplishes with an ingenious "three dimensional" interpretation of our individual consciousnesses endlessly flowing, algorithmic-like, data stream of biochemioelectric information.

Furthermore, this is most likely accomplished by an essential Awareness that is an intrinsic quality of the "Noumenon/Elsewhere" that births both the Primeval Singularity of our entire cosmos, and the subsequent experiential singularity of our "conditional" subject-object minds (by which it is known).

Hence it is an illusory "dualism" that independently focuses, as a matter of perspective, upon certain universal potentialities, by temporally, and therefore spatially, excluding all of the rest!

We are thus allowed to experience the essential Awareness that gave birth to the entire Universe, because it is simultaneously all around and within us, in the phenomenal manifestations of "Mother Nature". The universally Conscious processes of which, we are a living and integral part!

II

Take for example our leaf, at a closer range of focus we see a whirling galaxy of molecules, which then appears much like our familiar celestial "Milky Way", whose distant appearance is also a matter of scale. At this closer range of focus the independent form of the leaf is no longer perceptible.

Our familiar universe of elaborated forms and separate phenomena becomes unknowable, and it is impossible even to infer the independent existence of a "leaf"!

Evidently, in our rather limited view of our topsy-turvy universe, things are not always as they appear, at first glance. Indeed, we can "walk", which is really a process of losing and catching our balance from one foot to the other while falling forward into apparent space, only because our ability to pass through time as we fall is not obstructed!

Both space and time are always arrayed before us, and could even be obstructing our way, like liquid water does, as we push our way through them. But space nonetheless remains just as invisible to our senses as time does.

In a similar fashion, cause and effect are allowed to function within the range of foci provided by our senses. Which, in an enabling "negative" sort of fashion, includes the fourth dimension of space-time from which we perceive our tridimensional universe. Our perception is thus the "form and function" of the perceptual universe, in which we are enabled to move by what we experience as the "negative presence" of space-time.

The ubiquitous center of our Awareness thus experiences its ever expanding dualistic universe through the illusorily individuated consciousnesses of our body-minds. Thus we all experience the (selfsame) inflating universe, as if each of us is always at its very center!

Therefore we could as well say that our universe is essentially comprised of the extended potential of an energetic Universal-Consciousness. Which is experienced by an intrinsic presence of (pre-singularity) Self-Awareness, through the evolved vehicles of our subsequently emerging bodies and minds. Hence, we are also not at all incorrect when we observe that, "Wherever we go, we already seem to be there!"

Our dualistic consciousness effectively creates the invisibility of space by exclusion, because our individual consciousness cannot look into the dimension beyond itself from which our sentient awareness emerges.

Nor can our evidently "internal" sentient awareness, being essentially the projected (or involved) Self-Awareness of an independently noumenal Presence, see itself without the perceptive machinery of our body-minds.

This creates the seemingly limited and conceptually fragmented "positive" illusions of our intimately involved physical space-time existence, from an apparently spacetimeless infinitude of "negative" (non-physical) potentiality.

This is, in effect, the only manner in which a veritable "nothingness" can assume the semblance of a virtual "somethingness".

So, the next time that you walk by a tree, stop and take a closer look at one of its many hundreds of thousands of leaves, and you may discover that there is much more to even a "simple" single leaf than generally meets the eye! (See "Nothingness", pg. 557).

Secret Roots

An Astute Arborist Learns To Look not only at what he can see, but also, and perhaps more importantly, to look even more closely, for the unapparent and unseen.

During a career mostly spent in the quiet company of trees, I eventually discovered that a whole world of complex living connections exists, out of sight, down below the surface of the soil, where nobody sees!

It is this extensively interconnected, yet unseen, underground world which makes it possible for the myriad above ground forms of life that we do see to even exist!

This is the secret realm of roots, which supports, nurtures and connects the vast and varied above ground "Green Kingdom" of plants that spans our planet. It is an independently organic domain that is unlike all others.

Sustained not by life feeding upon other life forms, but instead, by a verdant *phototaxis* "response to light", that through photosynthesis almost "magically" converts the energy of light into the basic carbohydrate foodstuff that enables life to exist on our planet (see "Photosynthesis, pg. 563).

But people, like trees or icebergs, also reveal only a small portion of themselves, because the greater part of us too, is subtly secreted away from everyone's view! Nevertheless, there exists a vast and unseen subliminal world, which quantumly connects us to the entire universe.

It supports our apparent individual consciousness in much the same way that the secret root grafts and hidden denizens of the microscopic subterranean world connect and support an entire rainforest's ecosystem.[(21)]

Yet, even though we are all secretly connected, it may be unwise to place too much faith in the realities of other people, or you will most likely be disappointed in the end. For each person knows only the "universe" within his or her own mind, and yet most people presume that their private universe is shared externally by everyone else!

But by venturing out of sight, below our apparent universe of surfaces, deep within ourselves, we may discover an impersonal reality that exists even prior to our individual consciousness, which in its fundamental awareness, is shared not only by other people, but also by everything else in our vast universe.

Every manifest thing that exists is connected by this natural Awareness, in the same way that whole forests of trees share an invisible underground network of roots and *micorrhizae* (see pg. 509). Without this "inter" and inner connectedness, the forest, and indeed our entire universe, could not exist!

This is our secret "Garden of Eden". (At the root of which lies the hidden, self-Aware, inner universe of nature.). It is the force of Universal-Consciousness, the great *Mouna* "Open Secret". Which brings the universe into being and "orders the stars in the heavens."

But to sense its presence, we must be very quiet and still, as if we were anticipating the presence of deer in the forest. And after this secret Presence is known, there is the same delight and wonder as when a child first sees that there really are deer living in the forest!

So, if we must place our faith somewhere, consider that this may be the extraordinary place where it actually belongs.

Perhaps this is the deeper source of Shakespeare's inspiration, when (in Hamlet, Act 1, Scene 5) Hamlet rather wisely opined; "There are more things in heaven and earth, Horatio, than are dreamt of in your philosophy."

And although Shakespearean ghosts most likely do not exist, except in imaginary plays, has anyone ever expressed the wisdom of keeping an open mind any better; most certainly not I!

The Natural System

There Are Many Types Of Systems in our expansive universe, computer systems, mechanical systems, systems of thought, biological systems and even solar systems!

But here we will concern ourselves with the universal system of manifestation, which is quite likely the most fundamental system in all of nature.

All of the previous systems, save the one in discussion, are based upon the presumed material reality of the objects that comprise them, and of the space-time in which they appear to operate.

Only the last system, that by which objects, subjects and space-time become manifest in consciousness, is premised upon an entirely different paradigm.

I use the term *paradigm* "model" in our discussion of the universe's system of manifestation, because we must use language to communicate even that which is clearly beyond the capacity of language to do adequately!

But of course, any discursive (i.e. proceeding by reason) elaboration of a system of manifestation, which must be neither subject nor object, yet generates both, is necessarily only an inadequate model.

Yet, the fact remains that a paradigm is the best way to embody a complex system, which incorporates unfamiliar concepts and perspectives, in order to understand the familiar in a new way.

In other words, we must first understand our repurposed and new terms and concepts, and then use them to enrich our understanding of the unique and unfamiliar!

Therefore, please keep in mind that whatever is written, whether it is presented in the rigorous and exacting language of mathematics, or even within the ostensibly "enlightened words" of a purportedly "awakened" Master, are only symbols and signs poorly indicating the probable direction of any superlative "Reality"!

Indeed, science currently still bases its method of inquiry strictly upon the first premise, that is, the real or positive objective existence of objects and subjects, and then simply accepts, or overlooks, the quandaries (contradictory problems), which accompany this widely accepted positivistic concept (see "Positivist" pgs. 511, and 565).

I use the word "concept" pointedly here, because all that can ever be known of our person or universe are contained in our minds as concepts of thought, relative to some point of cognitive perspective.

By "thought" I mean the process that happens in our mind when it cognizes. "Cognition", here used, is fundamentally based upon our processing of the myriad electrochemical responses and the various underlying cerebral activities that support our minds and bodies.

In other words, it is largely the response of our *somatic* (physical) apparatus to its perceived environment. And by "environment", I mean to convey that which is generally presumed to have a real objective existence within space and time.

To be sure, some aspects of our cerebral activity apparently function independently of any sort of sensory input. Yet the entire psychosomatic instrument of our perception is nonetheless arguably just a small portion of our environment, which is external to the experiencer of its activities.

But, the overarching truth remains that it is quite impossible to know if any "environment" really exists externally at all!

I

Even within the often problematical present paradigm of empirical science, physics tells us that our entire universe is comprised of a complex gradient of energy, existing within a flexible field of space-time.

Therefore we can find an area of agreement between our two models. For both systems admit the presence of some sort of energy and the space which must be there for it to occupy. In other words, "nothing" must exist for "something" to occupy its vacuity, and allow for any apparent motion or change to occur!

Science, in its modern development however, inherits the age-old Hellenistic philosophical dilemma of *ex nihilio* "from nothingness", which continues ever onward as a result of its pragmatically nuanced position! (See "Pragmatism", pgs. 511, and 566).

This is to ask the rather perplexing question; "How did something appear out of nothing?" When common sense tells us that, as the pre-Socratic philosopher Anaximander long ago observed, *ex nihilo nihil* "from nothing, comes only nothing". And so the formidable wall of objectification (mistaking thoughts for objects) is soundly struck again!

But this is precisely the point at which it should occur to an earnest enquirer, that something must be wrong. However, what is "wrong" is so fundamental that even the most intelligent person quite easily overlooks it!

It is here that we begin our metaphysical inquiry into nature's system of manifestation, and how the system likely works. But, because we will achieve our understanding in part by introspection and pure reason, the ardently objectivist scientist, who relies entirely upon an empirical method to test his hypothesis, will most surely protest!

This, despite the fact that his ever changing theories of objectivism inevitably fail to explain adequately a system that includes his perceptions as part of its functioning.[(22)] (See "Forces, Laws and Causality", pg. 534).

Indeed, when our universe is *not* perceived, or if it is merely a perception, it may or may not even be there! Which effectively renders an empirical objection, into a moot point. Yet, what we inexactly call "history" apparently began at some time, when something evidently appeared out of nothing, and change began to occur.

But, what is history? Is it the record of the supposed chain of causation that began when some kind of "something" suddenly appeared? But, as already noted, something cannot exist without nothing! So what could possibly be knowable of a spacetimeless monistic "That", which somehow extended itself into an apparently diversified existence as our universe?

Necessarily, it is neither something, nor nothing, and could also not possibly be any combination of the two. Some theoretical physicists, like Stephen Hawking, have called it an "Elsewhere" and speculate that it is what is there, or rather, neither really "there", or "not there", beyond the edge of our rapidly expanding universe.

Indeed, some cosmologists even believe that our universe will eventually expend the energy that brought it into existence, when it suddenly somehow "banged" into being.

Functionally, the dispersal of this primeval energy through causality, is what is generally presumed to create the demonstrative history of our cosmos. But if this is so, then when the energy of the Big Bang is finally "expended", the universe will somehow end, by somehow simply ceasing to be!

But what is historical change, if not motion? Therefore a good question to ask is, "What is the essential nature of motion?" Above the subatomic level, it could even be argued that the entire history of the cosmos can only happen as it must!

In other words, any apparent potential of quantum randomness quickly disappears, even if only in our minds, into a causally ordered progression of "most likely" space-time bounded events. And moreover, evidently quantum gravity, as well as our consciousness of gravities phenomenal manifestations, is central to the simultaneous "inside/outside" motion which is necessary for any event to transpire in our cosmos! (I.e. ...evidently, both "inside" and "outside" of our minds).

II

However, any purportedly orderly progression of expressed infinite potential in our universe, rests upon a very disorderly foundation! Because, on the subatomic level, the expression and interaction of energy eludes our cognitive habit of objectification into the comfortable "somethingness" of our familiar subject(s) and objects.

Our mind does this automatically, in order to compare our experiential objects to one another and retrospectively to our customary subjective point of perspective, in order to gain an objective understanding of their apparent existences.

That is, any supposed "object" is assumed to have certain dimensions, which occupy a particular place in space-time, relative to our centralized point of perspective as their apparent subject.

But contrary to our more common sense, even relatively well known subatomic particles like the familiar electron will simply not follow our orderly intellectual rules!

Indeed, in order to predict the location of an electron, or any other particle smaller than an atom, the mathematics of statistical probability have been adopted and adapted (to the difficult purpose) by science.

As a result, while we can accurately describe the location of a given subatomic object, we cannot do so in regard to the speed of the somethingness of which the inferred object is ostensibly made!

This uncertainty infuriated scientists like Albert Einstein, who was moved to remark, "G-d does not play dice!" I am not certain if anyone really knows what he meant by "G-d", but Einstein knew that something was amiss and probably had some sense that these quandaries were artifacts created by whatever was missed, or incorrectly assumed at the time to be true.

Einstein wrote in his memoirs that when he was a teenager he wondered, "What might it be like to ride upon a beam of light?" (See, "Einstein's Original 'Speed of Light' Thought Experiment" pg. 528). From the happenstance of a youngster's musings, he later formulated his revolutionary theories about energy, matter and space-time.[23]

But ironically the very quantum of lepton energy, the photon, which Dr. Einstein used as a mathematical constant, was later discovered to possess the same quantum uncertainty that is apparently ubiquitous within a universal system that objective based science nonetheless hypothesizes is both objectively real, and materially substantive!

Not knowing the full extent of the quandaries cause, since cyclotrons and instruments capable of measuring on the subatomic scale were not yet invented, Einstein spent the rest of his illustrious life trying to tie up the now frayed ends of physics with a Unified Field Theory; an elusive "Grand Theory of Everything".

Particle and theoretical physicists are still trying to tie up the loose ends of modern physics, while physics, as we have known it, continues to unravel into the extensive quantum probabilities of an apparently fathomless uncertainty!

Yet many physicists are still valiantly attempting to salvage traditional physics, as the "atomic onion" that use to be the reliable foundation of our assumed objective reality, is being relentlessly peeled away, a quantum particle layer at a time. But they are often increasingly confounded as they learn more and more, about less and less.

III

Einstein probably never dreamt of quarks and charmed quarks, let alone the plethora of other subatomic particles that have been recently discovered! (See "Einstein's Posthumous Muon Proof", pg. 529).

And indeed, the invisible subatomic universe of energy particles may be limited only by our ability to create new instruments to expose them. No one currently knows if either or both assumptions are entirely true! However, many odd behaviors, in addition to only "probably" being there, have recently been observed.

Among these mysterious behaviors it has been discovered that even apparently distant particles can somehow interact with one another. These peculiar patterns of behavior indicate that a very different set of rules seems to apply as we approach ever closer to the fundamental level of the universe that we can perceive.

For example, energy components which are very small, are at distances which are very close, or are moving at very great speeds, behave as if they have a choice and sometimes even seem to decide upon a behavior which best suits the situation![24] (See "Totality", pg. 515, and 593).

This intrigues and puzzles physicists greatly, and is forcing them to redefine the terms which they use to explain why energy and space-time act as they do, on this more fundamental scale.

Thus, the objectivist is presented with the evidence that the assumed objects, and objects that are also "subjects" in his extroverted view of the universe, don't seem to abide by the orderly rules that are necessary for his traditional assumptions to be true!

Some very conservative scientists have even speculated that there are indications of a consciousness of sorts at work, within the quantum energy objects that comprise our known universe (See "Objectivist", pg. 559).

But logic asserts that the reverse must be the case. That the qualities and behaviors of quantum objects reveal instead that they are, in a sense, much like the projections of a Consciousness that is universal in scope and, from our dimensionally limited perspective, indirectly constructed of the same.

Moreover, this apparent conflict is most likely beginning to arise because physicists are finally beginning to approach the levels at which energy objects are initially manifested within Universal-Consciousness.

Albert Einstein, it seems, may have been closer to the truth of the matter when he imagined what it would be like to hitch a ride on a photon! Perhaps if he had turned his inquiry in the other direction, or intuitively inward, he might have become an awakened sage as well as a brilliant scientist?

Although I am also not entirely certain that what some astrophysicists are now calling the "Cosmic Perspective", might actually (upon closer inspection) be quite similar to the "Way of Enlightenment" that was once widely taught in the Far East by the historical Buddha.

But for us to gain a better "toehold" on understanding our universe, we must take a different approach and explore space-time to a greater depth, as it relates to the endless probabilities of phenomenality.

We will "listen inwardly" together, to reveal the mute but eloquent primeval poetry of life that lies hidden within our steadfast friends, the trees.

Inside the Phloem of Time

WHAT IS TIME? Is it an ever moving stream? Or is it perhaps some mysterious and invisible force? An ineluctable (inescapable) current, emerging from an unknown future, and ever sweeping past our bright present, as it slips away into an ever darkening past.

Carrying with it bits and pieces of our apparent present, in its relentless passing by. Carrying them forever away, as the fleeting moment of an unprovable "now" dims, and slowly disappears from our sight, around the unforgiving corner of our forevermore?

Is time our senseless enemy, a dark and indifferent force of nature, which not only sweeps away the lives of those we love, but cruelly strips away the tender leaves and living buds from our helpless tree of life, in its relentless passing by?

'Till one day our life tree stands, naked and alone, stripped down to the very "phloem" of our being by the ever passing stream of time! In due course, perhaps even destitute and lonely, bereft of dignity and meaning, hopeless and undone, do we not then topple senselessly into the indifferent coldness of time's ever moving stream?

Or is time like some vast and unseen "Titanic-like" iceberg which, as we presently sail by, takes from our life vessel bits and pieces, slowly scraping us away, until we are finally worn down to a mere nothingness. As we are impaled away by bits and pieces on the cold, sharp icy shards of an uncaring time?

Perhaps we only stand upon life's distant banks and watch, as the river of time rolls indifferently by? In the end leaving us with nothing left to say but; *Sic transit gloria mundi!* ("Thus passes the glory of the world!")?

But unlikely this, for the apparent subject within us is either moved, or ever moving, within the roiling waters of an apparently forever moving time! Yet, must we not have one foot planted firmly upon some unmoving, if not distant, shore to note time's relentless passing by?

Did G-d create us in the past, and so we move and grow, unfolding time from past to future, as along time's way we resolutely go? Or do we all emerge somehow, by mere force of happenstance, entirely upon our own?

But, then again, perhaps mysterious time flows backwards? Moving in arrears like a story already written, and just waiting to be told? Which in the telling, that is our living, carefully unfolds, and then refolds our present, back into our future's past?

Or is the present built up, like some evanescent (vanishing) skyscraper, upon invisible past floors, soaring ever upward, into G-d's illusionary space-time skies?

Indeed, with sufficient reflection, the gift of time, like many aspects of our dualistic cognition, is soon revealed to be, as they say in Greek; μια ευλογία και κατάρα! *Mia evlogía kai katára* "both a blessing and a curse."

But, perhaps it is entirely beyond our kin, to understand the mystery of time's relentless passing by.

Yet, friend or foe, be it as lofty and difficult to climb as snowy Mount Everest, or as difficult to reach as the deepest, darkest trench in the bottom of the deep blue sea, because we are but mere putty in time's ever ticking hands, we need to know, *mï Deüs*, how we need to know! (See "Phloem" pg. 510, and "Outer Cambium" pgs. 510, and 562).

What Indeed is Time?

"Time Keeps On Slippin', Slippin' Into The Future"... lyrics from "Fly Like an Eagle" by The Steve Miller Band, published by Capitol Records in 1976. Throughout time artists have tried to capture time's essence in song and poetry.

But recalcitrant time stubbornly resists every attempt at any such definition. Because, whether we try to capture time's essence in a lyrical phrase such as *tempus fugit* "time flies", or even more precisely with מהר הזמן *hazman mahair* "time hurries", time still depends upon three things, a past, a present, and a future.

Nonetheless, by the time that our senses and brains process any apparently present temporal event, 'though it is likely the only present that we may ever know, it has already slipped away into the past!

Yet the past it seems, may not truly even exist at all! Except as it is stored away, to be imprecisely remembered (and thus, slowly changed). Being itself composed entirely of the ephemeral memories of a rather questionable present, which has already passed away!

Moreover, while the future in a fashion "exists", it does so only in our expectations of what may yet come to pass. So you see, all three, "past, present and future", enjoy only a conceptual existence.

And as goes time, so too must space![25] But where then has all of our "space-time" gone? It seems that we have chased the somethingness of it quite away, into a very questionable existence!

But if in the end *tempus omnia revelat* "time tells all", and everything appears to be nothing much at all. (Because reality is apparently as much notional as it is factual). Then again we must ask, "What, indeed, is time?"

Does it not seem clear enough yet, that our cognitive experience of space-time must also be occasioned by the ever flowing river of the Force of creation and life, flowing through, and as, our everyday minds? (Not convinced yet?).

Volume and Duration

OUR OBJECTIVE UNIVERSE, of three apparent dimensions, most likely owes its existence to the creative act of a Universal-Consciousness. Indeed, as we have already seen, and will further explore, how else could our minds both "be" and work? Furthermore, the basic stuff of which our universe is largely made, is evidently "space".

Not empty space, but a veritable "Superspace" that is comprised of (for want of a better term) what we will reluctantly call "Universal-Consciousness". Which also contains every possible thing, in an ever expanding expression of its inherently endless potential, and only appears as empty space to itself (through us).

Because it cannot see itself through us directly, since it is the infinite source and center from within which all seeing, or apparently "individual" sentient awareness, begins and ends.

This is why apparent volume, or space, within our individual consciousness, can only be cognized along with our sense of duration or time. But our perception of space-time has a four-dimensional existence that is only relative to the other three dimensions of our perception. Which represent the various objects of consciousness that appear unbidden within our minds.

However, space unextended as volume is immeasurable and infinite. And as such, is probably as close conceptually as our individual consciousnesses can ever come to understanding our own essential basis in an independently spacetimeless Awareness that is essentially composed of a pure presence.

Yet, that which has no dimension and occupies no space is immeasurable, and is clearly either beyond, between, or prior to the incessant flow of involuntary experiences that goes on within our individual consciousness!

It is only from this basis that our emergent consciousness is able loosely to apperceive its own dimensionless existence in the fundamental primeval Awareness of the universe.

Moreover, it is also most probably an Awareness that was not only present before our universe was, but that is also present before and between the conceptual objects of our consciousness, that are spontaneously extended in an apparently sequential space-time.

Therefore our symbolic, algorithmic-like individual consciousness is a conglomerate process of evolved biological consciousness which, when taken altogether, comprises the apparent manifestation of our familiar phenomenal universe within our relatively "illusionary" dualistic consciousness. But what then is "phenomenality"?

Evidently we must investigate even further yet, and so we must continue on with our rather curious woodland inquiry!

Our Enduring Sense of "I"

THAT WHICH WAS NEVER BORN WILL NEVER DIE; but the body-mind by which it experiences (our) phenomenal birth, will, and when it does, it goes nowhere! It simply disperses like waves on a pond.

On one side of our Möbius-strip-like Noumenal essence we are an endless ocean of Universal Consciousness, and on the other side we are the floating "personship" of the evolved body in which our phenomenal "individual" consciousness operates (see "Möbius 'String' Paradigm", pg. 553).

But both sides eventually deconstruct, and then disperse as energy and waves of consciousness that eventually distribute the memories of who we were back into the deeper impersonal Universal-Consciousness, from which everything experientially emerges. While the "energy-string" side of things eventually absorbs and redistributes the "form" side of Universal-Consciousness (that contains our personship).

Both of which are temporarily brought together in our experiential body-minds. Wherein the universe of our life experiences is witnessed in a spatio-temporal fashion by the Noumenal "Elsewhere" substrate that supports and underlies our evidently conditional, but truly illusory, *individual* sentience.

However, before we can progress much further in our inquiry, it now becomes imperative that we finally summon the courage, and uncommonly "common sense", to allow ourselves to first admit, then to understand and accept, that when our body-mind is not present, there is also no way to experience anything whatsoever!

But this does not mean that "we" are not present, since evidently (as the substrate of the universe) we always are, right along with everything else on the sub-quantum level of pure Consciousness-Force, in our ever expanding Primeval Singularity!

Yet evidently, when we utilize this paradigm to understand our existential situation, only a great void remains, as the source of all things! But this "void" is simply what our mind is incapable of understanding, because it is the Source of its own sentient awareness. Not ultimately, but fundamentally, we are always simply *Tat* "That" alone.

"Heaven", evidently then, is also within the experiential realm of "Nevereverland", where we also have no choice but to live! This is because everything that is known within our consciousness, remains within our consciousness, and disperses just as our body does when we interpret its final dissolution as vile.

Naturally, as we age our body declines and our mind soon follows. But curiously, the sense of "I", of being "who" we are, remains ever the same! Nonetheless, as our body declines it also begins to show it, and it eventually becomes like some alien, bothersome thing. Now we no longer want to identify with it, but we have been in the habit of doing so for so long that it begins to "drag us down" too!

But the sense of I, of being who we are, remains the same as it was when we joyfully identified with it as a hope filled child.

And in the end, even the Force of Universal-Consciousness that underlies the entire universe ends, and all that remains is what was really there all along, the absolute and Supreme Awareness, which we wordlessly understand to be our enduring sense of "I"!

Goodbye

Our Little Lifetime "while" comes to an end, either suddenly, or slowly, with a sad little "goodbye". Either way, our final little goodbye is resolutely, finally and most simply put, just "Goodbye!"

Indeed, perhaps our loss of (the love of) "Be-ing" is the greatest loss of all, because when our consciousness goes, the entire universe quite suddenly comes to an abrupt end, with scarcely a sigh to mark its passing by.

Nonetheless, nothing has really changed (in a substantial way) at all! Where could we possibly go, being only an ideational inference from whatever is going on in our minds? Still, having the transient identities of our persons and universe(s) suddenly stripped from beneath our existential "feet", leaving us to walk alone upon the ephemeral clouds of an impersonal Consciousness, is hardly a very comforting thought at all!

Yet we eventually will wake up to the fact of our own non-existence, be it sooner, or later. So, "goodbye" eventually comes to us all, either suddenly, or slowly, and evidently there is never really anything that we could ever do about it!

Therefore all that we can do, is to "stand our ground" on the צור *Tzur* "Bedrock" of our essential being, as the Supreme.

Most astute Jews don't generally suppose that they will ever actually go to either a "heaven" above, or to a "hell" below, because they know that both exist only in the inferential universe of our hominid mind. But on the other hand, very few of us truly understand the eternal G-d that is always at the very center of it all!

And just like everything else in our shared universe (of the human minds'), our sad little "goodbye" at the end, eventually becomes our happy "hello" again! But don't make the mistake of thinking that the apparent person who eventually says "hello", is the same illusory person that once sadly said "goodbye"...

Tat Tvam Asi

Disillusionment is what the world most fears! Yet there is no difference between our transitory "persons", and the more durable universe(s) of our experience. But the universe of our experience is also purportedly comprised of other people. Moreover, each of these ostensible "persons" (that appear within our inner universes) is also comprised of thought, in "his or her" own head, or in that of another.

All is thus of *thought* comprised! Although everything that is (illusorily) "internal" is also thus essentially composed of nothing but ideation. Yet apparently "That" (by which *everything* is known) is not. And evidently *Tat tvam asi* (i.e. "This thou art").

But this is also where our universe of space-time experience allegedly comes to an end, and the impersonal Universal-Consciousness, of which the cosmos at large is composed, begins.

And apparently beyond this knowledge nothing more of real substance can be said or known, and yet only by the eternal presence of "That", by which even this is known, can anything at all (by anyone) be "known"! Yet, if this is so, how could this "That", which we ever are, ever even be "born" at all!

The Unborn

Naturally our psychological selves fear our ostensive "deaths", and subsequent "physical" disillusionment! But the Life within us is fearless, understanding without words that it is an independent and eternal Force of diverse and impersonal potential "Beingness". This alone is our *Nisarga* "Natural State".

But once we fully discover that we are always within it, we resume our true "unborn" state and can no longer convincingly communicate our revelation of impersonal Universal-Consciousness to anyone who still believes that they are somehow living outside of it! But this then fairly begs us to ask the question, "Is anything at all then even real?"

Which is the Real?

Again The Question Arises, "Which is the real"? But by now it should be almost apparent that force of habit, reinforced by consensus of opinion, creates in us the strong tendency to approach all understanding from a positive "Exoteric-Objectivist" viewpoint (see "Exoteric" pg. 507, and "Exoteric-Objectivism" pg. 533).

Generally, without pause or reflection, we almost automatically assume that an objectively material universe is the "real", and of course should, even experientially, always come first!

It is difficult to realize, and then to hold onto the realization, that there exists absolutely no independently objective universe, and that there could never be an "object/subject" (i.e. our illusory ego) to experience it in any case!

Even if it were possible for some sort of material universe to exist independently of our consciousness, it would be impossible either to prove or disprove its existence. It could as well be said that this is because Universal-Consciousness, operating as individual consciousness in our body-minds, is necessary to cognize our existence. Because its presence is also clearly necessary to the "presence" of every apparent object within our consciousness.

The material presence of a psychosomatic mechanism is also clearly necessary to the presence of our binary minds. Which serve as platforms for consciousness to function in the deeper presence of the universe's essential primeval Awareness.

Therefore, positing the independent existence of a material universe is as fruitless as trying to prove the wholly independent existence of an immaterial consciousness! (See "Psychosomatic", pgs. 512, and 568).

This is because any such effort erroneously attempts to divide apparently contiguous, but actually interdependent natural processes, and then compare the resulting mere appearance of their implied differences![(26)]

Moreover, every aspect of our universe is most likely a manifestation of the same essential substance. This substance is in essence a universal uniformity of Consciousness and Force, or more succinctly, a Universal-Consciousness-Force, or simply "Universal-Consciousness".

Simply put, all phenomena, which comprise the appearance or manifestation of the universe within our essentially illusory individual consciousness, are only apparent to the primeval Presence (i.e. "Elsewhere") that is there to experience them!

But our phenomenal experiences inevitably happen not in our evident "present". For they can only occur after our psychosomatic instrument generates a *vritti* "experiential quantum" of apparently individual consciousness. Which requires a certain quantum measurement of space-time (see "Vritti" pg. 515, and "Planck Time" pg. 564).

The objects of consciousness that represent our persons and universe(s) are thus created at an incredibly rapid pace, an experiential quantum at a time. Like a "paint-by-the-numbers" picture performed by Universal-Consciousness as it operates within our subject-object minds. Therefore, if you think that the universe is real, then for you it is. And if you think that it is not real, then for you this could as well be so!

But by not participating in thinking at all, you may come to see our incredible universe directly as it is, nature's extended (or evolved) space-time system of energetic Universal-Consciousness. Which inevitably includes both you and me as it's apparent or virtual "subject/objects".

This is to say that we are merely implied "subjects". Because our subject (illusory "ego") selves must also be the objects of our experience, in order for us to be cognizant of them as they automatically operate within the wondrous workings of a sublime, yet necessarily consciously implied, Universal Manifestation.

But if this is really so, and everything is just Consciousness, then why bother investigating anything at all!

What Goes Around

THE BUDDHA AND THE SAGES have said that whatever exists has always, and will always exist. But since everything is observed to change, and to do so until it becomes unrecognizable as being whatever it once was, whatever could they mean!

Today's science tells us that everything reduces into elemental "atoms". And that atoms are reducible into subatomic particles – and various forces – which reduce even further into "strings" of pure energy.

But even these strings (of pure energy) eventually reduce back into the all-inclusive spacetimeless Singularity that suddenly gave rise to our energy string and space-time comprised universe. Evidently nothing can be either less, or more, than "That".

Therefore all that we can say, with any certainty, is that it (i.e. the "Singularity") contains the infinite eternality of our (as yet to emerge) universe. But to do so it must be in a "spacetimeless" condition.

Of this essential, yet truly strange "non-stuff", not much can be said. Except that it must be present in some spacetimeless fashion, even below the level of strings, in order for anything to exist at all!

Apparently its only "principle" is its own ever present impersonal Presence, as our universe's pre-singularity "Elsewhere"; upon which everything depends!

But if this is so, then (disturbingly) outside of our decidedly "personal" conceptualizations of an essentially impersonal universe, even "meaning" is itself quite meaningless!

Meaning

BECAUSE A FLOWER or a sunset has no intrinsic meaning, is it any less beautiful? Our life opens and closes with the presence of our consciousness, like a sunflower, with the rising and setting "Sun" of our inner sentient Light, for as long as our body and mind are present within the universe. Thus, while our lives may well be meaningless, they are nonetheless extraordinary living examples of our Universe's "performance art".

But if this is so, then do we not share a connection with our fellow creatures in which the same universal light of Awareness must also be shining? And does it not then follow that they too must be worthy of a reasonable measure of human compassion? Thus the Jewish Rabbis of antiquity compassionately reasoned *"Thou shalt not seethe a kid in its mother's milk."* – Shemōt (Exodus) 23:19.

Contented Cows

THE YOGI'S CLAIM that innocent cows are often; "A good deal more worthy…", if not at least more *peaceful*, than most (authors and other sorts of) men!

And speaking as a one-time hard working cowboy, from nights spent amongst contented cows in the lime green mountain meadows of the Santa Cruz Mountains, under gentle starlit skies, I must agree!

Indeed, if the late great Jñani Sage Ramana Maharshi is to be believed, evidently even his faithful milk cow was (in the end) just as Enlightened as he! (See "Cows", pg. 523).

(With gratitude to the Calif. NRCHA - National Reined Cow Horse Association - for promoting, and preserving, the grand old "American Cowboy Way" – since 1949!)

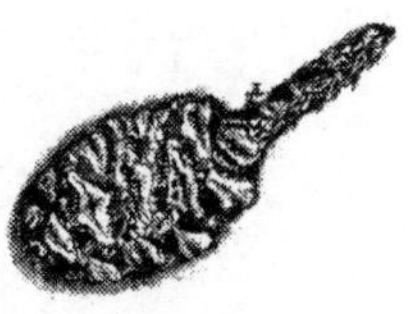

In Lovely Golden Light

BUT IN THE VERY MIDST of our forestland discussion, almost unnoticed, the lovely golden light and slowly gathering misty chill of evening has at last arrived. And it seems like a fortuitous time to light our campfire, boil some water for a bracing cup of hot coffee, and prepare our evening meal.

As I consume a simple dish of piping hot "chili frijoles" I can't help but notice that even the plainest faire often seems like "haute cuisine", when you are hungry in the wilderness!

And as the nighttime stars begin to shine and twinkle in the ever darkening sky above us, until the 'morrow, I wish you bliss, sweet dreams, and a gentle good night's sleep!

Santa Cruz Mountains
Northern California, USA

CHAPTER 2
"WESTERN RED CEDAR"

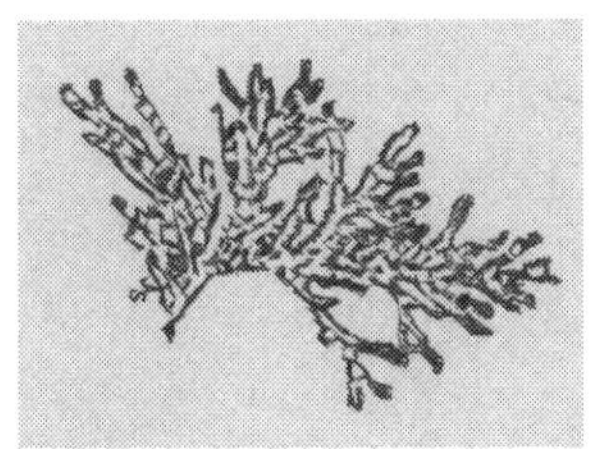

Dream Time

THE RUGGED ABORIGINES of the Australian Outback, and the diminutive Bushmen of the African Kalahari, speak of an ancestral Dream Time when men and animals could talk directly to one another. Everything was known and understood without the need of any language or teachings at all!

It is said that time stood still then, as it still does, for the giant Australian forest gum trees, the Blue Gums *Eucalyptus globulus,* White Gums *E. viminalis* and Royal Gums *E. regnans*, and the High-African-Plain Baobab trees *Adansonia digitata*. For these are among Mother Nature's most special trees, and the doyens say that they grow, with their roots intermingling still, amongst the ancient alluvial soil of the dark primeval "Dream Time" world.

For even now, the wizened older trees can recall a time, when men and trees once openly conversed. It is whispered amongst the gray haired elders that around the world, a few remote, uncut ancient trees still wait, to guide the sons of Adam back home again.

But this can happen only when humankind finally becomes tired of its bemused and painful wandering, inside the parched and desolate "Outback" of our waking dreams. For the oldest gums and baobabs still hold the ancient dreamtime memories of our primeval Tree of Life. And ever growing in the real Now, they steadfastly guard the pathway back to the deepest taproot of our inmost being.

Because life's taproot ever mingles with the dark, primeval life-force soil that is the provenance (origin) of our life's collective ancient Dream Time. These steadfast elder trees still faithfully hold the ancient occult knowledge that it is from this dark dreamtime soil that the great tribe of living minds sprouts forth, into the sunny daytime dreams of our everyday world.

But they will only share their deeper secrets with those who have not forgotten how to listen, to the quiet ancient wisdom of Mother Nature's primeval forest trees.(27)

(In respectful memory of the irrepressible Australian Naturalist, Steve Irwin).

Questions

When We Sleep, are we not everything within our dreams? And when we wake, is there really any difference? Perhaps sleep-full-ness and wake-full-ness are really only polar opposites of the same illusorily bifurcated (divided) consciousness. Could it be that we are in fact an ever present Awareness that has simply fallen behind knowing itself directly, in an eternally real and ever present timeless moment?

Have we perhaps simply lost our way, by habitually comparing this realer present with its immediate reflection in our short term memories, and thereby inadvertently imbuing its reflected image with the qualities of an imaginary realness?

Is it possible that this is how we lose our oneness with reality? A reality that automatically unfolds, but is truncated into an illusorily objectified and obscured diversity by our own subliminally spontaneous habitual actions of objectification? For are we not "It" and It us, eternally One, and no illusory "other"?

Might it be that our "fall from grace" is actually only an experiential disconnect from our *nisarga* "natural state" (pg. 510) of aware Oneness, into a reflective dreamlike tridimensional universe, only ideationally divided into an apparent experiencer experiencing an ideational experience?

And does it not then follow, that such an obscuration can only transpire, pursuant to an illusion? A mistaken perception, which artificially divides our consciousness into merely phenomenal, and not actual, "subjects and objects" of experience?

But how could a percipient causal subject that is itself an illusory phenomenal object, possibly ever be a truly volitional subject ego? Indeed, is any sort of "reality" then even *real* at all!

When we unwind the profound mystery of our own consciousness, even just this tiny little bit, it seems much more likely that even more of it remains unknown.

This may be, in part at least, what the Buddha was really getting at when he described our relationship with the universe at large.[28]

These questions therefore do seem worthy of our further consideration. But we will use an esoteric perspective from a rather well known, but at times tragically misunderstood, Middle Eastern source this time.

Torah, Talmud & Kabbalah

THE TORAH, TALMUD & KABBALAH represent the unique striving of a people for wisdom, and a deeper understanding of the physical and moral implications of the great mystery that is our universe.[29]

No doubt, great truth lies therein, but over time's long passing, as these teachings slowly spread around the world, they have collected accretions, like numerous barnacles clinging to the hull of a grand old ship.

Indeed, far too much has already been said by so many! So I seek here to strip away the plethora of words, until only the bare roots of verity (truth) remain, to expose the basic truth that in a very real sense, *homo mensura* "Man is the measure of all things".

In truly understanding this, we may even come to understand "all things", and can perhaps be done, once and for all, with all the rest!

בראשת
"Genesis"

I Am

אנ׳ מתקיים

Consciousness moves!

שה הכרה זזה!

Because consciousness moves, there is space.

הכרה זזה, יש מרחב.

Time appears within consciousness to measure space.

זמן מופיע בתוך הכרה למדוד מרחב.

Objects appear because I am the subject of space and time.

דברים מופיעים כי אני הנושא של מרחב וזמן.

I identify with my objects, and illusory egos instantly appear.

אני מזדהת עם חפוצים שלי, והאניה דמוני שלי עכשיו מופייע.

I suffer illusory experiences, and there is Universal suffering.

אני סובל ניסיונות משלים, ויש סבל עולמי.

שמות
"Exodus"

Suddenly there is only "I"

יש רק "אני"

No other remains!

אין אחר!

Illusory suffering fulfilled, I have at last come for my Self.

כאב משלה מושלים, באתי סוף כל סוף לעצמי שלי.

My time of wandering and seeking Self has begun.

הזמן שלי לנדוד ולחפס את עצמי התחיל.

ויקרא

"Leviticus"

In not singing, all songs are sung.
בלי כל שירה, כל השירים שרים.
By not worshipping, all worship is done.
בלי כל פולחן, כל עסות ייה נאשת
No longer seeking, the sought is found to be I!
כבר לא מחפש יותר, הנחבע מתגלה שהוא "אני"!

במדבר

"Numbers"

I am finally at rest.
אני עכלין דומם.
And because I am motionless, space disappears.
כשאשני מפסיק כל תנועה, מרחב נעלם.
In tranquility, there is no motion to measure, and time disappears.
בדומייה אין תנועה למדוד, וזמן נעלם.
No longer a subject, I have no objects, and dualism is no more!
באין יותר נושא, אין לי מתרות, ושינות לא קימות יותר!

דברים

"Deuteronomy"

No illusory egos remain!
שום אגו דמיוני לא נשאר!
The universe vanishes.
התבל כולו נעלם.
Suffering is no more.
סבל נעלם.
Perspective now true,
פרספטיבה עכשיו נכון,

I remain, but now there is no me.[30]

אני נשאר עבל עכשיב אין אותי.

Based upon the inspired teachings of Isaac Luria, and the late great Buddhist scholar, "Wei Wu Wei."

In loving memory of my dear Hebrew teacher Batya Eshel, too soon taken from this world by cruel סרטן *sartan* "cancer.

(See "Midrash", pgs. 510, and 551).

Religion

BUT PERHAPS I am being too hard on religion? Because there remains a profound need in civilized human society for meaningful rituals, and the reassuring continuity that is provided by time honored traditions. People need ceremonies to honor the major events in the ongoing cycle of their lives.

Birth, the coming of age, marriage, and death, for example, are all important "milestones" in our lives that need to be recognized. In the case of death, ceremonies are invaluable to help us with the process of grieving, and to provide a benchmark for the long and measured process of our emotional healing.

Without a doubt, the practical and emotional support provided by community is very important to all of these events. However, many people are understandably "turned off" by oversimplification where the deeper questions of our life are concerned.

As a result, I have met many people who can simply no longer believe in the traditional religious myths that are often still being propagated and presented as life's ultimate truths!

But while they may no longer believe in a literal messiah, resurrection, Armageddon, or even heaven and hell, they nonetheless continue to see the value in tradition, ceremony, and ritual for the individual, family and community.

Religion, for instance, teaches social responsibility, familial and civic duty, as well as providing socially important moral and ethical training.[31]

And indubitably, simple, pure and innocent faith is also a beautiful and sometimes even psychologically redeeming thing, and is most certainly of great value where our happiness is concerned, when it is balanced by reason!

In the main, it seems that the good done by religion fairly balances out the evil that it has done, in the name of many gods. But generally, only where religion ends, does true metaphysics begin (instead of an unlikely and wholly presumptive "spirituality"!).(32)

Indeed, the fully awakened Masters and Avatars, who are so often *apotheosized* "elevated to godhood" by religion, are nearly impossible to understand until the recondite nature of their original message is somehow understood! (See "Avatar", pg. 504).

Not many feel the need to work their way through the theological tangle of most exoteric religions to reach this more essential level. Many more are dissuaded by the superstitious taboos that are so often placed upon independent truth seeking.

It is especially for the sake of these few that I am laboring to provide a little taste of this deeper wisdom; a wisdom that only appears upon its surface to be at odds with the many *exoteric* "outer" teachings of religion.

Ancient East Indian Vedantic tradition and even Madame Blavatsky's rather imaginative, 1875 American "Theosophical" movement for example, provides an answer of sorts to the apparent esoteric/exoteric dichotomy, when they proclaim; *Satyan nahst paro dharma!* "There is no religion higher than truth"!

In other words, religion flows forth from the inner compulsion to search for our truer, "deeper" Selves.

Thus, the essence of all religion is really the search for our own essential nature, and once finding it, becoming stabilized therein. All of the various religious traditions are based upon concepts invented during this search.

But silence reigns when our true Self is at last rediscovered, and these intermediary metaphorical concepts are finally surrendered! This is the blissful silence of the abiding ملاس *Salam,* صلح *Sōlh,* or שלום *Shalōm* (i.e. Peace).

This is our essential Nature, and the final truth upon which the allegorical nature of all religious teaching rests. (See "Advaita" pg. 504, and "Vedanta" pg. 515).

Death

IN ADDITION TO SPACE-TIME, Mother Nature's system of manifestation, phenomenality, and religion, there is yet another important topic to discuss, the ever waiting specter of our own alleged "deaths". No lengthy discourse is required, for death is unfortunately quite familiar, and will soon enough come to us all.

Yet no topic, save the painful possibilities that may occasion our supposed death, is capable of striking such fear into the hearts of the unaware! For indeed, "survival fear" (the instinctual fear of death), its causes and post death possibilities, seems to be shared, in some degree, by all living things.

As my father, who rarely talked about his World War II experiences, once said to me, "Son, there are no atheists in a foxhole"!

Virgil, wise philosopher of ancient Greece, once pithily declared, "He who fears sleep, fears death." If you can accept his dictum with equanimity and also accept whatever next comes your way, then his rather stark advice may be "good enough", as far as it goes.

But if you believe that it does not go nearly far enough, then further inquiry into the mysterious matter of our inevitable so-called "deaths" is clearly required!

In the historical domain of human illusory subject/objects, fear, imagination, superstition and purposeful manipulation have created an aura of unnecessary anxiety where our so-called "death" is concerned. In point of fact, an almost universal aura of fear and uncomfortable mystery surrounds the ominous topic of our death.[(33)]

And while this is most certainly not an indictment of the American funeral industry, due to the widespread acceptance of these ominous sorts of unexamined ideas and resulting psychological pressures, people are often willing to spend exorbitant amounts of unrecoverable funds (and even go into debt!), in order to insulate themselves and their loved ones from what seems to be the irrevocable reality of cruel, uncaring death.

Sadly, they may even be pressured, by the more unscrupulous "merchants of death", into purchasing an expensive and elegant casket, when a simple pine box (or even a ceramic urn) might suffice!

But let us be brave enough to dismiss compassionately the unspoken taboo that surrounds our suppositious deaths and press on with our inquiry! Ironically, as unnecessary as this anxiety may seem, once exposed for what it is, it can serve as an impetus with the power to impel a deeper search for more reliable answers to the dark mystery of our rather indistinct deaths.

Evidently, the urge to find a way out of our apparent life and death dilemma is often just the beginning of a life changing "spiritual" journey. But other than the obvious eventual breakdown of our somatic instrument, just what exactly is death?

I

To begin with, it seems fair to say that, *"only that which is born will die."* And while trees are lucky that they don't have to give a thought to such things, where we dualistically intelligent human beings are concerned, all that is "born", and could ever be *born*, is our concept of a body and person, an "entity". But in order for our consciousness to conceptualize, our body and mind must first be present.

These two, our psyche and our soma, are thus inseparable aspects of our instrument of perception, the psychosomatic instrument that is our ostensible (outwardly appearing as such) "body-mind". But unlike a body, whether real or not, a concept cannot die.

It is either one collection of thoughts, or another, but it can never be other than the thoughts of which it is comprised. Because while our birth can exist as a concept, a concept cannot exist as our birth. In other words, it cannot exist in pre-thought actuality, whatever that actuality might "actually" be!

The same is true of the supposititious concept of our "death", and indeed even that collection of concepts which generally passes the phenomenal bar of acceptance as comprising the objective reality of our life.

But such a claim clearly deserves a better explanation, and it is really quite simple, because even our familiar person and universe exist only as concepts within our minds.

Therefore, we can accurately define our experience of death as just what it is, and all that it is, an ideation! Furthermore, within the obverse collection of ideations that is our experiential life, in order for anything to appear conceptually, a displacement in space-time must occur.

II

Our very first thought, of being or existing, is also the birth of all subsequent ideations, because what we experience as causality is actually the displacement of one thought concept by another, in our ongoing conceptual constructs of space and time. Hence, cause displaced in space-time will appear as its own effect.

This is to say, that our eventual death appears to us as the space-time effect of our own apparent birth. However, our birth, death, and whatever comes between them, actually occurs within the ongoing events of our each and every thought, which together comprise the entire manifestation of the universe, within our consciousness.

All of this conceptual busyness of objectification begins with the initial thought that that which we experience as "I" exists and happens only in a timeless present moment. "I", must therefore also be witnessed in the eternal present, which being timeless is also immeasurable, and is therefore spaceless as well!

All that could ever "exist", or be born, live and die, is a conceptual existence, which can enjoy only a dream like reality. Ergo (therefore), understanding the truer nature of our supposed birth and death seems essential, if we ever want to really understand and accept whomever, or whatever, it is that we truly are!

III

Is it not apparent then, that what we generally accept as objective reality is only a concept that describes the realistic appearance of our conceptual dreams, to we conceived dream characters who repeatedly appear within them?

If so, the grand cycle of our polar (i.e. having two sides) life/death, is really just a momentary illusion appearing within a subliminally inferred, and dreamlike, conceptual space-time.

The "fine point" here is that all of this can appear only to a witness that stands outside of space-time, and therefore also outside of any concept. So, you see, only that which is conceptually born, conceptually dies. But the true Witness of everything is always, in essence, the eternal Supreme Witness, which is the Divine, living and dying conceptually as both you and me!

Consequently, since we are always in essence the eternal Witness of the universe, what remains to be feared of our clearly conceptual deaths? But of course the body vehicle that supports our phenomenal minds, which generate and contain these concepts, will eventually break down.

Then it too will soon dissipate back into the self-Aware energy substrate from which it emerged for a luminous, but brief experiential time. (Within the familiar subject-object illusions of our individual consciousnesses).

Everything eventually returns back to the impersonal Universal-Consciousness of which it is comprised. Because Universal-Consciousness and the ever present Awareness from which it emerges is, quite fundamentally, if not yet demonstrably, the sum of our "Real" Universe.

Thus, the rumors of our supposed deaths, like Samuel Clemons terse retort when he read the false newspaper reports of his own death …"has been greatly exaggerated!"

All else is only the changeable phenomenon of veridical Reality's ubiquitous manifestation, being reflected within our dualistic subject-object minds by Universal-Consciousness.

But even our overarching Universal-Consciousness, which continually entertains itself with the ever changing wonders of the Universal Manifestation within our illusorily "individual" consciousnesses, eventually returns back to the pure and absolute underlying Supreme Awareness from which it once quite explosively emerged!

In the end, only the subtle foundational awareness of the Supreme Witness of our sense of "I" remains. So what then is the relationship of this impersonal subject "I", with the personal object I, that I so often mistakenly conceive that "I am"?[34]

> *"To fear death, gentlemen, is no other than to think oneself wise when one is not, to think one knows what one does not know. No one knows whether death may not be the greatest of all blessings for a man yet men fear it as if they knew that it is the greatest of evils."* – Socrates.

Amongst the Redwood Trees

Since The Early 1980's, there has been much ado in spiritual circles about two simple English words, "I" and "Am". Each one by itself seems not to amount to over-much, but when you put them together, in just the right way, there is apparently some real "magic" in their combination!

Let me explain, for each word is really quite old you see, and at one time, soon after they were first spoken, a very long time ago, they also had a rather exotic sounding third companion, *Tat*. These three words, put together, comprise the Sanskrit phrase *Tat tvam Asi* or "I am That".

But this concept is more precisely represented by the laconic (terse) mantra *sah-Hahm* "I Am"[35]; (see "sah-Hahm", pg. 512). "Not such a big deal", some might say.

But the phrase pops up again, closer to home this time, in the biblical palindromic sentence construction *"Ehyeh Asher, Ehyeh"*, or "I am That, I am"[36]. This is usually translated from the biblical Hebrew into English as "I Am, that I Am" ("that" with a small "t"). (I.e. a palindrome – like the words "mom" and "pop" – reads and means the same, in either direction.)

This gives the phrase a certain inscrutableness, that while interesting, is incongruent (does not necessarily follow), and is not at all in keeping with the character of its meaning, and probable early Indus Valley origin.

In any case, it seems only fair to point out that the phrase is written using the term *Ehyeh* to indicate the Hebrew future tense. Therefore, a literal translation into English is, "I will be that I will be", which is in the form of a "polite demonstrative" (requiring an external frame of reference).

When given this emphasis, superficially, it does seem to be a simple "declarative" (a statement or an exclamation).

However, while having G-d describe His existence with a terse anthropomorphic, "I Am because I want to be!" (Which is in keeping with the oftentimes self-contradictory exoteric Jewish, Christian and Muslim interpretations of the Bible). It is definitely at odds with the ancient esoteric meaning of the phrase! (See "Anthropomorphic", pg. 504).

Even if all things were equal, which they linguistically are not, it makes no sense for G-d (Yahweh, or Allah, etc.) to at once reveal and conceal "Him" or "Herself"!

That is, unless you are fond of Talmudic *pilpul;* a "tedious inquiry" that often yields illogical outcomes, and sometimes even proceeds from a questionable premise. (Such as the historically obscure phrase, "How many angels can dance on the head of a pin?").

However, if the phrase is translated, "I am *That,* I am", it is then in accord with its likely historical provenance, and principal esoteric meaning.[(37)] (See "I Am That I Am", on pg. 545).

It should also be noted that Biblical Hebrew has no commas, periods, or other noteworthy grammatical markings, and that a pause or emphasis is generally implied and eventually recognized by force of repetition.

Indeed, vowels are almost entirely lacking, and like the former, are inserted by familiarity, and contextual implication. This feature creates a formidable *mnemonic* "memory" obstacle for the inexperienced Hebraist, as any Bar, or Baht Mitzvah Jewish preteen can readily attest!

In other words, the latter translation allows for a proper descriptive focus on the subject of the phrase, which then brings forward its most probable meaning (see "Bar Mitzvah", pg. 505).

I

The second portion of the phrase, "I am", reveals an inner connectedness within consciousness that is apparently both intrinsic and necessary to any sort of Divine manifestation of our phenomenal universe.

That is, the first awareness of being is the "creation", or birth of man. It signifies the genesis of an only apparent individuality within the actual monism that is the veritable oneness of the Supreme. Which is in keeping with the Jewish mystical tradition of Kabbalah, Christian Gnosticism, and Muslim Sufism (see "Gnosticism" pg. 507, "Kabbalah" pg. 508, and "Rumi" pg. 448).

Moreover, it is also then in keeping with the phrase's probable origin. This is important, because great harm has resulted from the ideational divorce of a primeval principle of impersonal Consciousness from our universe, and the subsequent attachment of it to an imaginary deity.

In point of fact, there is very little real difference in the imaginary exoteric Christian, Greek, Hindu, Jewish, or even Muslim god(s)! Because these supposedly knowable "gods" are actually only, sometimes tragically harmful, appendix-like projections of our primitive human psyche.

Indeed, history bears witness to the unspeakable harm that the followers of fundamentalist religions, devoted to primitive superstitious deities, have sometimes wrought upon those inclined to follow another god, philosophy, science, or religion!

If we follow the latter interpretation, the sense of "I am" represents the first thought around which our experience of the universe is created. (I.e. By the addition of other thoughts and concepts into the secreted cache of our ever-growing long term memories).

Nonetheless, buy backtracking through the maze of our constantly accumulating thoughts and concepts, back to our split mind's principal generative thought, "I am", we may eventually maneuver into a position that allows for an almost complete disidentification with the virtual illusion of our personal individual consciousness.

And subsequently, a puissant reidentification with "That". Which is the actual reality of its impersonal Source, within Universal-Consciousness. However, it must also be noted that this cannot be accomplished by our intellect(s) alone!

Unimpeded, straightforward emphasis upon the essential oneness of all things is found in Indian Advaita, Chinese Taoism, Ch'an Buddhism, and Japanese Zen Buddhism. But because all religion begins with consciousness, the same essential teaching is also found in the Jewish *Kabbalah,* Muslim *Sufism*, and Christian *Gnosticism* (and with diligent inquiry, as the - late - brilliant Joseph Campbell discovered, amongst a very many others!).

Thus, like all objective things, even the esoteric roots of our subject/object selves' religious instinct, inevitably initiate from within the sentient experience of the primeval singular Awareness by which they are known.

When you walk into an undisturbed forest the first thing that you might feel, if you are calm and receptive, is the profound harmony that is the true nature of the Supreme Witness of the birth of our sentience in the notion of "I am".

From salamanders to trees, everything in varying degrees is "sentient". Every part of the forest knows that it exists. Everything within the forest knows its place and fulfills its destiny in harmony with its essential locus, as the ubiquitous Witness of the knowledge that "I am".

This locus is also ours. And if in quietness we allow it, nature may eventually remind us of our place within the harmonious orchestration of the great "Tribe of the Living", which is at once all around and within us.

"The two most important days in our life are the day that we are born, and the day that we find out why." – Samuel Clemons.

Beneath the Foggy Mountain Dew

Tonight, A Ghostly Gray Fog Rises Gently Up, from the not-too-distant Pacific Ocean, and slowly extends its cold gray tentacles inland. But it is once again being caught, and held tightly, upon the outstretched limbs, and needle "leaves", of the surrounding towering "old-growth" redwood trees.

Slowly coalescing there, it finally forms into little liquid islands that soon overcome the surface tension that holds them in their superficial places. And soon they begin to spill over, and drip a drizzling *pit-pat* of clear water onto the rainfly of our gray domed nylon tent. Which is snugly nestled on the forest floor below.

(From my childhood I have wondered what great secrets might lie hidden, snuggled away beneath this ephemeral "foggy mountain dew"!).

Certainly for a great many millennia before our tent made its momentary appearance on this lovely wooded mountainside, the redwoods have been quietly, and quite cleverly, watering themselves with this ghostly, foggy mountain dew. Thirsty giants, they require thousands of gallons of water to sustain themselves from day to day, and the rolling gray fog faithfully accommodates their prodigious watery needs! (See "End Notes", pg. 492, no. 102).

Quite recently, Berkley Tree Biologist Todd Dawson discovered that redwood trees actually absorb the majority of these innumerable "fogdrops" directly into their canopies. During the dry, Mediterranean climate like, warmer California summer months, this not only provides crucial water but also eliminates the work required, and precious energy expended, to lift the water against the force of gravity (often hundreds of feet) into the trees' lofty canopies.

But when trying to understand just exactly how this foggy wonder happens, it seems quite pointless to ask, "Which came first, the dewy fog or the thirsty rainforest?" For is this ancient hoary mist birthed either by the ocean, or by the land? Or does it spontaneously rise up into the moisture accommodating warmer air that floats above the nearby sea, because of the chilly offshore Pacific Ocean current that moves steadily Southward from the great frozen Arctic North?

This is a deceptively simple question, but it is nonetheless also a query with no end. For all things earthly are parts of a grand and interconnected "web", which at its fundamental basis is a subtle network that quantumly connects with everything else in our universe! Indeed, natural science may investigate "what" is happening within this cosmic web, but the essential "why" of the happening generally escapes the attention of even the most diligent researcher.

But those who do notice, recognize that a kind of "Gaia" consciousness seems to gracefully weave its way throughout all of nature. Everything, they notice, seems to be connected together in an apparently living and mysteriously aware *über* "super" system. But of course this substrate of cosmic "Awareness" is of a different order, more fundamental than, and even prior to, any sort of consciousness; at least in terms of our familiar conceptualizations! (See "Gaia", pg. 536).

Yet, by following the conceptual threads of an apparent causality, from event to event, what is happening on its surface, can be somewhat understood. However, this complex natural web is virtually endless, because each event is also connected, in probably inestimable ways, to other related, but often concealed, interconnected events.

So to understand what we can of nature's essential "whys", we must take a different path into the wilderness of the great unknown. Indeed, the energetic "threads" of this connectivity in nature branch off like the roots of an immense, cosmic sized tree.

But they are invisible, quantum sized roots, which nonetheless eventually stretch out through the fabric of space and time into every possible direction and dimension. (Until all of space-time is quite literally filled up with the energy of countless, nearly simultaneous events!). And although these events may appear to be contiguous, they actually overlap, like the circles of a Venn Diagram, but in a kind of enigmatic Möbius strip sort of fashion (see "A Möbius 'String' Paradigm of Space-Time", pg. 553).

The apparent seriality of any universal causality is thus quickly subsumed (included) by the actual interconnected simultaneity of our multiverse's countless events, whose discreteness is apparently only a perceptual illusion. But each event is also its own center, as it must be, in order to extend its necessary dimensions over enough duration, for it to either exist, or be noticed long enough for anything to happen at all!

Thus, all events, at all levels, possess their own identity or center as they must, in order to transpire, and since every "thing" is also an event, everything is evidently at the center of its own universe![(38)] (Dimension, remember, is the occupation of a certain amount of space by some particular amount of quantum presence, extending in differing directions, over a presumed finite measure of time).

Further, it would seem that to accommodate all of this rapidly multiplying activity, happening within three constrained dimensions, that a conjoined "fourth dimension" of space-time, which also contains the former, must expand with an ever increasing size, at an ever increasing rate! Or are we to believe that everything is happening simply because the concurrently increasing space-time either allows, or demands it?

I

At this point, perhaps we should begin to ask ourselves; "Are we then to believe the objectivist scientists, who postulate that the universe began from nothing, and will eventually expand endlessly back into a dark oblivion?"[(39)] But is this any more difficult to believe than the exoteric "Creationist", who claims that the world was created by an anthropomorphic god in just seven days? Certainly, the first hypothesis seems as unlikely, if not as absurd, as the latter!

The attempt to reconcile either position with common, or even logical sense is far too cumbersome. Therefore we have no recourse but to cut away the extraneous chaff by using Occam's Razor, which instructs that, "the simple, which accords with reason, is most likely to be correct, among competing hypothesis".

Therefore, both perspectives are rejected, as we continue to dig, searching deeper into the darker soil of probable reality, to arrive at a more likely, and perhaps even "testable", exposition of the basic experiential truth of things.

II

On the most fundamental level, each event is a quantum of quanta (amount of consciousness energy). This is because matter is a form of energy too, and time is our three-dimensional measurement of an energy event, as it expands through the fourth dimension of our experiential space.

We have no other choice, but to note change in this way, since we cannot otherwise directly sense in any fashion, the fourth dimension. It is most likely that the durability of any subatomic or quantum energy expression is also connected to the manner of its participation in the web of three-dimensional causality.

That is, the more durable, or some would say "dense", that the energy expression is, the more likely it will come to occupy enough space-time to enter into the tridimensional web of causation, and the longer it will appear to be there!

In other words, there appears a quantum, *vritti* in Sanskrit, or *nikkud* in Hebrew, which is the smallest possible point of pure, Consciousness-Energy. This Consciousness-Energy is the subtlest form of energy as it emerges into dimensionality. But it does not "pop" or "bang" suddenly into existence from nowhere! It manifests from where it ever resides, in the ostensible fourth-dimension.

This is the portion of our own indispensable witnessing Awareness that is indirectly accessible to our individual consciousness. In Sanskrit it is called the *Nisarga* "Natural Awareness" by which it is also simultaneously aware of itself, or self-Aware. Or if you like, by being intrinsically aware, it is then simply the pure Awareness by which we are simultaneously aware of both the contents of our consciousness and of ourselves as our consciousness' ego experiencer!

III

Dimensions, 1st, then 2nd etc., are apparently built up by ever increasing quantum layers or membranes of Consciousness-Energy, until finally reaching the three-dimensional subatomic level, where they become subject to the sequential causality in space that we experience as time. The familiar chain of events then appears within our consciousness.

It should be remembered, however, that even these so-called "dimensions" exist only as creative measurements within our minds; that unique expression of impersonal Universal-Consciousness that automatically appears within our individually manifested psychosomatic instrument(s) of perception.

The foundation of this natural system of beingness, insofar as we are concerned in this enquiry, is its apparent manifestation as the psychosomatic phenomenon of our "person", and the subsequent conceptualizations that emerge within our minds. Which are then assembled into our ongoing experiences within space and time.

But do not be misled into objectivizing our body-mind, because it is really the biological equivalent of a cybernetic "avatar". It is the conglomerate subject concept of our dualistic minds, by which Universal-Consciousness enjoys the vivid and compelling experiences of our essentially "illusory" individual consciousnesses.

Thus, the true dividing line between our experiences of inner and outer, and subject and object, remains unclear because it too is just another artifact of our dualistic perceptions! To ascertain more fully the inner connectedness of Universal-Consciousness, which links everything together in our minds, we must step outside of the familiar and look from a "right angle", in the further direction of the fourth dimensional "in" side of everything.

This perspective of withinness contains our third dimension, within an essential fourth dimension, as an in-"tuition"; in Sanskrit this action of manifestation is called *prajna*. This is the reason that it is difficult to comprehend how the manifestation, of our person and universe, can be at once within and outside of our minds; it is all a matter of perspective! In other words, from which dimension are you looking? (See "Prajna", pgs. 511, and 566).

From this further angle, it is apparent that space-time, as we perceive it in constant serial motion, as events apparently connected in this movement by causality, is a complete fiction! Moreover, everything that is connected to a time concept is evidently also incorrect in its essential incompleteness.

And when the essential, but nonetheless illusory time element is removed, space collapses, and our perception of a three dimensional universe is taken up into the apperception of a fourth dimension (see "Zeno's Paradox", pg. 597).

Thus, our Enlightenment could be said to be like waking up from a dream. But the Awakened have said that it is only seeing directly, or *in*-seeing, what is already completely there within our ever-present substrate of primeval Universal-Consciousness. Hence, any true "Awakening" can only be subsequent to our final remerging into the primeval Awareness that is the source of our sentience, through its extension as Universal-Consciousness.

Science may tell us "what" and "how", but only we can discover within ourselves *"why"*. Yet I must caution you, it is important to remember that the Universal Manifestation is not at all as I have just described it. If it were, it would only be another concept and could not contain even the slightest trace of any real verity!

IV

It is crucial that we understand that our essential Awareness cannot be conceptualized, it can only be realized directly as our own true and natural state. There is no practice or discipline that we can do to attain it, because it is already exactly who and what we already are, and we can only obtain whatever we are not!

But, perhaps I am also writing this to those who have already bravely tried, even mightily, to "awaken" themselves, and finally have wisely given up trying? Or to those who may have even tried to do so with the disingenuous guidance of a spiritual con artist, or the empty promises of a well-meaning but misguided "guru".

Those sorts of disingenuous and selfish people that are generally hungry not only for money, but for the power to deeply influence other people's lives as well! These are the false guides who are actually quite lost themselves and don't even yet know it, or more often don't even care! Yet without conscience, they continue to sell the tainted water of misinformation to people that are already drowning in a river of confusion!

To you who may have sadly suffered in this way, even after trying so very hard, I must reluctantly say; "I certainly know how you feel, because the task is not only difficult, it is patently impossible for *anyone* to do!" But to better understand what probably went wrong, without "throwing the baby out with the bathwater", please consider' for a moment, just "who" is trying to awaken "whom"?

Is it not apparent then, that our illusory little "character-self" (i.e. our so-called *ego*), which regularly appears in the Universe's lucid dream, cannot possibly step outside of the durable dream that it is a part of? Which, in any case, is not really even its dream at all!

Albeit (even if) the real Divine Dreamer may seem to be presently experiencing the dream as if it were? In other words, no matter what technique may be employed, if you unquestioningly believe what the considered mind tells you, you will most likely be disappointed in the end!

Instead, we must not give away our executive powers to others, or strive in this particular regard to do anything at all! While also being aware that purposefully doing "nothing", is still doing *something*! Some of humanities Awakened ones have advised that a way of "nonaction" is to simply allow ourselves to be what we really are, by ceasing to act and think of ourselves as what we are not.

Therefore, understanding the implied temporal actions of the imaginary person, who persistently appears before our ever witnessing Self, is evidently quite essential! One rather wise Master has said that all that is needed is a "shift in awareness" that can come about not by doing anything, but by clearing up all of our doubts and misconceptions, which keep us from realizing that we are already fully awakened![40]

Apparently it is only this kind of "nonaction", the achievement of which is really a durable shift in our perception and attitude, that presents a remedy of sorts. But in the end, all of the awakened Masters attribute our essential metanoesis to an act of grace.

In other words, Universal-Consciousness spontaneously does whatever may be finally best accomplished in our successful "not doing", or more accurately in our not "not-doing". Because this is the doing of neither something, nor nothing. And it is therefore something that no one else could ever do for us, and that we evidently could also never even do for ourselves! (See "Non-Action", pg. 557).

Quite a Ride on a Windy Day!

The Sun Is Now Brightly Shining, and golden shafts of light gently touch the earth, in ever shifting glowing patterns, between the leafy penumbra being cast down from the forest canopy high above us. A living ceiling of green stretches out in every direction as far as the eye can see! And as patches of powder blue sky peek shyly through the leaves of green, the leafy bower above us is slowly transformed into the verisimilitude (likeness) of a colossal, natural "stained glass" ceiling.

This massively verdant arboreal *ventana* "window" is cordially supported 'round about, by the numerous giant ruddy columns of ancient, tranquil redwood trees. In the midst of these forest giants there is a sacred silence. The thick sienna colored bark batting, which clads the giant sequoia's trunks, fits like a snug fireproof overcoat. All the while absorbing the ambient sounds of the forest, as it quietly protects the massive Methuselah that is safely snuggled away beneath its rough, reddish-brown folds, through the passing vicissitudes of the ages.

Naturalist John Muir and wildlife photographer Ansel Adams likened the quiet stately redwood rainforest to a peaceful "Cathedral of Nature". In this forest cathedral, we find ourselves, quite serenely, in the very best of company amongst the chocolate brown Newts, irascible Gray Squirrels and sapphire blue Stellers Jays (mountain "Mocking Jays") that make this emerald green redwood rainforest their permanent home.

The air here is crystal clear, filled to overflowing with chemically fresh oxygen from the steady photosynthesis of countless evergreen needles, and numberless deciduous leaves. Indeed, there is a certain ambiance in the air within a redwood forest that evokes the comfortable feeling of being home. Perhaps it is the visceral remnants of some ancestral memory, of a not too distant time, when it was far truer than it is today?

Through a chance break in the extensive leafy bower above, the top of a distant redwood giant suddenly appears, slowly swaying in the sprightly coastal breeze, which so often blows, just above these ancient American redwood trees.

It is an intermittent breeze, striking only those trees strong and bold enough to poke their heads up, and into the warm sunlight that still shines brightly, high above the relentless shade of the dense rainforest canopy below.

'Though from afar the swaying seems slight, from experience I know that the distant massive tree's top is really rocking back and forth perhaps a half-dozen feet or more. It is truly amazing how flexible such massive trunks can be, many of which are greater than six, eight, or even a dozen feet in diameter! To any creature in the redwood treetops, it can be quite a ride on a windy day!

Anyone who has climbed there is likely smiling and nodding as they read this, for we share a rare experience that leaves its expansive mark upon you, like an astronaut stepping out onto the far side of the moon. And like the dark side of the moon, the high redwood canopy conceals a unique and marvelous hidden world (but perhaps this is an excellent topic to explore on a less windy day?).

It is always fun to muse and reminisce, but alas, it seems that every picnic also brings its own supply of bothersome "ants". For the sonorous bellowing of a noisome distant radio now quite suddenly diminishes the redwood rainforest's antediluvian peacefulness!

The excited staccato cheering of some (quite likely inebriated) sports fans accompanies the loud radio chatter. As it echoes upwards through the ghostly white trunks of the birch trees that grow on the slope below our little mountainside camp, from another campsite located somewhere in the canyon far below us.

I

The awakened East Indian Sage Nisargadatta Maharaj once ironically observed, "Consciousness is hard to bear, so people keep their minds occupied with something every minute!" I am as fond of sports as the next guy, but all that I can presently think is; "Really! It's certainly obvious that you guys care deeply about whatever the score might be! But must the entire quiet mountainside be made aware of it too?"

Indeed, the quiet mind in its peacefulness seems to invite a greater awareness that we are in fact not living, but are "being lived". The greatest Masters, our only real authorities on the "whys" of life in this regard, say that this Natural Awareness, which is living our lives as us, is "That" which in impersonal reality, we always are. Otherwise what we assume to be our very own personal life, is really only a momentary figment, appearing within our (only illusorily) "personal" minds!

It has also been said that besides learning the answers to life's "whys", other than achieving a quiet and peaceful mind, there is precious little else that we can do. Metanoesis (see pg. 509) will "sooner-or-later" come, they say, to the "worthy", and by *worthiness* the aforementioned seems to be the plain sense of what they are saying.

"Oh happy surprise!" For the irreverent electronic cacophony, and accompanying verbal mayhem have quite suddenly ceased, and quiescent peace now reigns again in the forest primeval. Nonetheless I simply cannot restrain myself from exclaiming out loud into the (now quiescent) mountain air, "Ah, golden silence!"

Trees and Threes

HAVE YOU EVER NOTICED, from a certain distance, that most conifers begin to assume the shape of a narrow upward pointing triangle? This pointed shape is called *fastigate*, and the wider shapes of other trees, *deliquescent*. The discerning eye notices that many things in nature, besides trees, often come in "threes". But this should come as no surprise, for is the universe that we sense, not sensed in three dimensions?

Hence, the up and down, back and forth, and in and out of things, is apparent to us everywhere! Indeed, the essential three of our genetic "Tree of Life's" evolved cognition are, *perceiver*, *perception* and *perceived*. And like removing the supporting trunk below the canopy of any living tree, removing any one of these "basic three" of our genetic "Tree of Life's" critical cognitive faculties renders it not only dysfunctional, but perceptually quite dead as well!

So, does it not seem passing strange that we apparently cannot perceive that very aspect of time's three dimensions that presumably contains our "real" tridimensional space-time universe. Which we nonetheless experience as if we too exist within it? (I.e. in the fourth dimension). All that we can ever truly perceive of time is the anticipation of a future space that is already in the past, both of which possess no substantial reality, since they exist only in our imagination, and in our memories.

But our memories and imagination are effectively "unreal", because they are substantially nothing more than the conceptual musings of our mind. Put another way, one thought is actually about the same as another, and both add up to nothing (of any real substance). Yet this is not to say that our thoughts may not be filled with whatever meaning that we think they might contain, or that they have no real importance either!

Moreover, without the presence of the "I" to perceive our thoughts, they have no existence whatsoever! Indeed, it is the perceiving of the *thinking* of a thought that imbues it not only with meaning, but with existence as well. Therefore, the independent personal reality of our thoughts is evidently their intrinsic nothingness, but for the "somethingness" which they seem to become, in the presence of a truer impersonal Reality.

Yet if this is so, then to what space-time reality could we possibly be referring? Why, the presence of our own impersonal and veridically non-conditional "Reality" of course! That's because the reality of our own essential Presence is the only experientially verifiable reality that there ever really is. This is also to say that the purported reality of our present experiential moment cannot be within the sequence of time that we perceive.

This is because, even prior to the essential delay for the cognition of our every thought, the time sequence that we belatedly experience (as our "present" moment) evidently originates from within the ever inaccessible, yet always contingent (and therefore always in-the-future) "present" reality of the universe's foundational Universal-Consciousness.

This is what we then mistakenly assume is our very own personally substantive future! But, of course, it could also then not be within the ideational present moment that we commonly perceive. (And please don't worry if you feel that you have not yet fully grasped all of this rather "slippery" space-time-stuff all at once. Because we will soon be revisiting it, in a much more direct fashion!).

But the (Supreme) Presence that is intrinsic to our every experience is in reality prior to the individual consciousness that contains our experiences within space-time. Which is even prior to the Universal-Consciousness that allows it to appear to be there. And it is this constant and ever ubiquitous, spacetimelessly impersonal "Presence", which is also the superlative Reality of us all!

And so you see, the triangular shape presented to our senses by a tree, could also serve as an "arrow" that points us in an inward direction. Which eventually leads us to a better understanding of our natural world. But it is also a world that only appears to exist tridimensionally all around us. As we in fact view it not only "internally", but also universally, from within an ideational fourth dimension. And moreover, we are quite likely even aware of our viewing of it from an independent "fifth"!

Therefore, must we not also then be viewing it from the before, between and after side(s) of our objective thoughts as well? Indeed, upon closer inspection, at first curious, and then even curiouser becomes the everyday perception of our astounding universe, and our surprisingly not-so-prosaic "persons" as well!

Camping Out

Perhaps The Real Charm in "camping out" is precisely why so many people take such little pleasure in it. It is the utter simplicity of the experience. Living is divested of its complications, 'till only the bare bones of potential verity remain. "When I'm hungry I eat, and when I'm tired I sleep!" Such was the simple advice of a "Realized" Zen Master, when he was asked how to find the ultimate truth.

What he is really saying to us is that our *Satori* "Awakening" is not something to be found on a mountaintop, or to be acquired by some arcane spiritual practice. Satori comes upon us of its own accord, if it will, in the everyday process of living our lives. But of course preparing ourselves, by making ourselves available for it, is also an essential part of Satori's natural process.

It is, however, essential to understand that the apparent "we" can do nothing to compel the process of our own awakening! But if it is so that everything we experience is an illusion due to its incompleteness, or because it exists only as a "reflection" of a deeper reality, then it may also be true that everything in our lives is in fact an indication, poor though it may be, of a far greater impersonal Reality!

In this way, every moment presents an opportunity for Satori to reveal itself, within the unceasing flow of experiences that comprises our lives. We have only to live our lives with the same simple directness that is the real charm of camping. Because being in tune with nature's Taoistic charm is no doubt an indication that we are living correctly.

For it is the stirrings of the intuitional *in*-seeing of things, which evidently presages the final revelation of Satori, into our little reflected sequential "waking dream" moments of space-time. Because only this mirrored moment, where we cognitively dwell, contains the possibility of sparking an immediate awareness of our phenomenally obscured, yet spacetimelessly real, "present" moment.

This quandary also has no real answer, just as the experience of any subsequent Awakening cannot really be explained! Therefore, when a Zen Master is asked, "What is the meaning of Zen?" he might simply reply, "The Cedar tree growing in the courtyard."[(41)] Because the tree, of course, like everything else, is really "growing" in our heads!

"There is an ecstasy that marks the summit of life, and beyond which life cannot rise. And such is the paradox of living, this ecstasy comes when one is alive and it comes as a complete forgetfulness that one is alive. This ecstasy, this forgetfulness of living, comes to the artist, caught up and out of himself in a sheet of flame; it comes to the soldier, war-mad in a stricken field and refusing quarter..." – The Call of the Wild, by Jack London, 1903.

Ignorance and Bliss

Most People Happily Pursue Their Lives, ignorant of the fact that it is they who are being "lived", for the most part unaware that they possess no independent volition. Then one day something happens and they become discontent.

Possessed by a feeling that there must be more, they become convinced that something is missing from their lives. Then the misery of searching for whatever "it" is, begins. Until finally, often years later, and usually in utter frustration, they stop searching for it!

It is only then that the opportunity begins to exist for the missing component of our blissfully undivided Reality to come searching for itself within us. But once this process of transformation, by the shifting of the locus of our awareness back into *it*-Self begins, in this too we will soon discover that we also have no independent volition!

Then one day, our locus shifts. This is the sudden inevitability that is the realization of our truer Self within the eternal Present, which is the fulfillment of our complementary Self-ignorance.

Everything else is eventually discovered to be nothing more than a persistent illusion, which automatically transpires when we are using our brain's evolved faculty of pattern recognition to try to make sense of our inferential realities incompleteness.

Thus, the evident ignorance of our true Self, never truly exists! Simply because, our veridical "Enlightenment" never ceases to be That (Supreme), by which our apparent ignorance is perceived![42]

But ignorance itself is rarely simple, and since tomorrow is a day that memorializes, in a cautionary fashion, the disturbingly dangerous and persistent ignorance of war, it seems only fitting to observe it together, in nearby Memorial Park.

Memorial Day

We Americans, cherishing our freedom as we do, wisely choose the noble eagle to symbolize what we hold most dear. (Indeed, this too is why, long before America was even dreamt of, that a brave proudly wore his eagle feathers for all to see!)

And today, as I watch an eagle soar up into the sky, "wing surfing" on the thermal currents that rise up in great transparent waves from the sun warmed redwood forest below, I feel a swelling pride in the precious liberty that we Americans strive to live by.

Beautiful Memorial Park, now embracing us with its serenity, is dedicated to the brave Americans that have given their most precious possession, to defend the priceless concepts of liberty and justice for us all.

For these two basic principles of freedom are also the reflections of humankind's highest nature, and are thus the very presence of inviolable truths within our thoughts and actions, in this, our universally shared "Waking Dream".

Yet inevitably, in our strange, mutually shared universe(s), there will always be "good guys" and "bad guys". For such is the peculiarly "polar" nature of our shared experiential world(s)! On the consciousness stage of our apparent universe, sometimes the actors may change places over time, but it remains nonetheless essential to be one of the good guys![43]

And although our "good guys" sometimes have to fight to protect the innocent, and to preserve justice and truth, we must also never forget one of the most important of war's grim lessons that; "The ends should never be used to justify the use of unjust and inhumane means!"

I

In pursuing our conflicts, whenever unnecessary brutality prevails, we diminish our own freedoms and precious liberties. Indeed, we insult the honor of the legions that have died to protect what we would, in anger or ignorance, unwisely give away! These truths are universal in nature and thus are the property of all humankind. But, as wise Cicero of Rome once trenchantly opined; *Inter arma silent leges* "Among arms laws fall silent."

But on this Memorial Day, while laboring over my writing, my eyes are beginning to fill with involuntary tears, as I recall the steadily growing loss of my military brothers as they were swept away, one-by-one. (To be forever lost in the dark and bloody waters of the ever escalating Vietnamese Civil War.).

How sad that each subsequent generation of Americans seems to have its very own war to attend to! In retrospect, because the price of war is so dear, we simply cannot afford to surrender reason to fear, nor allow ourselves to be manipulated into any unnecessary conflict.[44] When we ignore the hard won lessons of the past, we do so at our own peril!

Indeed, all around me, like the steadfast ghosts of the lost warship Alice Marr, I can almost feel the restless spirits of the fallen men and women, which this place seeks to honor. I can almost feel their outrage, their powerful love of family, tribe and country, and their profound concern for our welfare. And as they try, perhaps, to find some way to reach my ear, it is that portion of my heart that all decent men share that ever honors freedom and justice, which cannot help but heed their warning cries!

And so I speak now for those brave souls who would, but are no longer able. I speak for them with all the hope that as an American I can share, that we will always hear freedom's call for a return to justice when it's needed and thus preserve our most precious liberty.

For liberty, justice and freedom are not unlike the stars that continually shine above us, beyond the intervening fog, on even the darkest of dark nights. And 'though we may never actually "touch" the stars that perennially shine above us, it is by their bright light that we must find our way back to just and honorable actions, from the darkness into which we might unwisely tread. (At the peril of our own precious liberty, and even perhaps, the executive foundation of our humanities reflectively sentient "soul"!).

For our most important battles and meaningful victories are not won by the "force of arms" alone, but also by the *honorable* ways in which we use whatever powers that we may come to possess in our shared, cognitively virtual world(s).

But despite what anyone may ignorantly claim, it is never too late, to build a better America, to resume our role as a respected world leader, to partner with other nations to save our world, and to achieve a more peaceful and prosperous tomorrow.

We must never allow it to become too late to fulfill our social duty to the future generations of humanity. Who will also need to work towards making real the elusive human dreams of ubiquitous prosperity, enduring freedom, and lasting peace for all. We must never allow it to be too late! But we also must not wait (before we finally act), for perhaps we already tarry for too long…

II

There is a certain grim wisdom that eventually comes to a warrior, if they survive long enough to learn it! As my *TsaLaGi* uncle Leonard taught me, when I was still a child, a Cherokee Indian Brave wears his eagle feathers with pride, because they are earned with honor, as well as with steadfast courage.

But the heady (and unfortunately, not that uncommon!) toxic combination of pride, fear and fanaticism is easily confused with truth and honor. Nonetheless, and perhaps as a result *even more so*, life must always be honored over death because, "War knows no end but death", as wise Plato long ago observed.

And as almost every veteran warrior, and West Point graduate knows, even if we were to try with all our might to adhere to Augustine's "Principles of Just War", when the battlefield need for selfless courage is finally done, the relentless violence of war's memory can only through honor's "sacred way" finally attain some measure of lasting peace.

Every seasoned officer will also tell you that the purpose of the military is not to "wage war", but to restabilize and return human society to the normalcy of peace, when for some reason the peace has been disrupted. And moreover, they also will agree that while acts of valor should always be honored, that war itself should *never be glorified!*

Wars have almost always been fought for control over land, food and possessions, and in essence, rarely over the ideologies that are offered up for their justification. Indeed, wars, like America's shameful Vietnam fiasco, that have been fought over mere ideologies, have proven to be especially despicable in their brutality. But today, humanity stands on the brink of our own destruction if we cannot redirect our present errant course.

At the very heart of our dilemma lies the "quiet-but-puissant" struggle that has been going on between religious superstition and science for several thousand years. This is the conflict between reason and superstition that began with the rise of sentience within our species ever evolving human mind. Nowhere is this more evident than on the present battlefields of the Mideast, where the probable last chapter of the conflict is now being fought.

It is a "war" that humanity will likely win as a more enlightened civilization takes hold around the world, but at what price? It is up to us to use our power responsibly, to minimize the suffering whenever, and wherever we can!

Courage is laudable, but honor is sacred, because it reminds a combatant of their humanity, and retrospectively even has the power to heal the stubborn inner "wounds-of-war", that no one else can see (and which the "experts" are now rather superficially calling "PTSD"!). Moreover, only honor has the power to shield (somewhat) a veteran warrior from the inner ravages of war. Because, only the truth of honor can protect the inner innocence by which we know not only that we are, but also "Whom" we really are.

Losing this, we lose ourselves. Indeed, throughout all the apparent days and changes of our lives, the principle of "identity", our essential presence within, remains constant and unchanged. And this tender essence is exposed to every experience within the consciousness by which it experiences its own existence within the universe.

But until we experience our own so-called "enlightenment", and then fully Awaken to the fact of just "Whom" it is that we truly are, we will continue to remain trapped within the dualistic subject-object grasp of our ephemeral consciousness, imagination, and memories.

So it is important to understand that only when the heart of a warrior honors peace above violence, will his actions on the field of battle be steadfastly honorable, rather than most often simply "brave".

Therefore, I must conclude this inevitably insufficient eulogium of valor, with an aging warrior's simple, but cautionary prayer; "May G-d forever bless America, that it shall ever be the 'land of the free', and the 'home of the brave'. But may we always balance our reaction to violence, and the sometimes necessary, if not always saddening, 'Call to Arms', with honor, justice, and *compelling reason*. And never forget *Officium, Honorem and Patriae!* (Duty, Honor and Country) – Amen

"The nation that insists on drawing a broad line of demarcation between the fighting man and the thinking man is liable to find its fighting being done by fools, and its thinking by cowards."

– Sir William Francis Butler

(Please see, "Piae Memoriae", pg. 600).

Tao Te Ching, Ch: 31

"Weapons are the tools of violence; all decent men detest them. Weapons are the tools of fear; a civilized man will avoid them except in the direst necessity and, if compelled, will use them only with the utmost restraint. Peace is his highest value. If the peace has been shattered, how can he be content? His enemies are not demons, but human like himself. He doesn't wish them personal harm. Nor does he rejoice in victory. How could he delight in the slaughter of men? He enters a battle gravely, with sorrow and great compassion, as if he were attending a funeral".

~ Lao Tzu ~

(Dedicated to the brave citizen IDF's (i.e. Israel Defense Forces) of *Medinaht Yisrael* the "State of Israel", who are daily forced to do what they must to protect the innocent. But who most often do so not only with courage, but also with the heavy burden of a compassionate heart).

Taking and Giving

THERE IS A "GIVE-AND-TAKE" to things in nature, a natural rhythm that determines when it's right to give, and when it's time to take. There exists a quiet harmony, a balance that is thus maintained. In trees you can see it in the uptake of water and nutrients from our earth's life giving soil.

On their "up" side, for example, trees give moisture and oxygen back into the air from which their leaves have taken in carbon dioxide. And by giving back moisture and oxygen to the air through the *stomata* (pg. 514) of their leaves, trees take water back in again through countless tiny root hairs that grow below the ground, on their "down" side, where no one sees.

From millions of miles away in space, our bright sun gives out light that is taken in again by *palisade cells* (pg. 510) in the far away leaves of earthly trees. The solar energy thus given is taken in, but it is soon given back out again, transformed through the green photosynthetic "prism" of chlorophyll into the carbohydrate food stuff that sustains earth's trees. Green plants everywhere take in sunlight and carbon dioxide, and give back into nature the food and oxygen that sustains the life that encircles our planet.

But the life that nature gives is soon taken back in again, as the expired forms which once sustained it give their life material back into the "star-dust" soil from whence they came. The enriched soil then gives back its myriad disintegrated life forms as nutrients to sustain the life of earth's trees, plants and fungi. And the endless natural cycle of "give and take" is then, for a time, complete![(45)]

I

There is a harmony in nature, a balanced rhythm of repeated change, through a system that operates by giving out and taking in energy at the right time. Yet nature does not own a clock and knows nothing of our human concept of "time". Which is but a poor mental mimicry of the harmonious cycles and orderly rhythms that we observe operating all around us within nature.

The circling of our planet around its life giving sun, we take to be a "year". The rotation of our earth balances its own essential gravity (and much more!). And we take its rotation from sunlight to shadow to be a "day". A day we then, for more anecdotal reasons, presume to be twenty four "hours". Which we can arbitrarily now do, because the turning of the earth, which we have taken into our minds, has been turned into a "concept", and in consciousness we can now do with it whatever we please!

This part of nature's rhythm has been converted into insubstantiality. It has become a portion of our inner illusionary, conceptual universe. A "universe" that may or may not even approximate reality! Indeed, except through us, nature knows nothing of the cognitive illusions of consciousness that are constantly operating within our minds. Which we nonetheless take into our trust as being the "reality" of a wholly physical universe.

When we act upon our thoughts we give back energy into nature's earthly system, which our bodies connect us to. But we rarely do so in harmony with nature's subtle rhythms, and may even attempt to force the system to adjust to the selfish whims of our imaginary (conceptual) universes.

And then, there is no end to the "mischief" that is done! We are ever so connected, and the experience of independence from the grander "System of Nature", of which we are but a tiny portion like a single cell amongst the trillions within our own body, is just a fiction!

Thus, independent existence and volition are actually the mental equivalents of independently operating mirages. They are merely persistent cognitive illusions that are created by the automatic *psychosensorial* "mind and sense" actions of cognition that define our tiny (humanly somatic) portion of the overarching System of Nature. Ironically, the entirety of the natural system that we personally partake of exists as a virtual phenomenal manifestation within our binary minds.

But it does so only through the auspices of the impersonal Universal-Consciousness of nature. And although nature's entire system may seem to be individually ours, we are only simultaneously aware of our subjective selves and our objective universe because of the witnessing presence of the universe's subquantum substrate of intrinsically self-sentient Noumenal Awareness. Indeed, this is what enables our apparently personal body-mind systems to function, and so, to exist experientially!

If this seems at all hard to fathom, it is simply because "Mother Nature's" impersonal Universal-Consciousness exists in a superior dimension, as well as within the three dimensions of "Her" mental manifestation.

Which are the occult boundaries of the spontaneous perceptions that define our phenomenal universe.

Nature's independent Universal-Consciousness thus creates an endlessly "streaming" cascade of realistic and seemingly "personal" experiences. Which it then experiences through us, as we cognitively move through the fourth dimension. But we cannot see nature's implied "fourth dimension".

So we must sense its implied dimensional facets, of inner volume, apparent separation, and causal connectivity, by measuring what we experience as invisible space, with the serial and therefore apparently "causal" motion of time.[46]

Indeed, from beginningless beginning, to endless end, the totality of nature's system, it seems, is the function of an independently involved, or involuted, Universal-Consciousness!

II

So perhaps we should avail ourselves of the opportunity now, to re-familiarize ourselves with nature's own Universal-Consciousness. For it is only by essential Awareness's observing through our body-mind constructs, which have evolved through the agency of its extended potential as impersonal Universal-Consciousness, that we can even incompletely sense *anything* at all!

And we do so quite "thoughtlessly", as if the automatic subject-object processes of sensing alone were our very own personal presence of being! Nonetheless, it is this arrangement that almost magically imbues our inner conceptual universe with a powerful, albeit illusionary, sense of reality.

For the rhythmic cycles that sustain our body, as an aspect of nature's overall system, will all too soon cease their function. Its purpose fulfilled, our body will then quickly disintegrate, and give itself back into the earth and universe from whence it came.

Our mind too, and all that it contains, will also soon dissipate back into the space from which it temporarily emerged, within the traveling charade of our conceptual time.

Every thought and concept, our personality and all that we have learned, will all abruptly cease when the psychosomatic instrument that projects our thoughts, comprised of images appearing on the inner "screen" of our mind, quickly disintegrates upon its death. But in truth, our body-mind and all of its concepts, which together encompass what we experience as the reality of our self and universe, never truly existed at all!

What will remain is what was always there, in the fundamental reality of nature's own Universal-Consciousness. All that can be said of it, is that it is evidently an impersonal and Universal-Consciousness, which has emerged as if uncalled from within an absolutely supreme Monad (i.e. an all-inclusive Oneness).[(47)]

But this fundamental Presence remains ever transcendent to our transitory dreamlike participation in the workings of nature's multi-universal totality and wondrous universal cycle that manifests as if it were our personal psychosomatic manifestation. Apparently, such is the inscrutable "will" of G-d. For it is only in the apparent juxtaposition (ideational union) of Universal-Consciousness and our limited individual consciousness that the notion of a "god" even appears!

Therefore, if you must attach the idea of a "deity" to something, it is evidently best ascribed to the Supreme witness of Universal-Consciousness, rather than unwittingly to the deific delusions of the illusory subjective instrument of Universal-Consciousness' dualistic experience (i.e. our ego).

Indeed, if such a thing as an absolute G-d exists it must be this Supreme witness, which is always transcendent to any ideational "god". In the final analysis everything eventually resolves back into the Supreme from whence it initially emerged (as the universe's primeval Singularity) and then extended *it*-Self into the temporal "god play" of Universal-Consciousness.

Thus, to become "fully" aware of doing so while still in our body, is to become fully "Awakened"! Such, it seems, is our true condition, and our truer deeper Nature. But how then, do we surrender our deeply ingrained habits of misdirected identification? To help us better understand, Swami Ariananda unexpectedly arrives in his ochre robes and reddish brown *rudraksha* "prayer beads", to share his insights with us. (See the "Supreme", pgs. 514 and 585).

A Campfire Discussion with Ariananda

Ari: "The concept of surrender is a hard sell to most Americans, who are a brave and strong people that have struggled hard to win and to keep their freedom. The very sound of the word 'surrender' evokes a feeling of repugnance and defiance. But the spiritual concept of surrender could help us to understand better the mysterious nature of our of lives. How would you explain its true meaning?"

Ariananda: "Surrender is understanding and accepting without reservation your true life situation".

Ari: "And what might that be?"

Ariananda: "You, as impersonal Universal-Consciousness, have identified with many body-mind organisms, over many recurring lifetimes. Thus the ego, that is, the false self, has been created and recreated, over and over again"!

Ari: "Is our situation hopeless then?"

Ariananda: (Laughter) "Oh no, of course not! What fun would there be for the Divine in that?"

Ari: (Also laughing) "I see your point, but is there nothing to do, or should I ask, what happens next?"

Ariananda: "When you finally inhabit a suitable organism, you may begin to suspect that your current 'ego-body-mind' construct is not your real Self".

Ari: "OK, so now you've become somewhat aware of your dilemma, is there more?"

Ariananda: "Most certainly! Now you begin to search for who you really are".

Ari: (Smiling) "You're not going to make this easy for me are you?"

Ariananda: No response.

Ari: "Please explain the process by which you might discover who it is that you truly are".

Ariananda: "Eventually you realize that you are not your ego, but are instead the impersonal Consciousness that has created your ego by identifying with its evolved human body. The mind turns inward, but the ego still believes it is the Seeker".

Ari: "You have described the condition of most of the 'spiritual' people that I know".

Ariananda: "Yes, but the process of disidentification does continue at its own pace, until the seeker finally realizes that he is not the one that is doing the seeking, because the real 'Seeker' is the impersonal Universal-Consciousness".

Ari: "So, you are telling me that the independent seeker is not even real?"

Ariananda: "Exactly so, and eventually, or suddenly, you will decide that there is really no point in seeking; at this moment, of incipient enlightenment, you begin to disappear".

Ari: "So, when you give up the search, you at last begin to have a chance of understanding what it is that you are looking for?"

Ariananda: "Correct. This acceptance is surrender, and finally accepting that complete Awakening cannot even happen in this body-mind, is its end result".

Ari: "But after all the effort, this seems to be a sad realization".

Ariananda: "No, what is sad is not accepting the truth. The resulting consequences of which are either the sadness of continual disappointment, or the desperate, but ever hopeful, acceptance of a delusion in lieu of true Enlightenment!"

Ari: "I understand".

Ariananda: "Clear intellectual understanding leads to this point. But true understanding emerges only with Divine Grace, after you fully comprehend that there is no one to be concerned about it".

Ari: "I think I understand. Are you saying that gaining an intellectual understanding of our true-life situation is important in that it prepares the way for Enlightenment to appear, when it will?"

Ariananda: "Essentially, understanding totally, impersonally, comes only without an individual comprehender, by re-identifying with our impersonal Universal-Consciousness".

Ari: "Ah! I see what you are saying. With acceptance and a clear intellectual understanding that our individual consciousness is just an illusion, which can never really attain our essential Consciousness, then true Consciousness, which is universal, now has a chance to emerge"?

Ariananda: "You have my meaning, and in this process nothing has changed but your perspective".

Ari: "So, this is the metanoesis of your point of view?"

Ariananda: "Exactly, and you have accomplished nothing, but the 'Lila' has now come to an end with this Awakening".

Ari: "And this 'Lila' is the creative 'god-play' of Universal-Consciousness?"

Ariananda: "Yes, it is a Sanskrit word for the concept you have just described. Another term is 'Maya', which describes the power of Lila to hypnotize Universal-Consciousness into believing that it is the body-mind, and that the experience which it is having, is reality".

Ari: "And so you become trapped?"

Ariananda: "And so the witnessing Supreme, that is experiencing the universe through impersonal Universal-Consciousness, is tricked into believing that it is confined in a body, as you".

Ari: "Thank you Ariananda

Ariananda: "You are most welcome Ariay."[(48)]

Author's Note: Swami Ariananda is using the term "Enlightenment", as a synonym for "Awakening".

Sentience

The Late Great Dr. Alex Shigo, forester, arborist, researcher, teacher and friend of arborists and trees everywhere, once confided in me, "You know, Ari, when you touch a tree, it 'touches' you back!" Alex Shigo was a beloved toucher of trees, and a consummate scientist whom, I firmly believe, came to understand that even trees are somewhat sentient too! But now that I have openly made the claim that everything from plants to people are sentient (and most likely every apparently inanimate object, in a fashion, is aware as well), I had better say something to substantiate it!

Dr. Shigo used to advise, "First define your terms, then everyone will know what you are talking about, and only then can you present your ideas". So just exactly what do I mean, when I say that I believe in the essential sentience of nature, in all of its expressions? And please bear in mind that I am not implying some sort of pantheism (nature worship), or supernatural mysticism.

Sentience, the awareness of being consciousness, is what happens when Universal-Consciousness begins to operate the operating mechanism of an organism. This Consciousness is the impersonal force behind the manifestation of everything in nature and could also be used to describe what nature is.

Therefore everything in nature is sentient to a degree that is based upon the evolution of its operating element. On one end of this "rainbow-spectrum" of sentience are inanimate things such as a stick or a stone, whose sentience is rudimentary and solely based upon the fact that it, like all manifested things, is essentially made of numerous subatomic "strings" of energy quanta, all of which are comprised of the same "One" Universal-Consciousness!

Or, as the ancient Hebrew Sages once quite revolutionarily declared; *Adonai Eloheinu, Adonai Echad*, "the Supreme our G-d, is the supreme One". And evidently the supreme G-d that they referred to as *Yahweh* (syn's. *Parabrahman, Deüs*, the Supreme, etc.), is both transcendent yet immanent, and always Divine. Yet paradoxically appears within our imperfect divisional reality, as both the sacred and the profane! (I.e. Yahweh or Allah, etc., is the primeval Witness of both Universal, and personal consciousness).

At the other end of this "rainbow-spectrum" of sentience are we human beings, whose operating element has finally evolved enough to become an instrument through which Universal-Consciousness can become more fully aware of itself.

This state of initial "Self-realization" is the condition of so-called "Enlightenment", which knowingly dispels the darkness of our erroneous psychosomatic identification. But this condition is still a state of mind, and it will either end in a reification of Self-forgetfulness, or it may bring on the complete body-mind disidentification of our full Awakening.

Have I lost you? Then, just in case, let me succinctly explain that this deistic paradigm may not be quite as "far out" as it at first sounds! This is to say that there cannot be a total disidentification with our psychosomatic instrument of subject-object experience (i.e. our so-called "body-mind"), because our living organism still operates.

What does change is the locus of our subjective perspective. Our point of view *re*-becomes "impersonal" with our Enlightenment, and veridically "universal" with our subsequent *full* Awakening! Our "body-mind" organism is then seen to be a part of our personally veritable, but cosmically impersonal, virtual "Waking Dream".

But is this really our true-life situation?(49) Probably, but accepting this as an irrefutable fact, without more information, still seems just a bit too perfunctory!

Of Jays and Squirrels

Last Night, Another White, Misty Fog rolled inland, from the nearby Pacific Ocean. And this early morning, the trees, shrubs, and forest floor all around us are again wet with its watery condensation.

But no great mystery this, for once moved inland, by the faithfully recurring clockwork rhythm of a daily coastal breeze, this foggy mist soon begins to cling (like crystal clear liquid glue) to the coastal redwood rain forest's numberless needles, and legions of leaves! And eventually it falls back down to earth again, in a shower of gentle "fograin". The cold fograin that is now dripping steadily down my back, and soaking my poor uncovered head!

After drying my cold wet hair with a fire-warmed towel, I summarily don my brown "Stetson" Bufallo-felt (waterproof) cowboy hat, and begin to tend our stubborn little "Indian style" campfire, trying to coax it back to drying warmth and fiery life.

In the meantime, a midnight blue Stellers Jay, sporting his distinctive black crested "executioner's hood", squawks and flaps, while chasing an annoyed and loudly chattering squirrel through the smoke filled redwood branches that stretch out and disappear into the foggy sky above us.

But then I am unexpectedly showered with a deluge of more cold wet "fograin"! As the unhappy Gray Squirrel scampers higher up through the cold wet thicket of slippery twigs and branches, desperately trying not to fall as he seeks a safer haven. 'Though discomfited by the sudden icy shower, I also find myself quite unable to repress laughing out loud at their arboreal antics. "What a show!" I blurt out to the surrounding trees, since (besides myself) no one else but you and they are here to hear me anyway!

But I am suddenly seized by a rather curious and unexpected "train of thoughts". Spiritual "Seekers", like jays and squirrels, seem to be of two different sorts. There are those who use methods and techniques, trying to improve or inspire themselves into enlightenment. But then there are also the rare others, who come to recognize that their persons (and personal universe(s) are only an illusion, or at best only a partial reality.

The second, I think, have the advantage, not unlike the aerial mocking jay that is chasing the arboreal squirrel away. Some call this advantage "humility", because all credit is then wisely given back to the Divine. (But in all fairness to our poor annoyed squirrel, problem-solving *corvids,* like crows and jays, are generally much smarter than most other creatures, large and small!).

Perhaps this is what the unknown comedic Sage meant when he (or she) quipped, "Angels can fly, because they take themselves lightly!" In any case, most tree trimmers, who like the arboreal squirrel must climb trees to earn a living, sometimes secretly wish that they could "fly" too. But perhaps with enough understanding, and complete faith, one day we all might do just that!

Yet evidently, until that day comes, only our ever evasive witnessing "Who" knows how. Indeed, this unknowable *Who* is the only one that ever truly does know (anything at all)! But whenever I ask, this fundamental "Who" of ours, either "how" or "why" this could possibly be so, Who remains trenchantly silent about the whole affair! (Piae Memoria - John Davey, and Millard Blair).

Illusions and Delusions

OUR PERSON AND UNIVERSE are but an illusion, while "nostalgia" is clearly a delusion. It is little more than a dream within a dream, tightly wrapped within the conundrum of another illusion! Yet the illusory universe that contains our "person", is also by its illusory person contained. But if so, then is no one ever really "present", to wake the dreamer from his convincing dream?

But while our illusory persons rather quickly pass away, even the less transitory universe (real or not) that seems to contain them does not last forever! And while it seems intellectually obvious that our consciousness is unreal, it is not at all obvious to our ego's common experience that its contents are not! Hence, it is difficult to discern "reality" from delusion.

The essential problem lies neither in our illusory universe nor within our ephemeral "persons". It only exists when we are looking backward when we should be looking forward, and even more so when we mistake a statement for a question.

Ironically, the only question that stubbornly defies even the most diligent scrutiny is, "Who wants to know?" Which begs the question; "Who is there to be overwhelmed with either amazement or "Enlightenment", when it is finally understood"!

Pale Suns and Bright Moons

TODAY, WHILE GAZING at the pale orb of the sun, its brilliance diminished now by a heavy fog, I have a last "preparatory" sort of thought (for the novel notions that we will soon encounter on the less travelled pathway that lies ahead), which I will leave you with for now, in the manner of a simple metaphorical paradigm.

In trying to imagine the "shape" of a life, I am unexpectedly inspired by today's pale sun, and last night's bright full moon. And in trying to imagine the shape of *all* life, I see near infinite "moons" attached at nearly infinite points upon the surface of a single, perhaps immeasurably large, "sun".

But, while trying to imagine the shape of the "Superverse" being manifested by Universal-Consciousness, which contains the possibilities of all life, I must think of this solitary sun and its companion moons as hollow ספהירות *spherōt* "spheres".

With a near infinite number of various sized hollow moons, attached at a nearly infinite number of places upon the surface of the hollow sun. (In perhaps as many places as there are subatomic strings within our universe?)

The hollow moons are attached on the inside and on the outside of the hollow sun, so that each moon passes through the hollow sphere of every other moon.

Moreover, each sphere is also contiguous, because it is offset by the minimum distance of the tiniest possible subatomic particle into a dimension that is at "right angles" to its neighboring sphere, and what eventually results is the conundrum (puzzle) of a "hollow" solid sphere!

If the eternal present lays in the center of each sphere, then the sphere's shells are comprised of every possible moment, offset from the eternal present into an only experientially evident past and future. Past and future are therefore contiguous with the present, and as such, a single sphere represents a potentially eternal present life!

From this perspective, it becomes apparent that each near infinite point upon the near infinitude of every sphere is, in kind, the potentially eternally present center of every other sphere. That is, our ubiquitously aware, true primal center is located in an eternal "Now". Which actually comprises the totality of all the *spherōt* (i.e. spheres, dimensions or "planes") in which Universal-Consciousness manifests an endless realm of potentialities, within our deceptively illusionary individual consciousness(es).

Thus we have a conceptually correct, but of course in actuality insufficient, metageometric multi-spherōt representation of our experiential multiverse. (Or of G-d's "Creation" in our minds, if you will.). I cannot say, with any surety, what the utility of this "Thought Experiment" might be. But it is nonetheless compelling that a quasi-geometric representation of that which we cannot see, can be envisioned within our imagination, even if its full function and true dimensions cannot (yet) be fully grasped!

But perhaps you may yet discover some use for it?[50] (See "Forces, Laws and Causality", pg. 534.

"Two roads diverged in a wood, and I took the one less traveled by, and that has made all the difference". – An excerpt from "The Mountain Journal", *The Road Not Taken* by Robert Frost, first published in 1916.

The "Geometry" of the Big Bang

IN THE BEGINNING, most people find it difficult to imagine the geometry and sudden inflation of the Primeval Singularity and Big Bang events. But it need not be overly difficult to understand.

Just imagine that suddenly, from a spacetimeless, and therefore almost indescribable "Elsewhere", a tiny sphere, for want of a better descriptor, of pure "Awareness Force" appeared (i.e. because we can only say whatever it was not!).

At first it was so small that it had only one dimension. Yet within this tiny sphere of pure "Awareness Force", the energy and accompanying spacetime of the entire universe was already present!

But because everything was *suddenly* within it, with nothing to stop it, it inflated like a balloon, instantly becoming thousands of magnitudes larger in size. And with nothing to stop it, it did this even much faster than the speed of light!

Nonetheless, even though it had now acquired three more dimensions (including space-time), because it was still immeasurably dense (much like a Black Hole), even though it was also immeasurably hot, it was also still completely dark. This is because it was so dense that even photons of light could not escape from it!

But we reckon that around 330,000 years later, it had expanded enough in size, and cooled down enough, so that light could now escape from it! Therefore, wherever we look out into the universe with a telescope, we are actually looking back in time, toward the moment that light was first released from within our early inflated universe.

To envision this, imagine that the inflated ball of visible glowing hot energy, which we just described, still lies at the "center" of our universe. And that our visible universe simply continued to expand outward from its inflated super-hot mass. Until it eventually became large enough to generate our galaxy, sun, solar system, and finally, our home planet, earth.

Hence, when we look outward from our planet, if we look far enough in any direction, we are actually looking back at the leading edge, of the inflated "center", of our already expanded universe. (Which is remarkably still expanding, at an ever increasing rate!) But since it was also all that existed at the time, and there is nothing visible beyond it, it also appears to be a sphere that surrounds us in every direction that we look!

Vrittis and Strings

Just To Be Very Clear, before we can with any confidence proceed, we need to thoroughly comprehend that "vrittis" (the basic constituents of consciousness), and "strings" (the basic elements of energy), are fundamentally one and the same. In the very first "Planck Moment" of the Big Bang there was a perturbation in the sudden presence of impersonal Universal-Consciousness, as its constituent Energy emerged from within our universes (original) Primeval Singularity.

And almost simultaneously a subquantum fluctuation in space-time occurred. Which then gave rise to googolplexes upon googolplexes of potential subatomic strings that were suspended within a perfect subquantum "fluid-like" membrane of Universal-Consciousness Energy (See "Googolplex", pg. 537).

This is the essentially "limitless" Energy of Universal-Consciousness. Which always (spacetimelessly) underlies, and thoroughly permeates, the quantum energy "sea of strings" from which our boundless Superverse is constantly emerging. And through which it also (forever) remains, both quantumly and Consciously interconnected!

And because it exists in a "limbo" sort of state, apparently before, between and beyond all of the merely physical laws of our overarching Superverses coalescing Multiverses, it is still endlessly extending itself into lattice-like dimensions of endless parallel Universes!

Moreover, the apparent "space-time", in which all of our Superverses post quantum events continues to transpire, then quite simply, goes on perhaps forever! Emerging (as it does) from within our universes incredibly hot, and apparently unstoppable, rapidly inflating Primeval Singularity. Because, with nothing to stop, or even slow its progress, the foundational Primeval Singularity of our Superverse is forever expanding, at an ever increasing FTL (faster-than-light) speed!

Nonetheless, the Divine's all-inclusive primeval "carnival cloud" of cotton-candy-like perturbation strings (which are comprised of Cosmic Inflation energy) remains invisible, as we approach its spacetimeless Source. Because it is still so incredibly dense, that even photons of light cannot yet escape from the powerful grip of its intense gravity!

Cotton Candy Clouds

THIS EVENING IS DIFFERENT than most others, in the enchanting forest primeval, because the usually foggy gray sky, that seems to stretch out endlessly above the redwood rainforest, is now remarkably clear! And as the flagging reddish evening sun begins to sink slowly into the distant blue Pacific Ocean, the "robin's-egg-blue" sky that now stretches out overhead is becoming dotted with small, fluffy, pastel pink, "cotton-candy-like" clouds.

And as the evening sunlight slowly fades, first the pink clouds, and then the light blue sky above us assumes the same tired, muted tones, as it slowly soaks up the spreading grayness like a gigantic celestial ink blotter.

Then, as the cheerful colors of the evening sky begin to slowly fade away into gray, the first bright stars of the emerging night now begin to twinkle and cast forth their jovial bright light.

High above us, the sky's evening gray finally fades even further away, into a fathomless nighttime blackness, and the crystal clear heavens above are suddenly ablaze with the light of countless scintillating stars, glowing planets, and distant sparkling galaxies.

On this rare clear night in the oceanside Santa Cruz Mountains, it becomes quite clear that our entire planet is little more than a "microscopic" blue speck, adrift in a shadowy celestial sea.

But by now I am not only tired, but am also thoroughly humbled by the dark and seemingly endless wilderness night! So now seems like an opportune time to turn off our sputtering Coleman® lantern's wavering golden light, and bid the primeval forest around us yet another well-spent "good night".

(Dedicated to the Silicon Valley visionary, Elon Musk, and to the future colonization of Mars by the courageous astronauts of the "Space X" Team. And perhaps of equal importance, to the perennially youthful magic of Pink Cotton Candy, and to "idle dreamers" everywhere!).

Destiny

Morning Has At Last Arrived, and our time here amongst the redwoods is nearly at an end, at least for now, and soon I must "break camp" and drive back home. I wish I knew, but in truth I really have no idea what I am returning to. Can we ever really know what awaits us, even though everything has already been causally "written" within the potential of the early Universe's own ספר של חיים *Sefer shel Chayim* "Book of Life".

Albeit, despite the fact that the variable process of so-called "Natural Selection" is the Universe's principal author of the great tribe of the living! But of this much I am certain, when a body enters this world, everything that is going to happen, or not, will of a certainty transpire, just as it was always destined to do! (See "Natural Selection", pg. 555).

The only liberty that we in all likelihood ever truly possess, is to take the "inward" direction. Otherwise, we are evidently only circling around the rim of a potential reality!

Because the only change that we can ever truly make, is whether or not we will identify with our body-mind, or with the Noumenal Witness of the impersonal Universal-Consciousness that ever experiences our body-mind's antics. As our clockwork cognitive mechanism incorrectly assumes that it somehow controls the moving hands of its own destiny!

The Sages of the Talmud wisely put it thus, "All is in the hands of heaven, save the fear of heaven!"(51) But whatever could "heaven" be, if not a recognition of our deeper spacetimeless connection with the ever present Noumenal Supreme Witness, from which the universal substrate of impersonal Universal-Consciousness emerges?

For this is evidently not only the ultimate "sum", but also the Primeval Singularity's ultimate source for all consciously manifest things as well. But perhaps it is only natural to at first fear that which is attained only through the dissolution of our illusory individuality. For indeed, only the deathless can ever fearlessly pass through "heaven's" gates!

Hence, some of the best advice that has ever been given to an unsure humanity, was uttered by a clearly frustrated, but nonetheless determined Winston Churchill, in a speech meant to inspire the beleaguered British people, during the relentless Nazi bombing of Brittan during the Second World War; "Keep calm, and carry on!" And evidently, it worked!

On Knowing

It Can Be Said that knowing the "Nothing" that knows nothing about not knowing is what needs to be known. But inevitably it seems, no matter how much we twist or turn our thoughts and words, all that we can ever come up with is just another deficient descriptor of the indescribable! Indeed, how can "Nothing" possibly know some- thing, or any- thing, let alone if only "No"- thing is ever really there? And if only nothing knows nothing, only nothing is actually ever known!

Alas, such circumlocution is simply "beating around the bush" of objectivity, where nothing is known, except by inclusion of its opposite pole. Thus the question naturally arises, "To and by what does our knowing know, and know that it knows?" This is to say that, "To and by what does our Self's knowing of inexistent objects appear?"

No Exceptions

In A Word, whether we understand it or not, there is nothing before or after this apparent world, but our own indescribable Self. Were it not there, no experience would be possible, no exceptions!

(Pea Memoriae, My "Brothers in Arms" Chuck and Pig,
kind Mountain Charlie, and the unforgettable "Gypsy
Jokers", wherever you guys are…).

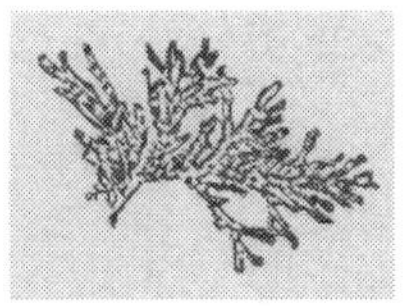

Santa Cruz Mountains
Northern California, USA

Chapter 3
Quercus "Oak"

Risk Assessment

"Here I sit under a clear blue mountain sky, hemmed
'round about by massive redwoods thrusting up
to meet the occasional cloud wafting by. Beauty
abounds here, yet even that small lifeless limb I spy,
dropping from its lofty perch on high, might deal
such a savage blow, 't would cause a man to die!"

~ From the field notes of an anonymous Arborist ~

One Of The More Important Duties of an arborist is "risk assessment". Because trees are not only useful, necessary and beautiful, but they are also often big, tall and heavy. Indeed, when large trees unexpectedly drop limbs, or occasionally fall over, something, or someone, could get hurt! Knowledge and experience in tree biology, decay fungus, insect pests, hazard recognition, science based tree care, the law and several other related disciplines are all necessary to fulfill effectively the arborist's duty of tree risk assessment.

However, just two elements are unquestionably fundamental to the necessity for tree risk management, trees in proximity to people, and by extension, to what they value. From this elementary juxtaposition "risk" emerges and the principal work of an arborist begins.[(52)]

It is a matter of wisdom, born of experience in life, to understand that deep meaning is often masked by simplicity, or sometimes shrouded by the seemingly obvious. Therefore an astute arborist learns not to dismiss too hastily the deceptively plain, or even the apparently simple!

In fact, the deceptively simple formula, hazard + people = risk, accurately describes many crucial occupations and essential work functions of any organized society. A doctor, for example, manages health risks, while a policeman manages the risk of crime, and a soldier that of war. The occupational success of such critical professions benefits the continued survival of the individuals that collectively comprise our modern world's more advanced societies.

But a deeper, occult (hidden) imperative, like incipient (early stage) decay lurking deep inside a tree trunk, is the presence of survival fear, which results from believing that we are only a sensing body with a thinking mind. Unhappiness and suffering are the inevitable result of this erroneous belief.

This is true for the individual, and also for any society comprised predominantly of illusory individuals who remain ignorant of their veritable existence as impersonal Universal-Consciousness, and of their essential nature as the Universe's immortal core of transcendent Awareness.

It is an essential Awareness that, like the vital heartwood of a living tree, always lies within the innermost *Hridaya* "Heart" of its own ever-expanding Primeval Singularity (see, "Hridaya", on pg. 508). But what is the nature of this recurring pattern in life, and what, if anything, might it signify?

I

It seems as ineffectual for an arborist to inquire, "Which came first, the climber or the rope", as it is for anyone to ask, "The acorn or the oak?" Yet however simple in appearance, this is neither a bona fide insufficient, nor an irrelevant question. For it reveals a profound quandary, which nonetheless seems quite fundamental to our dualistic perceptions and common cognition! Indeed, a comparison of opposites is "how" we think. But how we think has nothing to do with the basic Awareness, by which we understand that we are thinking.

Renee Descartes opined, *Cogito, ergo sum,* "I think, therefore I am". But if we are not aware that we think, our cognition then becomes quite impossible! This is because our purported "being" or presence, provides the "place" or center of perspective from which our cognition operates. And our dualistically beleaguered consciousness ultimately derives its implied meaning, as well as its apparent existence, therefrom.

Interestingly, psychology, the "study of the human mind," refers this necessary function to an "ego". Our purported "ego", however, could not possibly be the true locus of our Self-Awareness. But it most certainly does seem to be the operative center of our "split" (i.e. subject-object) mind.

As such, it possesses the necessary subjective sense of identity and personality to function in an apparently diversified objective universe. It is clearly a necessary, functional aspect of our individual consciousness.

Yet most of us see ourselves as much more than mere function(s)! For we are not, and cannot ever be, whatever appears before us in our field of experience. This includes our illusory egos, which by definition could not possess any truly separate identity.

Moreover, this is inexorably so! Because our ego is actually only a collection of thoughts, concepts, assumptions and such that are loosely held together within an ideational space-time that exists only in our phenomenal memory.

Hence, because phenomenal "subject/objects" (i.e. our ephemeral egos) demand a noumenal subject in order to be seen, we are then most likely the impersonal self-aware Awareness, by which we understand that we do exist!

I once asked my late dear friend Ken Brown, a gifted electrical engineer and computer scientist (and the youngest son of a bone fide Rocket Scientist!) "Will computer science ever succeed in achieving artificial sentience?" He answered, "I don't think so Ari, but we probably will succeed in creating a computer that not only thinks much faster than we do, but it will also have much more memory capacity and intelligence than us, per se."

Future computers will, no doubt, be able to do everything that we do and probably much more, and not simply faster, but considerably better as well! But it is unlikely that strictly mechanical/electronic computers will ever be able to admire a beautiful *puesta del sol* "sunset", write a distinguished novel, compose an artful song which "speaks to our human hearts", or fully experience wonder, or love. Perhaps it is unlikely, but also not impossible!

In 1980 Professor David Cope of UC Santa Cruz wrote a computer program that was based upon his own minds method of musical composition, rather than the obtainable A.I. methodology of the time. The program works by uploading all of the available compositions by a composer, which it then analyzes and uses to create a new composition that is in the style of the composer but does not copy any of the music that is in its database. The result is a lovely new composition that is so much in the style of the composer that it could have been composed by the Maestro himself!

As it turns out, by simply copying the creative processes of our brain, rather than trying to reinvent what nature has already accomplished, he may have found a better pathway to A.I. (artificial intelligence).

But this should come as no great surprise, since Nature's evolutionary path is so rich with improbable events, that "impossible" might not even be in Her creative vocabulary! But, should we succeed in creating Strong A.I., now being a "lesser god", we may soon become irrelevant, if not unnecessary, unless we are destined to become "cyborgs"![(53)]

II

We seem to possess an uncanny awareness that transcends the local biological machinery of our bodies and brains, and which is also quite likely responsible for the conglomerate (perhaps even "non-local") phenomenon that is our mind. However, our mind is not only intellectual and cognitive (able to think), but it is also affective (emotional), and intuitive.[(54)] And remarkably, it also seems as if we are able to somehow "stand back" and observe our mind's operation without actively participating in it.

This indicates that our Awareness is not a mere "ghost" in the bioelectrical machinery of our body-minds, but that it most probably exists in another dimension as well! Its locus is most likely within a fourth dimension, or perhaps even a fifth, since we also perceive the fourth dimension of space-time operating as if before us.

It also seems likely that this, not strictly local Awareness, is not only independent of our psychosomatic instrumentation, but that our localized body and mind are dependent upon it to be what we understand and commonly experience as being "human".

It is improbable that we will ever be able to build a computer that can be connected in this intimate fashion to the entire universe, unless it is also somehow intimately connected to us. Which could only happen independently through the quizzical property of "quantum entanglement", within an artificially intelligent "Quantum Computer" (see "Microtubules", pg. 550).

But the perception of opposites, our apparent "alpha and omega" of experience, which are so basic to our dualistic consciousness, curiously, may actually be that of "complementariness". Or perhaps more specifically of complementaries, which are the likely binary conceptual products of our conditional subject-object perspectives.

In other words, we see the universe as we do largely because of the way in which we sense and cognize it. Moreover, it is evidently absurd to believe that the universe does not extend far beyond the boundaries of our sensorial, cognitive and affective abilities to perceive it!

Thus the ideation of opposites seems basic to our assumption of temporal causality. However, by fully admitting our apparently contingent existence within the continuum of an infinite and eternal Universe, Multiverse, or perhaps even "Superverse", the seeming opposites of our common experience are enabled, or allowed within consciousness, again to be One. As they are indeed, within their primal actuality of essential Reality!

This is to say, that the apparent opposites of our experience are in fact the opposing ends, or complementaries, of a single aware universal continuum. A "Consciousness Substance" that is comprised of living Consciousness Energy, and might even be best described as being a "Universal-Consciousness".[(55)]

III

No matter how engaging that they might seem to be, the "isms" and "ans" of religions, philosophies and nations also inadvertently (conceptually) divide humanity. Which we then must (rather ineffectively) strive to reunite, or at least keep from trying to dominate and kill one another!

The telltale attachments of our sleepy anthropoid intellects, that are still generally in unwitting service to our false ego "self", inevitably brings with it all of the blessings and curses of complementariness. The cause of this is embedded deeply within the dualistic nature of our individual consciousness, which must constantly cognize by division and comparison.

But the solution for humankind's problems cannot be found by merely attaching more suffixes, infixes, or prefixes to nationalistic or religious conceptualizations, or in the inhumane actions that so often accrue when they are! For indeed the function of mind in this way is only a הבל הבלים *havel havelim* "vanity of vanities, as *Kohelet* the "Preacher" (i.e. King Solomon), most eloquently observed, so long ago.(56)

But rather than a paean of hopelessness, is his refrain perhaps not merely the lamentation of a wise poet for the unbridled foolishness of man? For perhaps the hope of our true freedom is freedom from the Grand Illusion (called "Maya" in Vedantic Sanskrit, and "Malkuth" in Kabbalistic Hebrew), which divides, in experiential concept only, that which is always One.

It is the universal play of יהוה *Yahweh* comprising the consciousness of man, which arises "as if uncalled," from the actions of our body-mind and all that its emergent ego consciousness contains. Or, if you will, it is *El elyōn* "G-d on high", the creations ultimate Manifester, which inevitably includes its space-time co-creator, man.

This is to say that It is the one transcendent G-d and all that is contained within "His", but also nonetheless also "our", phenomenal human consciousness.

Thus, being transcendent yet immanent (indwelling), the purely sentient Divine thus comprises both the alpha and the omega of the manifest, and the infinite potential of the unmanifest. In other words, the Divine is always an all-inclusive "One" *(Adonai echad)!*[57] (See "Yahweh", pgs. 516, and 596).

(Piae Memoria, consummate Tree Trimmer, Consulting Arborist, talented Artist and fellow Cowboy, John Brittan. We all miss you John, *Vaqueros*, and *Paniolos* too!)

A Sotto Voce

Consciousness Is Clearly Necessary for our every experience of person and universe, but just what is our consciousness? Our personal experience is that it seems to place us within a physical universe when we are awake, and into a dream universe when we are asleep. But must we not also be conscious, to know that we are dreaming?

Yet, in deep sleep it seems obvious that we must not be conscious, because we have no memory of it! Hypnosis, however, reveals that even the most fundamental elements of our consciousness, including the so-called "unconscious", are automatically memorialized as they are subliminally tucked away into the subconscious depths of our minds.

Therefore it seems quite likely, if not mandatory, that memory is an essential ingredient of everyone's consciousness.[58] For whenever we dream, perceive, or cognize, we build imaginary universes of concepts, which linger for a while in our memory; sometimes even after we awaken from a dream. And surely imagination is our minds method of creating, and recreating, the thoughts of which this continual stream of representative concepts is comprised![59]

We have nearly all had dreams that seem so very real that when we finally wake up, for just a moment, it seems as if we are dreaming and that the dream which we were just having is the real! Indeed, it is very difficult, if not impossible, to determine whether or not what we so fervently believe is "real", may actually be nothing more than the equivalent of a persistent waking dream![60]

I

All that we can ever know with any certainty is that both universes, our dream universe and our waking universe, exist only within our consciousness of them. Our cyclical lack of consciousness also compromises our ability to render certain and trustworthy opinions about the reality of our experiences.

All that can be confidently established, is that consciousness and unconsciousness seem to be the polar opposites of the normal range of our experience. Yet it remains impossible to establish reliably, the independent reality of either state! But then, how is it that they do come to exist, real or not, within the limited range of our cognition?

It seems obvious that sensorial and cognitive information is continually processed into the ideational stuff of our experiences. What is less obvious, is the important roles that our memory and imagination play within this ongoing experiential process.[(61)]

Our mind seems to be forever busy, reminding us of its presence with an endless stream of cognized images, words and feelings. Indeed, our experiential universe is literally comprised of them, but of what are they comprised? They are most certainly the objective phenomena of our subjective experience, and are therefore comprised of the same consciousness. Which displays them as apparent objects to its subjective aspect (our illusory egos), within an ideational, sequential space-time framework.

Therefore, we cannot accurately determine if our ongoing experiences of person and universe are real, or merely the equivalent of a subjective dream! Indeed, all that we can reliably infer is that we, and the reality that we seem to occupy, are apparently a rather large collection of conceptual, imaginary thought forms. All carefully arranged with lightning speed into a sequential memory stream of subject-object consciousness.

To match this high speed level of intimately interconnected cerebral activity, a conventional computer would require a memory capacity of 2.5 "petabytes", and a trillion-bit-per-second processor! (I.e. 1 petabyte = 10^{15} bytes, which is one million gigabytes, or one thousand terabytes!).

Our imaginary experiences are thus continually, and automatically, generated and recorded within our consciousness, until the death and dissolution of the psychosomatic machinery of the body-mind that generates them. But what then makes our waking, and sometimes even sleeping dreams appear to be so real?

Obviously consciousness must be present and functioning, but this alone does not constitute "sentience" any more than opinion constitutes "science", and it is sentience (knowing that we know) that lends a sense of reality to everything that we experience!

Sentience is only possible in the presence of an Awareness that is independent of whatever consciousness may be experiencing within our body-minds. But there also has to be a spacetimeless gap of unconsciousness, which is necessary to differentiate our apparent subjects and objects within our ideational space-times.

And it is apparently necessary because both our subjects and our objects are essentially only thoughts, or conceptual objects that exist only within our consciousness, and therefore cannot share either the same inferred temporal, or imaginary spaces.

Unconsciousness is therefore necessary for our consciousness to differentiate conceptual subjects and objects within a variable space-time.

Thus both consciousness and unconsciousness are apparently necessary for us to distinguish subject from object (i.e. experiencer from experience), and object from object, within a conceptual field of malleable space-time.

Our body-mind first generates the conceptual objects of our experience, then immediately experiences them subjectively, by subliminally (below the limits of our consciousness) inserting gaps of unconsciousness between our experiential subjects and objects.

This is the cognitively invisible spatiotemporal system of our current hominid mind, which we automatically interpret as the incessant flow of an invisible space-time.

The subjective aspect of our binary consciousness simply cannot be conscious during the imaginary expression of its cyclical polar opposite of complementary unconsciousness! In other words, our conceptual expression of dualistic consciousness necessarily takes, then holds in our memories, the form of a temporally elaborated "subjective/object" to return to (i.e. our illusory "ego").

This is also true of every object of our consciousness, including our minds conceptualization of unconsciousness, which doubtlessly transpires through its faculty of imagination. Both consciousness and unconsciousness are therefore fundamentally ideational images "in array", and are thus essentially (only) imaginary.

Yet, in order for them to exist, they must do so within our experience of the subjective pole of the underlying consciousness in which they exist. In other words, there is a difference between the total lack of dualistic consciousness, when our brain is "dead", and its purportedly "unconscious" state when it is alive!

Therefore our dualistic minds enjoy a "symbolic" existence, which is wholly contingent upon the presence of an independent Witness. That is, they exist as ideational opposites, like the contrary arms of an imaginary seesaw.

Hence, when one aspect of our conceptual consciousness is ascendant, its complementary polar opposite expression is inevitably either repressed, or suppressed; including that of an only symbolic "unconsciousness" as well.

This repression of consciousness is what provides a conduit that allows our independent space-timeless Awareness to experience itself as an apparently "mortal" subject/object. Thus we confoundingly cannot see that we exist only symbolically within a presumptive individual consciousness that coalesces (forms) within an ideational, and temporally bound universe![(62)]

II

If the entities of our common experience are imaginary, being sequentially composed by consciousness of "consciousness objects", which are actually dimensionally expanded aspects of an immanent yet transcendent Awareness, what could the character of this independent primeval Awareness possibly be! All that can be confidently said, that is not likely quite "wide of the mark", is that it is neither a some-"thing", nor a no-"thing".

In other words, any such "supreme" Awareness, although present within our consciousness as its necessary Witness, cannot be either a "something" object, or a "nothing" object, and could also not be a "something" subject, nor a "nothing" subject. Being neither an illusory objective subject, nor any sort of an object within our consciousness, it could also not simultaneously be both, but must certainly contain them both within its necessarily intrinsic Self-Awareness!

Therefore it can only be pointed at by saying that it is a not "not-nothingness", and also a not "not-somethingness".[(63)] If this wording seems unusual, it is so for a good reason. That is, it at least linguistically removes the dualistic problem of implied opposites, which will never do as a descriptor of that which is descriptively "nonpareil" (beyond comparison)!

Evidently, without being supererogatory (superfluous), this is about all that can be said of our essential nature, and perhaps even the saying of this is too much.

So, let it simply be said that our intrinsic yet independent Awareness is "That" by which we know that we exist and can thereby, in this very moment, understand the words that we are reading, while simultaneously knowing that we are also separate from them.

Because we cannot simultaneously see and "be" whatever we see, since a subject could never be an object, nor an object a subject, except in a purely unadulterated *illusion!*

Yet our essential self-Aware presence, like "lightening in a bottle" certainly cannot be dualistically contained. And like a contrary existential puzzle that is missing some critical pieces, it cannot even be completely understood within our binary consciousness. Hence, it must appear within the domain of our independent Awareness, in a sentient fashion, in order to be perceived as existing.

Because, even though our foundational Universal-Consciousness cannot see itself objectively, since it is also the ideational space-time in which cognition takes place within our consciousness, it most certainly must be aware in some fashion, in and of itself!

Which is to say, that it is "infused" with the immanent presence of a transcendent Awareness, which is being "reflected", so to speak, within our phenomenal body-mind consciousnesses.

But if this is so, then perhaps the most that can then be accurately said, to succinctly summarize our current "symbolically embodied" situation, is that, like the opposing sides of the black and white Taoist "Yin/Yang" symbol ☯, our essential Awareness is always phenomenally transcendent, yet always remains noumenally immanent.

Our reflected deeper Awareness is the interior *sotto voce* "soft [small] voice" that we occasionally experience, and may even come to understand quite well! Yet even familiarity with our deeper intuition does not necessarily equate with a true understanding of our arcane, impersonal Divine nature.

III

If we truly, fully understood our intuition and the deeper Divine Source of our awareness, then many aspects of science, technology, the acquisition of power, pursuit of pleasure and certain other acquisitive human endeavors, which are born of our conceptual understanding, may be in danger of becoming trivial!

For intuition is a metacognitive awareness of the omnipresence of our essential, but essentially "passive", primeval Awareness. It is therefore likely an impersonal and ubiquitous Awareness which, of necessity, connects us from the "in" side of our universe's foundational Universal-Consciousness to all things conceptual.

A sudden and complete holistic understanding is thus the hallmark of any genuinely intuitive insight. However, it is only through a humble and "hearing heart" that the noumenal presence of the Supreme can make *it-*Self known within us.

But how are we to distinguish between invisible Noumenon and compelling phenomenon? And Solomon prayed; *"Give thy servant a hearing heart that he may discern the truth."* – from the Torah, 1st Kings, 3:9.

(Dedicated to my halcyon Soto Zen Instructor, Kobun Chino).

Phenomenality

PHENOMENALITY CAN BE THOUGHT of as being broadly comprised of three aspects, our birth, our existence, and our death. The one factor that all three have in common is their reliance upon "presence". Our birth could as well be said to be the beginning of our presence, existence the duration of our presence, and death the dissolution of our presence.

But for our presence to have any meaning, we must first be aware that it is there, which even in an accurate "Positivistic", "Objectivist", or any other fully reasonable sense, is what our ostensible "birth" actually is!

From this standpoint our birth begins when our mind becomes aware and subliminally identifies its awareness with the presence of the physical vehicle that grants it the phenomenal experience of existing. It is then that our illusory existence as a body-mind summarily begins, as a discreet "subject" to every other apparently external "objective" thing. Which then appear to exist as separate objects that are suspended within space-time.

As an illusory entity, the universe is always experienced with our person at its perceptual center. In other words, our presence comes into existence with the first thought that "I am", or that "I" exist(s). Other thoughts are then accumulated, and the basic concepts of our person and universe are built up.

These concepts are constantly reinforced by sensory input and our various and sundry associated thought processes. But the fact remains that all that we can ever know of the universe is purely conceptual in nature. Indeed, in order for us to experience anything, it must first be expanded in our minds for a sufficient duration to be cognized!

This expansion is the occupation in experiential space by the concept, and its duration of occupation in experiential space is time, hence "space-time". Space-time is thus the invisible stuff that seems to contain all of our ideational phenomenal manifestations. The objects of consciousness that represent the person and universe within our minds are built up dimensionally, and given their continuity of forms sequentially therein.

Our body-mind automatically generates a cognitive "spark" within our minds, which for a better clarity of understanding we will call a "quantum" of consciousness. Then with it, it serially builds up the dimensional objects of consciousness within our minds. By the exact placement of each thought quantum in a particular ideational location, within an infinitesimal present moment.(64)

Of course we are not directly discussing the energy quanta of quantum mechanics. But we are also not merely discussing the electrochemical activities of our bodies and brains, which are surely also an indispensable part of our experiential processes! However, we would also have no existence per se, but for the presence of a consciousness within our minds that interprets them as being actual, and in so doing, grants them an apparent existence.

The question, which naturally arises, is "Which, then, is the real?" But there is no "real" answer to this question, which innocently tries to divide an apparently contiguous (not touching), but actually interdependent natural process, and then compare the resulting mere appearance of implied differences! Like trying to separate an object of reflection from the mirror that appears to contain it, this ironically points out how very real that the experience seems to the unreal person who is having it.

(Hence it does not really constitute a bona fide substantive question at all!) The phenomenon of the motion of our experiential objects of consciousness through space-time, or "change", is thus manifested along with the sequential creation of our apparent person and universe, appearing within, or before, the dualistic subject-object "mirrors" of our minds.

Aspects of phenomenal change, such as our apparent identity and the negative distribution of entropy in living things (and the thoughts of living things), are indicative of this sequential process in which each apparent quantum must be different than the previous one in order to dimensionally exist.

In other words, insofar as its appearance within our consciousness is concerned, evident "entropy" (disassociation) results from our necessarily sequential space-time processes. Our mental disassociation occurs automatically as each thought quantum is replaced by a new and necessarily slightly different one, in the veritable timelessness of an eternally "present" moment!

Ergo, no two thoughts, leaves, or even snowflakes can be identical, fill the same time in space, or occupy the same space in time! This forever "present-moment" exists before our perception brings it into focus within our next sequentially present, but forever transitory, experiential moment(s).

But this all happens so rapidly that the imaginary cognitive passing of space-time seems to be a seamlessly smooth flow of actual, instead of inferred, constant change.

Indeed, this phenomenon is ubiquitous and redundant, for not only are no two snowflakes exactly alike, but when we are finally able to look and measure on the quantum scale with a neutrino microscope we will most likely be able to substantiate that no two atoms, or even any two of its constituent subatomic strings are exactly "identical".

The reason for this is simple, just like every-"thing" else in our topsy-turvy "inside/outside" universe, no two thoughts, or even their constituent quantum of consciousness-energy, can occupy the space of our phenomenal consciousness at the same moment of time.

Thus, to the surprise of many astronomers and astrophysicists, even the observable planets, moons and stars are also quite different from one another. Hence, their previous concepts of categorization no longer seem to fit in well with the rapidly changing facts that are being daily revealed by our constantly improving instruments and methods of observation!

Indeed, some researchers are even beginning to suspect that every observable thing is quite likely only a miniscule portion of an infinitely diverse continuum that (infinitely) extends not only microscopically, but macroscopically and perhaps even into endless parallel dimensions as well! (See "Cognition", pgs. 505, and 522; and "Neutrinos", pg. 556).

But because there exists a *jijimuge* (Jap. "interdiffusion") of consciousness and awareness, we can only attend to one three dimensional thing at a time. Therefore, in every apparent "where" in the universe, change seems to be constant. What we normally don't consider, is the essential part that the phenomenon of memory plays in the ongoing mental processes of assigning our ideational subjects and objects not only their apparent forms, but their proper spatiotemporal locations as well.

Consequently, it also seems like a topic that we should take the time to consider in greater depth (see *"jijimuge"*, pg. 508). But before anything at all can transpire in the universe, pure energy must somehow acquire mass in order for stars to produce the "stardust" comprised elements of which we and all earthly "material" things are made. But how could such an implausible thing ever come to pass?

The Gravitas of Gravity!

Gravity Is A Force which provides weight to mass, but is also an expression of organized energy, which has no appreciable mass and therefore no quantifiable gravity. Therefore it takes a tremendous amount of energy to aggregate and maintain enough mass for a particle of "matter" to form. On the atomic level this is indicated in the well-known equation $E=MC^2$ (I.e. the energy required equals the speed of light squared.).

Leptons, like electrons and photons, for example, carry a charge which is measurable and can initiate motion but cannot be fixed simultaneously in (both) the space and time matrices of a conventional four dimensional space-time manifold (see below).

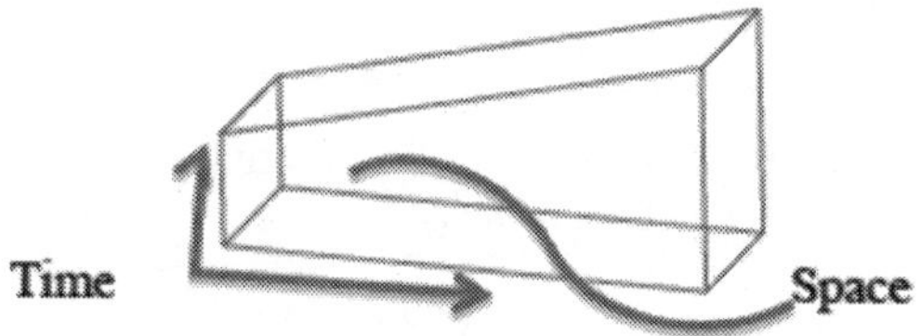

Observationally this appears as a particle wave duality. Ostensibly because energy is not exactly a "particle", which has mass and a definite spatial location, or a "wave" which has a temporal location, but no definite place in space.

But, this is a quandary that is not explained by simply calling it a "particle-wave duality", and where there is an apparent quandary, there is often in fact a lack of adequate knowledge.

If we render the universe down to its simplest terms two things remain, presence and the space-time necessary for its appearance. What is present seems most simply to be energy. Which nonetheless must retain its duality since to be present it must occupy a definite place in both space and time.

This apparent necessity is further reinforced by the multitude of events that have subsequently transpired in space over time. (After the primordial appearance of energy in a multitude of spatiotemporal locations, following the "Big Bang" expansion of our universes Primeval Singularity.).

To better understand this, we must first look at the basic conflict between Newtonian physics and Einstein's Theory of Relativity. Newton's equations accurately describe the manner in which the motion of bodies in space function over time. But they also assume that the basic force of gravity (which drives them) is spontaneously present and also not subject to temporality.

This is to say that gravity must be instantaneously present everywhere, regardless of space or time. For example, in the unlikely event that our sun were to suddenly disappear into a gigantic space-time anomaly, such as a humongous interstellar "worm hole", all of the planets in our solar system would careen off into space!

But Einstein discovered that apparently nothing in the universe can travel faster than the speed of light. Therefore approximately eight minutes would transpire before we would notice that our sun's light had gone out. Only then would the absence of the sun's gravity release the planets from their heliocentric orbits.

To explain this, Einstein unified gravity, space and time into a single geometry. Space and time were unified into a single space-time "field", which becomes warped by the mass of whatever occupies it.

This warping is what Newton thought was a discreet force of gravity which acted upon all bodies depending upon their mass. The greater the mass, the greater its attractive force (see below).

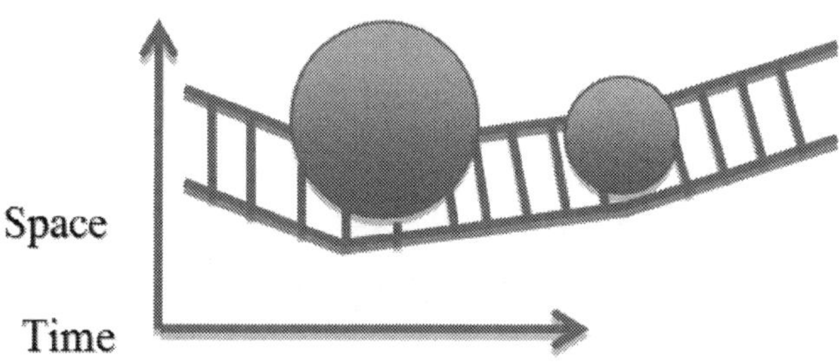

But Einstein's paradigm proved that it is the field of space-time which acts upon bodies as it is warped by their relative masses. It is important to remember that the field of space-time is constantly expanding and it is the mass of the bodies acting upon this field, which causes it to warp and appear to "pull" the bodies together.

It is also important to understand that space-time is not merely a two dimensional "fabric". Because it actually extends in every direction and encloses everything that the universe contains, out to the very edge of our ever-expanding cosmos.

As an extension of his theory of General Relativity, in 1916 Einstein also predicted the existence of "gravity waves". Gravity waves are perturbations similar to those that likely gave birth to the sub-atomic strings that underlie the emerging sub-atomic particle/waves that eventually generated the atoms that created the molecules of which our substantive universe is comprised. Initially, Einstein predicted, massive amounts of gravity waves were probably caused by the expansive force of the Big Bang.

But no one knew if they actually existed until their existence was recently substantiated by a newly developed powerful laser detection device (a "Laser Interferometer") as they undulated out and deformed space-time from the forceful interactions of two supermassive Black Holes.

This is most certainly a validation of Einstein's elegant theory, and moreover offers astronomers a possible means for probing even deeper into our evidently limitless Superverse. Yet it apparently also offers little assistance in solving the stubborn quandaries of quantum gravity interactions, which eluded Einstein because he did not yet have the temporally advanced tools at his disposal to detect anything on such a miniscule scale.

Thus, his theory of General Relativity (I.e. the ongoing relationship or "relativity" of bodies to each other in a clearly "malleable" space-time) was soon pushed aside when physics began to probe even deeper into the sub-atomic level! Here the rules of space-time seem to lose their predictive ability and the universe appears to be either fundamentally, or intermittently, chaotic.

Before the universe expanded, it was all contained in one infinitesimal point, a "singularity" in which negative entropy either did not exist, or did not yet exist. This means that the Primeval Singularity was either in a perfect state of order (i.e. Supersymmetry), or that entropy, or disorder, did not yet exist. A new physics was clearly needed to explain the resulting quandaries!

By Campfire Light

It Is Now Dark In The Forest, and a crisp gentle breeze rises up the mountainside from Pescadero Creek as it splashes blithely past a colorful array of intervening sand stones in the canyon far below our campsite. It chills the air, forcing me to inch ever closer to the comforting warmth of our campfire, as the *basso profundo* "deep bass" of a distant bullfrog reaches my ear. And as I gaze into the everchanging sea of our campfire's glowing red embers, I am compelled to wonder what the universe might look like from the Supreme's point of view.

But clearly, such a lofty perspective is impossible for we mere mortals to achieve! Yet, is it not likely that the objective universe, which we believe that we alone subjectively see, is also the Supreme's very method of perception? And by "Supreme" I mean the source of the Universal-Consciousness, which is responsible for its own identities large and small, and its own apparent actions, whether grand or minuscule, while acting at large as our entire universe!

Evidently this is so from the smallest subatomic energy particle to even the largest celestial mass. And whatever "space" seems to separate, contain, and allow for any apparent dimensional existence, or relational identity, within the spatial extension of an endless (but nonetheless illusory) "time".

But any experiential perspective must also be functionally constrained by the limits of the Supreme's body-mind instrument of illusory individual perception. If so, is it not then also true that our perception is essentially that of the Supreme. As it witnesses events dualistically through its dimensionally extended, evolutionarily elaborated, Universal-Consciousness, operating *as* "us", within our experientially individuated "space-time" fields!

Therefore what we see is also what G-d sees, through our temporal subject-object perceptions of the universe. Problems arise only when we forget that our lives are essentially being lived by the universe's ubiquitous impersonal Consciousness; the fundamentally "Self-sentient" basis of which, is likely the only veridical G-d!

Ironically, only when we begin to forget ourselves can we start to remember "Who" we truly are. But is there no simpler way to understand this?

Two Simple Statements

THE UNIVERSAL MANIFESTATION can be summarized in two simple, yet nonetheless equivalent, statements:

- UNIVERSAL *ABSENCE* (*A*) VIEWS ITSELF AS UNIVERSAL *Presence* (*P*)
- ABSOLUTE ABSENCE (A) KNOWS ITSELF AS ABSOLUTE PRESENCE (P)

1. *A* is equivalent to *P* except insofar as an apparent difference exists within Universal-Consciousness, as a point of perspective within our individual consciousness.
2. A is equivalent to P through the resolution of their respective Noumenal and phenomenal perspectives in the principality of the Supremes' foundational Awareness. Which is prior even to the Universal-Consciousness that precedes our individual consciousness.

Right Understanding

WHAT IS SEEN IS WHATEVER WE BELIEVE has been or will be, but never what simply "Is". Because what *Is*, is what we truly are! Therefore the present never actually exists, while the future does so solely in our imagination, and the past only within our memories.

Our every experience thus transpires only within the illusion of a conceptual space-time, and is comprised of dualistic concepts, which are subliminally assembled from our collected thoughts.

Whatever is experienced is therefore a fraud and all claims to any authenticity of existence, outside of our consciousness, is equivalent to so much hearsay! Indeed, in this regard, it is not the phenomena of our experiences that are important, it is understanding the true nature of our veridical Self, by which our every experience is known.

This is why Gautama Buddha taught that "Right Understanding" is crucial to our Awakening, or perhaps more aptly put, in the philosophical Sanskrit vernacular of his day; *Om Mane Padme Hum!* (see pg. 560). [Author's Note: "Pali" however, was the lingua franka of the historical Guatama Buddha.]

We could also say that as soon as we are present, the fourth dimension of space-time separates and brings into cognitive focus our future and past perspectives. In other words, perspective provides the mental "shading" by which our thought concepts are transformed into the apparent objects that seem to fill our consciousness.

And even if we do not yet know it, when taken altogether, these "thought objects" and the evidently "thoughtless" space-time penumbrae (shadows) that appear to separate them, actually comprise the entire manifest universe of our experience! For we can only be the Awareness by which the phenomenal objects of our body-mind consciousness are known.

Simply put, we cannot possibly be whatever appears to be "before" us, including the illusory subject/object ego that we observe experiencing the ongoing stream of our *vritti* comprised thought objects! In other words, we cannot be the phenomenal subject (i.e. illusory ego) that we see experiencing the mental objects that comprise our experiential universe. Because, upon closer inspection, it is revealed to be just another object of our experience!

Thus our experiential "present" is actually the moment that some particular aspect of Universal-Consciousness becomes separated within our individual consciousness. This is accomplished by its elaboration into an ideational emergence (i.e. a "dimensional" existence) for at least the minimal duration necessary for it to be cognized or manifested within our minds.

In the field of our experience we are represented by space-time, because we are the necessary living field of Consciousness Force within which the ideational objects of our consciousnesses coalesce. They appear within our dualistic subject-object minds, before (i.e. "in front of") our Divine Awareness, or so-called *Dharmakaya* or "Buddha Mind". Indeed, only thus do we "know" objects, as they expand into our dualistic or "split-mind" cognition, over an inferential space-time.

But we become easily confused by the compellingly personal nature of these continuous (and evidently contiguous) ideational experiences. And all too soon we begin to believe that we are the contingently sentient illusory ego reflection of the veridical occult "light" of our Awareness's essential Presence, by which all of our experiences are known! But if this is so, then how can the ineffable possibly ever become "effable".

Pointers

Pointers Come In Many Types. There are the old fashioned wooden pointers, like the ones with a soft black rubber tip that our teachers used, if you are old enough to entertain this memory, to point out words and numbers hastily scrawled in dusty white chalk on a large slate "black board". (Then there is the bright red laser "pointer", lying idly in my sock drawer, which I sometimes use to point out tree parts to my clients).

But in point of fact, science and mathematics are also "chock full" of pointers, because pointers can also be concepts! And these kinds of pointers are rather valuable, because when they are used correctly, conceptual pointers can direct our attention inward, towards the truer nature of our own consciousness.

"Pointedly", this is why even the most precise of languages, mathematics, is at once the most abstract. Because the very premise of number upon which mathematics is based, except by exclusion and contextual agreement, is actually impossible to define precisely! But it is the very abstractness of numbers that makes them so useful.

Perhaps the most telling of these pointers is also the most basic, that of an energies identity; and of course even solid appearing matter is actually energy in constant motion too! This type of "identity principle" simply points out that something, even an abstract conceptual something, must first be present in order for its existence to be acknowledged.

I

Numbers are clearly symbols, or conceptual place holders for other concepts, which are generally assumed to be actual objects, existing in a concrete external universe. Yet in actuality, every such assumed "object" is, always, only a mental construct. Its physicality, like the incorporeality of an assumed individual soul, is patently impossible to precisely determine!

In other words, whatever we experience could as well be nothing more than a dream. After all what is a dream, or even what we assume to be reality, but the experiential interpretation of an unremitting flow of mental symbols (i.e. *vrittis*). But the identity principle makes it necessary for even the symbols, of which a dream is constructed, to be constructed of *something*!

If we again apply the thought tool of Occam's Razor, nominally, this "something" could as well be nothing more than some sort of energy, or more likely, a Self-aware energy. Indeed, this may well be the simplest, most logical, and perhaps even the most elegant answer, to the stubborn qualitative quandary of "mind vs. matter".

And indeed, for all practical purposes, at least those useful for determining anything with any certainty, the "person and universe" that we commonly experience could as well be nothing more than a complex experiential analog, similar to the orderly symbolic conceptual world of mathematics!

This is because the abstract "Number Concept" can be used to electronically represent different identities and orders of concepts. Some of which may soon become the operative equivalent of the occult cerebral "Biorithms" that are currently operating our own hominid brains! (see "Algorithm", pg. 504).

II

Algorithms are nothing more than interrelated sets of "if A than B" types of instructions, fundamentally no different in their thoughtless functions than a mechanical mousetrap. Yet their equivalence is found operating behind the scenes everywhere in our universe, from the currently invisible subatomic quantum universe, to the very chemical and electrical processes that support life itself!

Eventually, they even accompany and support the emergence of our mind. These symbolic numbers contain hidden abstract logic that supports their definition, or "nature of presence", and allows them to represent increasingly complex concepts of number, which can then be used in a myriad of ways.

Hence we have "discovered" many types of numbers, from zero through "real", "natural", "irrational", "transcendental", and even allegedly "transfinite" numbers! The endless trail of numbers leads of course from zero to infinity, for one very good reason, and one alone, 'though it may be seen from a very many perspectives. For ultimately even zero and infinity are equivalent!

This is simply because neither can be reached by any sort of concept or percept! Hence, they will forever remain just what they are, purely "ideational" abstractions, which are borne mysteriously from within the peculiar dualistic shape of our mind's expression.

However, mathematics remains the nonpareil language of science in that its exactitude, while actually based upon an intrinsic inexactitude, well serves the purpose of science's empirical search for truth. Nonetheless, there exists a certain irony in the incertitude of the quantum field and the basic premise of identity upon which both we and mathematics stand![65]

III

Both mathematics and the science which it serves eventually point us inward, toward the conceptual cerebral machinery that acts as the "Metacognitive Processor" of our body-minds. Because whatever we experience, is done so within a psyche that is operated by nature's own evolved biological "Software", which is essentially comprised of thought and concept.

This is the brain's own O.S. "Operating System" that constitutes our everyday consciousness, and also limits our experiential realm. Nonetheless the intrinsic honesty of pure science and mathematics cannot be easily ignored. Even if, as they stand today, they are largely limited to a dualistic subject-object perspective that is trying to somehow look beyond itself!

But, this also means that both science and mathematics are excellent "pointers". Because everything in our universe eventually cancels out into the nothingness of zero and infinity, all that remains is the essential Awareness by which nothingness is known to exist, within all of its endless iterations of implied or symbolic states (of artificially dualistic "somethingness").

This means, of course, that the essentially unbounded potential of a "universal nothingness" is whatever appears before our own true nature. Which is the unadulterated sentient presence of pure Awareness that has been present in the universe since the perfect Primeval Singularity first appeared out of the absolute nothingness of a spacetimeless "Elsewhere".

However, this also disallows the self-refuting philosophical musings of nihilism, for the simple reason that there necessarily exists a principle of identity in order for anything, including non-existence, to exist at all! (See "Nihilism", pg. 509).

Fundamentally, this may amount to a nothingness that is neither corporeal (physical) or incorporeal (spiritual) in its essential nature. But necessarily, it must be aware in some fashion in order to acquire the identity by which it exists in our consciousness. In point of fact, the entire universe only "appears" when we begin to measure our sentience against the evident nothingness of our inferential interior space.

But this is actually an illusory "space" that only appears to be "before" us through the agency of our functioning hominid vehicles. Which are actually the Divine's own evolved vehicles of imagined phenomenal experiences.

Thus, we complete the universe when we spatially identify its apparent existence, and thereby create time within our experience of it. And as fantastic as this may sound, it is well supported by the rigorous mathematics of quantum mechanics as well as the exacting empirical method of science!

Ultimately 'though, only our veridical and ever witnessing Divine Awareness remains. But inevitably it does so from before, between, or beyond, our body-mind perspectives of both number and science. Therefore, both mathematics and science are very good pointers indeed!

However the existence, extent, and nature of our own essential Awareness cannot be fully conveyed or grasped by any combination (or effort) of our intellect, imagination, or emotion.

But this does not mean that being an ignoramus is of any particular value either! For apparently, our amazing mind does possess the ability to transport us to the very edge of ordinary reason. And this precipitous "edge" becomes a very good jumping off place into the *terra incognita* "unknown territory" of a more uncommon, and perhaps even superlogical sense.

This explains to some extent the utility of the Sages' unique kind of intuitional teaching, which can lead us to the edge of common understanding and eventually perhaps even beyond![66]

But if our experiences are merely the byproducts of an automatically functioning psychosomatic mechanism then what, if anything, makes our clearly ideational "us" so special?

Robots

LET US SUPPOSE THAT we will one day accomplish the artificial construction of a human-like body and brain.[67] Every cell, around ten trillion of them, will be assiduously (carefully) reproduced in all their various functions, by "nano-sized" computer controlled wireless robots shuffling around atoms and assembling molecules. In fact, this we can already clumsily do, by using a tunneling electron microscope!

There is no reason to assume that such a creature would appear to be any different from us in any way, even insofar as its various functions are concerned. It will look like us, move like us and think like us. It could even be engineered to require similar solids, liquids, and gasses as fuel and molecule replacement sources.

It might even be possible to have different "sexes" that are able to combine artificially engineered genetic material, similar to the human genome. And of course it would also need to incorporate all of the important chemical and mechanical differences that would allow a new replacement robot to grow (see CRISPR, pg. 523).

The new robot, like a human infant, would move, think, consume nutrients, and continue to develop in a fashion paralleling human development. Eventually an adult robot would emerge with the nano-encoded cellular instructions to age, and one day even die, exactly like a naturally evolved "human" being.

In other words, we have now satisfied all of the basic biological requirements of life! No one would be able to distinguish it from any other person. And if we were to conceal from our robot the fact that it is an artificial life form, there is little reason to believe that it would even know that it is not human!

I

It is also conceivable that the inner content of its experiences, the response to art or music for example, or in the recognition and appreciation of love, excellence and beauty, would be similar to us, nature's evolved human biological machine. It seems quite clear, in this "thought experiment", that there is effectively no difference between our mechanical robot and nature's human biological robot.

Moreover, neither robot necessarily hosts an incorporeal soul, nor is any such an improbable thing required, for its existence, its functioning, or for its continuation through progeny. Unless we are willing to play with words and impute the epithet of "soul" to the fuzzy experiential stuff that comprises the *qualia* "inner content" of our experience.

However, the fundamental necessity of "identity" remains intact within life, and artificial life, as well as within what is generally assumed to be the qualitatively different realms of our mental and physical experiences. Indeed, from the identity principle, the measuring rule of number emerges, and all else follows!

But it seems inconsistent to assume that at some point in the growing complexity of combinations, which are either built up through the processes of natural selection, or in our thought experiment by human agency, that an awareness should magically occur, by which the identity of any sort of form is then recognized.

Our robot, for example, could simply be doing everything that we assume identifies us as human beings, without any inner content of awareness. In fact, it is quite possible that our robot has even become in-a-sense "sentient", understanding that it exists, but still without any actual inner content of awareness, or of the symbolic algorithmic constructs by which it has achieved its apparent sentience. But this creates a conundrum, since the same can also be said of us, who are in actuality nature's sentient biological robots!

II

Nonetheless, this is the current belief of many neurologists, who feel that whatever is going on in our minds, is ultimately somehow the result of what is going on in our brains. They argue that we are not aware of most of the things that go on in our brains, which nonetheless function quite nicely in both running our bodies and generating our minds.

Awareness, however, is most certainly an intrinsic quality of "identity", just as identity is an intrinsic quality of "presence". In other words, whether you contend like Descartes that *Cogito ergo sum* "I think therefore I am", or *Sum ergo cogito* "I am therefore I think", the principle of our identity remains codependent with any sort of conceptual or substantial awareness.

This is to say that if something could nonetheless be present, it would be lacking in identity if the awareness of its presence is lacking. Indeed, it becomes impossible for any material or even wholly conceptual thing to be actually present, because both presence and identity are apparently only enabled by the pure presence of an independent Awareness.

Moreover, if Awareness is "present" it most certainly has an identity, even if this identity defies our brain's dualistic method of understanding, because it is incomparable in its intrinsic and all-encompassing monism.

And most certainly this Awareness is present even prior to the consciousness that eventually emerges. (Due to its accumulating "sentient presence" within the dualistic consciousnesses of both our mechanical and humanoid biological robots!).

Consciousness is therefore contingent upon the functions of both sorts of robot's bodies and brains. But Awareness is likely co-intrinsic with the basic identity by which our robots and their complex brains are known to exist. It is also likely that the fundamental principle of identity, which underlies any sort of subatomic existence, is co-intrinsic with the Awareness by which its presence exists!

And, of course, this foundational primeval Awareness exists even prior to our dualistic consciousness, apparently in much the same way that its aspect as the identity principle of unconscious form does. Therefore energy in the form of matter, while being necessarily "unconscious" outside of a functioning cognitive platform (either synthetic, biological, or an amalgam of the two) is not really in-sentient, because it is actually *pre*-sentient!

Yet paradoxically, the presence of pure Awareness must also be noumenal to the phenomenality of our identities, because they are opposite poles of the same primeval substance, which is neither corporeal nor incorporeal.

However this substance is closer in nature to the consciousness that emerges within the elaborated minds of our sentient robots, than it is to the insentient material from which their bodies and brains are constructed. This is because the subtlety of discrimination that eventually emerges within growing consciousness, emerges from within the intrinsic Awareness of the elaborated substances that allow our minds to appear (i.e. subatomic *vritti* "strings").

In other words, consciousness, rather than merely self-referential mechanical function, emerges only because of the co-intrinsic identity of the primeval Awareness that lies within every subatomic particle of which both types of robots are comprised. But it cannot emerge unless an intelligent robot is present, because it requires a specific physical platform to operate. Because, although intelligence and function can be engineered, apparently Consciousness cannot!

Therefore, the kind of Consciousness that has the potential to evolve itself into sentient cognition, within a sophisticated electronic computer, as well as within an average human brain, only appears because of the fundamental nature of the material of which our bodies and brains are comprised.

And mere unexamined group consensus, superstition, spirituality, quackery or habit of opinion does not alter this situation. (Any more than ignorance of the law will keep us from getting a speeding ticket if we are caught breaking the posted speed limit!) Because physics is the universal law of Nature!

It is to this substrate of essential Awareness that the Sages refer when they talk about our true "Self". Therefore it is not at all beyond the realm of reason to predict that humankind might one day create an entirely new sentient species!

But in the final analysis, is this accomplishment not simply the ineluctable force of Nature, which through a patient process of "natural selection" is giving birth to the fuller expression of its own primeval Awareness, rather than we ideational human beings successfully playing at being god?

Life and consciousness are thus apparently aspects of one another. However, if the universe and consciousness are indeed "Awareness based", then what of our subject-object based values? Are they simply arbitrary, or do they have even a modicum of intrinsic veracity in themselves? Since we order our lives by them, this seems an important thing to know if we are to gain a better understanding of ourselves; and so our sojourn into the great mystery of our consciousness continues!

Hope

"Lasciate Ogni Speranza Voi Che'entrate"! "All hope abandon, ye who enter herein!" This inscription appears above the gates of hell in the Divine

Comedy, authored by Dante Alighieri in the 14th Century. And perhaps Dante is, in a manner-of-speaking correct. For is not this hope fundamentally only a desperate prayer, which our illusory ego entity holds out against all reasonable hope? Hope for the reality and immortality of our ego's own illusory experiential self and its passing universe of ephemeral concepts!

Surely this unexamined hope, which wounds as much as it heals, must be abandoned before exiting the illusory infernal gates of the sometimes lovely, but always beguiling Maya, the "goddess" power of universal illusion.

For what could the king of hell, *Lucifer* the "Bringer of Light" possibly be, if not just a masculine Middle Eastern name for East Indian antiquities dear beguiling Maya? A symbolic deceiver, illusory Lucifer is simply the occult dualistic demon of partial truth's false light and matter's impossible cyclical destruction.

Hence, the fictional diabolical author of the fanciful living illusions and persistent memory myths of our supposedly eternal post-death torments, simply cannot be trusted!

But what is our deluded illusory ego, if not the servant of Lucifer and perhaps even the ideational charmer himself? Indeed, the faux light which our illusory ego entity brings to our consciousness is but a dark web of dualistic concepts.

A continual array of inevitably incomplete concepts that finally displaces simple reality with a powerful group illusion that all too often becomes the most hellish portion of our persistent waking dreams!

So-called "heaven and hell" are but the extreme polar opposites of our dualistic life experiences. And the egoistic hope that is born in the darkness of this false light, places us in opposition with our own deeper, truer Natures.

Therefore, only when we find the wisdom and courage to finally, "Abandon all hope!", can the gates of our self-created "hells" finally be breached. Because only then will we be free to enter into the truer heaven of reliable truths that an unencumbered mind and careful science might now reveal.

For this false hope, not to be confused with other sorts of better dreams, is actually our illusory ego's attempt to take charge and influence the experiential events of our lives, as they automatically unfold within nature, according to the inexorable will of Universal-Consciousness. This "will" emanates from within the intrinsic Awareness that precedes the identity of the presence of Universal-Consciousness, by which anything and everything exists!

"Heaven, purgatory and hell" are thus merely potentials within the expansive realm of Universal-Consciousness, and can only be experienced conceptually through our minds. Verily, all around us the world is filled with people who are unknowingly living in a solitary purgatory of imagination, and sometimes, sadly, even in their own private, emotionally disturbed "hells".

Which is certainly an unpleasant set of natural processes. But as such, they also remain processes to which the supernatural forgiveness of "sin" is simply irrelevant! Moreover, is this not especially (if not egregiously) the veridical case, when a simple visit to a competent Psychopharmacologist is much more likely to provide an actual cure?

I

Few are experiencing the transitory happiness of the involved or "lower" heaven, which is only available within our dualistic consciousness. Because happiness, as everyone knows, is fleeting. Fewer still come to know the "higher" heavenly *Shalōm* "Peace" of a persistent Self-Awareness while yet residing in their bodies.

The bigoted religious fundamentalist, for instance, who believes that everyone else that does not share his or her faith is going to hell after they die, has already unwittingly entered the gates of hell, all upon their own. Thus, no outside assistance is required. Nonetheless they may even claim that, "The Devil made me do it", when their erroneous assumptions inevitably lead to actions that have quite the opposite effect of their good intentions!

But there is really no one to blame but themselves. Because they do not understand that there really is no birth, and therefore no substantive suffering, or even final death. And that heaven, hell, and even their ideational god, exist only within the ephemeral and ever shifting concepts of their temporary individual consciousness. For it rarely occurs to them that this representational god is actually only an ideational idol, or a "Golden Calf", if you will.

Entrapped by the emotions created by superstitious fears, imprisoned by intellectual dogmas, and hindered by admonitions and obsessive rituals, they soon cannot escape the walls that they have erected in order to protect their imaginary souls from the devils within their own minds. Because the unexpected demon of their own illusory ego still holds them tightly in its grasp! How can they understand that their ideational soul knows nothing of their demonic fears, or their religious concepts?

And so when their body servant naturally expires, the Life Force and Awareness that animated them for a while, simply reenters the universal totality. Like a drop of glistening dew slipping from a leaf in the early morning sun, it blissfully merges back into life's covert "home". Rejoining the vast occult, sub-quantum, Primeval Singularity perturbation "waves" of Universal Consciousness, that float above the dark Elsewhere "sea" from which our universe explosively emerged around fourteen billion years ago!

But because life is ever a work in progress, very few discover the Universal Manifestation as the Divine likely intends it. For the innate universe is actually nature's גן עדן *Gan Eden* "Garden of Eden".[68] The awakened Sage, compassionately neither "dead" nor "alive", abides peacefully therein, moving unconcerned amongst the endless experiential mental constructs of the Manifestation, as if among fragrant flowers, without lifting a finger or even moving from his seat.

And when his, or her, days naturally end, their essence peacefully reenters the *Gan Eden* with scarcely a sigh, having never truly been anywhere else! Having rediscovered that their essential nature is the solution to the illusion of their mortality, what need has such an über man or woman for transitory hope?

But in this purported "hopelessness" lies the unlimited power of our entire universe, which continually manifests an endless stream of boundless, compassionate *sah-Hahm*. This is the veridical portion of our universal "Be"-ing, with the power to reach our deeper subliminal programming and free us from the considerable inertia that prevents us from confronting our long habit of wrongful identification with the lost and illusory thought constructed aspects of our virtual conceptual beingness.

Such a Sage has transcended both the universe and humanity. They may look like a "person", but while yet embodied they are actually the living מרכבה *merkaba* "chariot" of the Divine, as are we all!

As an initial result, of this greater understanding of our essential deeper Nature, we may mistakenly be tempted to try to deny the life that we now understand that we cannot change. We may even believe that this might somehow stop, or limit our involuted Awareness's spontaneous participation in the ongoing stream of our everyday consciousness. But this is an erroneous assumption that only leads to a rather disappointing, if not eventually painful, (and sometimes even literal) "dead end".

Denial

Surely Denial Must Be A "Sin", if such truly exists? The beauty of the dawn, the infectious smile of delight on the face of an infant, the first kiss and tender "miracle" of love. All of these wonders and more, which are carried in the bosom of that cosmic enchantress, "Mother Nature", only amplify the sorrow of those who, sadly, may never even guess at their profound loss.

For indubitably, there are those things, *sine qua non* "without which life loses its savor"! Therefore nothing essential should be given up, but the ignorance of our own true nature. Indeed, just as our life is not really an event of our own making, neither is there any real choice in its living, "giving up", or otherwise!

But we can cause much unnecessary pain to ourselves and others in trying to practice its denial. As Gautama Buddha observed thousands of years ago, unless you are fond of grief rather than enlightenment, asceticism is patently absurd!

Only remember the truth that you and your story are being lived by the Divine Universal-Consciousness, and finally abandon the very idea of its denial! Upon discovering this, the ancient Chinese Masters would say, *Zhòng shēng wúbiān shì yuàn dù!* "I vow to liberate all beings, without number!" But if a Buddha is not self-made, and we are all really just "Buddha's in waiting", is he (or she) somehow then "chosen"?

The Chosen

Yes, Jews Are The "Chosen People", G-d's and mine, since I became a גר *Ger* "Jew by choice", and by *Beyt Din* "Rabbinic Court", many memorable years ago. But what a weighty, if not presumptive, task to teach the already wise! So, if it pleases, my dear *Yiddishë* "Jewish" kin and brother *Rabbonim* "Rabbis", I must beg your pardon, but the אמת אלין *Emet-elyōn* "highest truth" apparently has no worldly preference. It surely embraces all who earnestly pursue it!

For is not the presence in our mind of a self-referential consciousness that is capable of generating our experiences of "person and universe", not also responsible for generating the concept within our mind of "god"? Surely you do not suggest that we should bow to a mere idolatrous ideation of the Divine! So where then is our true G-d, and do we indeed suffer for "His *or* Her" sake?

If we but have the courage to abandon our familiar conceptual space-time, within the indelible quantum "Akashic" record, we may even now experience the suffering of our brethren in the *gehinonim* "hells" of Birkenau, or a thousand other unspeakable places, as we painfully await our own destruction. Because such cruelly unmitigated suffering, surely, knows no space-time boundaries! Understandably, many have asked, "how can it then be, that יהוה *Yahweh* the "Supreme" has breached His own covenant with us"? (See "Birkenau", pg. 505).

But I will not waste our time placing this petty illusory "god" on trial. For wiser men, well-schooled in justice, have already tried this purely conceptual god, and found him not only culpable (blameworthy) but astoundingly irrelevant, if not confoundingly insufficient!

Indeed, his unprotected, innocent victims fairly cry out for exoneration from the lie that they should bear the liability for their own suffering and destruction! A phenomenon that often transpires when men try to explain why an anthropomorphic god would allow the indescribable suffering of the innocent, especially on such an unimaginably massive scale! (See "Consciousness and the Law" pg. 403, and "Akash" pg. 504).

For is this anything other than the abomination of a faulty reason that cheapens the sanctity of life and truth by unfairly blaming the innocent victims of unspeakable brutality for causing their own suffering? So, let us once and for all bury this petty conceptual god, as we strive to understand the truer G-d of Abraham, Mohammed and Jesus. But where then, and whom, is our G-d and how might we best live in accordance with the occult will of the Divine?

Such existential questions have occupied Jews since the time of Abraham. Indeed the tireless search and endless sacrifice for their truthful answers has arguably changed, in a meritorious way, the otherwise apparently wayward direction of the entire world! If just the earnest search for trustworthy answers to such questions has had such a universally beneficent power, how much more so must be its transformative power upon the sincere seeker!

But is Yahweh or Allah not the One, blessed be "His" or "Her" name, whose presence in consciousness grants understanding to whatever we experience?

If we must *daven* "pray" to something, perhaps we should worship the Divine presence that is within us all, by which we all understand that we exist. Because if we were bereft of awareness, our entire world would be denied existence! But in the greater presence of our minds witnessing

Divine Awareness, the quandary of our life and death, and even that of our existential "heavens" and "hells", is ours wordlessly in "grace" to understand, as the pranic breath of eternal Life begins to seek its own solution for mortality and suffering within us (see "Prana", pg. 511).

But what is, was, and shall be, will not change. Only the belief that we were ever born, or were actually involved in the machinations of an ever-turning *Chakrapani* "Wheel of Fate" (see pg. 521). Or that we can personally change what is already written in G-d's ever unfolding *Sefer shel Chayim* "Book of Life"!

However, this transformation cannot transpire until we first understand what our mind truly is, and how it subtly works. Otherwise we are in danger of becoming stranded upon the painful shores of our ever accumulating unhappy memories, as the river of life passes indifferently by.

Indeed, this deeper sort of understanding must be "first hand", taking place from our *in*–side, as our mind automatically accomplishes the geometry of its experiential "magic". Weaving together the concepts of our person(s) and our universe(s) with an ideational space-like "shuttle", and illusory time-like "thread", into the complex tapestry that is our experience of their ongoing manifestation in our dualistic consciousness.

But from our binary mind's perspective, only the tridimensional objective "knots" and dangling ideational "threads" of our experiential four dimensional universe can be seen. We simply cannot reconnect the ever divisive threads of the constant space-time bounded stream of apparently separate "consciousness-objects" that is being continually generated!

The "celestial tapestry's" indescribable beauty only appears when we look outwards from within the fifth dimension of our whole Mind. (Indeed, this spontaneous meditation, or "self-inquiry" goes on automatically, all of the time, eyes closed or eyes open, whether we are aware of it or not!)

But eventually, as we begin to fully awaken to it, we may even become convinced that what we see directly is the absolute truth. In other words, the "absolute truth" already lies directly before us! But it is a task for each of us alone to discover whatever this "directness" might truly be.

No one else can do this for us, and only a fool would believe that suffering and injustice can ever be explained away with mere words, ceremony or even with well-meaning prayers! Frustration, distrust and distain will ever be the final reward for any such calumny (discrediting) of reason!

But the *Sefer shel Chayim* "Book of Life" is only closed to the dead and wherever the *mirakuru* "miracle" of life remains, seeing the truth of being is always possible. For it is indeed the truth of the Supreme, for which we suffer, until we come to understand the deeper meaning of the Divine Covenant and so fulfill the command to "become a light unto the nations." – "Isaiah 49:6.

Yet until you understand, you may cry *apikoros* "sacrilege"! And I may even appear to suffer at your hands, while you – most likely – continue to conceptually suffer the unnecessary vagaries of an illusory, experiential world. But even if the Book of Life is always open to us, how can we be "reading" it without knowing that we are, and moreover, how in the world can "you" and "I" possibly be One?

In loving memory of my Christian father and older uncles who courageously fought to liberate our surviving Jewish people from the horrors of the Nazi extermination camps. And of my Jewish uncle, Aryeh Jacobi, who did the same, and then together with the help of my dauntless aunt Shulamite, heroically dedicated his life to help rebuild the nation of Israel. An honorable, if not sacred, undertaking which their children, and children's children, bravely continue to this day! You can hear his personal Holocaust recollections, along with those of the still living Jewish survivors, by visiting the website at:

https://www.youtube.com/shared?ci=k50hu6IDw88

אם תרצו׳ אין זו אגדה. ואם לא תרצו׳ אגדה היא תישאר!

Ēm tirtzu, ayn zoh agedah. Veēm loh tirtzu, agedah hē tēshaair!

"If you will it, it is not [merely] a dream. And if you do not, a dream it is, and a dream it will remain!" – An excerpt from, *The Jewish State*, published by Benjamin Ze'ev Herzel in 1896. (Please, see pg. 602).

Our "Book of Life"

We Live The Pages of our ספר של חיים *Sefer shel Chayim* "Book of Life", and with each passing moment we are subjectively "reading" the universe and ourselves into a symbolic experiential existence. But if this is really the case, then simply put, "How-in-the-world" could it ever be accomplished within the cognitive confines of our everyday minds?

In order to better understand our own common everyday *ex animos* "conundrums-of-mind", let's preform another "Einstein-like" thought experiment. Let us imagine (for a moment) that we are a two dimensional being. Because we can now only sense the universe in two dimensions, anything having to do with a third dimension (of height and depth) is completely invisible to us! This is akin to how we three dimensional beings sense the fourth dimension of space-time.

Now, to progress even further with our symbolic "thought experiment", let's imagine that our Book of Life lays open on some sort of table (that almost magically appears before us!). As a two dimensional being, we can only see the two dimensional surface of the three-dimensional "page of life" that is currently open before us.

Therefore, when we are moved by the universe past the third-dimensional edge of one "life-page", on our way to the two-dimensional surface of the next page, we would not even notice that we had just passed through it!

In a similar fashion, we are constantly moving through an invisible fourth dimension of space-time. But we are not cognizant of our journey through the fourth dimension, because it is forever beyond our brain's ability to sense. Indeed, the only way that we can even know that the fourth dimension is here, is by its persistent absences!

But we rarely question the mysterious manner in which everything before us seems to automatically change. Even though the method of change (like the energetic oxidation of dry campfire wood that is caused by igniting it with the flame of a burning match) may seem apparent to us, the mechanism of time's constant change remains stubbornly imperceptible. Nonetheless, change is always going on everywhere, within and without us, like a magician's crafty trick that is always happening right before our very eyes!

Now, imagine that (again as a two dimensional being) we are looking at the numerous pages of our Book of Life from the "side view" (of the third dimension). If we begin to flip through its pages, as if we were shuffling through a deck of playing cards, this is how the force of the Big Bang is experientially propelling us through the spacetimeless void of the pre-Singularity "Elsewhere"!

In other words, our entire Superverse (including the invisible spacetime in which it appears), which is comprised of all of the expanded and endlessly expanding multidimensional "parallel universes", is already fully present within our individual and collective Books of Life!

Albeit, even if we can only see a miniscule three dimensional part of it, our entire Superverse is also (always) multi-dimensionally present throughout all of space and time. Indeed, what we commonly think of as the "past, present and future" are also already fully present. Even if we cannot directly sense them all at once, because of the limitations that are imposed by our brains and senses.

Thus, as we more "three dimensional" beings cognize our own Book of Life within three dimensions, it becomes rife with intermittent episodes of apparent (four dimensional) nothingness! Which we can't help but notice in a serial fashion, as we flip through the four dimensional spaces of our own Book of Life's pages, in the temporal experience of living our lives.

But it is these spacetimeless moments (of pure Awareness) that provide an opportunity for us to rediscover that we are actually always (only) the Supreme! Evidently we are nothing, more-or-less, than the Supreme, experiencing the puissant distraction of its dualistic "selves" (i.e. our many illusory egos), within the constraints of three implied dimensions, wrapped in the space-time conundrum of an apparently invisible fourth!

But if we cannot see either the fourth dimension of space-time, or even the eternal pre-fifth dimensional Witness thereof (that we in eternal Essence most probably always are), how can we possibly ever know if our thought experiment is valid, or not!

Shema

Don't Worry, this is not for Jews alone, for truth is both the property and the concern of everyone. For are we not all equally the children of the Divine? *Shema* שמא is a *pa'al* "simple" Hebrew verb of the imperative form, it means "Hear!" As in intuitive listening to the highest Truth.

But this quality of attention is only possible for those who remember. Therefore **remember**, your apparently "personal" experience is actually G-d's grand adventure! And moreover, you are in fact only an infinitesimal portion of G-d's experience, in the totality of our astounding multiverse!

This is because the essence of all experience evidently lies within the presence of the Divine Awareness upon which the appearance of everything else depends. Hence, what many erroneously believe is "god", is only an idolatrous and "fundamentalist" notion of יהוה *Yahweh* (or الله "Allah", etc.) the true fundamental Supreme from which the entire universe is manifested. It is "That", *Tat* or *Sat* in Sanskrit, by which we know that we are and by which, "…all the stars in the heavens are arrayed!"

Indeed everyone, and everything, is fundamentally comprised of the selfsame Universal-Consciousness (i.e. *Elohim*). Which is an impersonal and Universal-Consciousness that is unaccountably and recurrently emerging from within the Supreme. In the East Indian Vedantic scriptures the Supreme is called *A-dvaita*, literally the "in-divisible" and omnipresent One, that in ancient Hebrew Scriptures is praised, *Shema Israel,*

Adonai Eloheinu, Adonai Echad! "Hear O' Israel, the Supreme our G-d, the Supreme is One!"

However, the "Shema" is only sacred to the extent that the Divine truth of the Universal-Consciousness contained within its words is actually understood! For it is only by embracing impersonal Universal-Consciousness fully, at its Self-sentient supreme Source, that its truth may again live within us, by the grace of Divine intuitive revelation.

Indeed, in many ways the primitive mind of humankind, of six or eight hundred thousand years ago, was much closer to the grace of the primeval שהורש *shoresh* "root" of the Universal-Consciousness, from which all life springs forth.

Like our present day animals, that have not yet constructed an elaborate synthetic world of concept and language, the mind of primitive man was not yet too far divided into an ideational subject and object. Humankind had not yet ventured too far (in mind) from the *Gan Eden* "Garden of Eden" and was still relatively innocent.

But from around ten thousand years ago, humankind apparently became almost hopelessly lost in the increasingly complex dualistic consciousness that had finally evolved within our progenitors increasingly busy collective minds.

Our ancestor's rapidly growing and unwieldy subject-object minds, brimming over with contrary concepts, ever growing subliminal pattern recognition, and puissant emotions, required an increasingly discrete, yet ever more powerful executive center. As a result, our species divisively "self-ish" human ego eventually made its appearance.

When conceptual "good and evil" finally entered the mentally divided world of humankind, some way to control our unruly ideational world was needed. The next step in the evolution of our hominid minds is recorded in the symbolic tales of tribal humanity everywhere around the world.[(69)]

In one Jewish tribal tale, in order for this event to transpire, Isaac, the innocent but illusion bound mind or "son of Abraham", must be securely bound and sincerely offered up as a sacrifice to the Supreme. This is to say, in an allegorical form, that it is important to return to an accord with our universal Source! (See "Abraham", pg. 503).

The Universal-Consciousness that gives birth to nature is also arguably "life affirming", and while it is unambiguously "impersonal" in nature, it is also relatively balanced, peaceful, and complete in itself. Therefore, it serves as the basis for justice and peace to appear within the reflective consciousness that is operating in the divided mind of modern man.

It is the foundation of humankind's symbolic world, and when this connection with our essential Source is lost, there is no end to the mischief that an unrestrained and imbalanced illusory ego might unleash upon the world! Reestablishing this accord is what religion is really all about (i.e. *religio*, in old Latin means, "To tie back").

But of course, the ideational threads which bind our ever dividing concept bound mind must increase in number and complexity just to keep pace, and a point of diminishing returns is soon reached. In other words, without a clear understanding of the inner meaning that esoteric traditions rest upon, the Jewish "Shema", Hindu "Advaita", Buddhist "Sutras" and the esoteric traditions of all religions are merely words.

And even prayerful words may soon become only more conceptual stumbling blocks, as the rather superficially understood "Bal Shem Tov" has so aptly pointed out! Consequently, quiet "inward listening", rather than constant immersion in study, prayer, and ceremony, is evidently of paramount importance (see, "Bal Shem Tov", pg. 517).

Only remember that it takes diligence, earnestness and endless patience to do this kind of *Kavanah* "Inward Listening". Listening of this order is more akin to meditation and introspection, or if you prefer "self-inquiry". But intuitively hearing, and fully understanding, what has already been revealed to our innermost hearts is finally a matter of Divine "Grace".

Therefore the attunement afforded by *Kavanah* is certainly not without merit, or honor. After all, everything, always, lies within the provenance of the truer G-d of Abraham that is ever behind, and within, the Universal Manifestation!

But, with all due respect to my Muslim *tios* "cousins", to whom, like many of my fellow Kabbalists and "Zen Judists" (i.e. Jewish Zen Buddhists), bowing is itself a form of respectful prayer that even non-Muslims should be respectful of), if so, then do we really need to grovel before G-d, as if the Divine were some kind of fickle earthly king?

Bowing

We Are Told That Bending is good exercise for our backs, but that stooping under a heavy load is bad. However, where "bowing" is concerned, our true healer is evidently the eternal One that delivers us from the crippling burden of our own self-ignorance. And once this weight is removed, we are told by the Sages that our Life-Force will sooner or later rise up to heal the illusions and subsequent delusions of our supposed births and deaths.

As we have already seen in our earlier inquiries this is most likely true, and even more compelling evidence awaits us further on. But since the verity of this claim awaits our personal verification until it comes to pass, it seems wise in the interim to bow before no other! Or perhaps, even better to "stand tall", with an unshakable faith in the redeeming munificence of our own Divine Source!

Here and Now!

Life *Is* A "Miracle" Of Sorts, it is what a Japanese Zen Roshi sometimes calls the *mirakuru* "miracle" of Consciousness. Because our body, it seems, is either composed entirely of tiny subatomic particles, wavelets of energy, or wavicle expressions of some sort of fundamental essence, or Universal-Consciousness (see "Wavicle", pgs. 516, and 596).

Indeed, the entire cosmos is thus likely composed of the same impersonal Universal-Consciousness that is conceptually contained within our only apparently "personal" individual consciousness.

The universe emerges quite suddenly from within nascent Universal-Consciousness, and Universal-Consciousness emerges from within the universe simultaneously. Because there is an essential identity of emerging Universal Consciousness with its evidently endless energetic expressions of infinite potential. Thus, the fundamental wavicle energy expressions, of which our universe is likely comprised, are actually the extreme distal (far) pole of undivided Universal-Consciousness.

But there is a continuum of universal Presence upon which both depend. When we merge into this Presence (of Universal-Consciousness), we enter into wordlessness. But thereafter, upon the event horizon of its cusp, every word uttered is from within the ineffable and ultimate "Source" (Shem Tōv) of all scriptures and universes.

Herein lies the root of the manifest consciousness that is our own Essence. It is also the source of all prophesies. But even this emerges from within the *Parabrahman* "Supreme" (see pgs. 514, and 585). Which is the root of the root of the Universal-Consciousness that consciously contains, and energetically expresses our entire universe, multiverse, and superverse!

People often misconstrue this merely to mean "infinite potential". But, even this rather imaginative conceptualization is still sorely lacking as an adequate descriptor for the *Parabrahman* (an ancient Sanskrit term that is also synonymous with our modern astrophysicists pre-Singularity "Elsewhere"). For the simple reason that "That", which is necessarily prior even to Consciousness itself, cannot be measured in conceptual terms!

In other words, the presence of the veridical "present" is always prior to both the universe's impersonal unitary Consciousness, and our effectively "illusory" individual-consciousnesses. But it is nonetheless also quite inexistent, without the fundamentally self-aware presence of a witnessing supreme Awareness.

So just where, in fact, can we say that the "present" might be found? Yet it is nonetheless always "here", and ever "now", and therefore always foremost in our dualistic minds![70]

But in order for anything to transpire in our minds, first the energy, then the space and time of which our mind is composed, and within which it apparently operates, must be "present" in some fashion. Yet it must also be able to change without losing its essential identity. Hence a good question is, "How does this curious arrangement function in such a manner that space and time seem to be ever at our disposal, while both we and the objects that space and time seem to somehow "contain" are also apparently at its rather indifferent "mercy"?

Fractal Bridges

REMEMBER EINSTEIN'S WELL KNOWN COMMENT on Heisenberg's Uncertainty Principle, "G-d does not play dice!" But perhaps a more contemporary combination of modified "Chaos" and "Fractal Geometry" theories could be used to create a concept bridge between Einstein's and Heisenberg's widely dissimilar models of our universe?

However, unless you are by trade a mathematician, a scientist, or perhaps a computer programmer, you're probably not very familiar with either fractal geometry or chaos theory.

But all that you really need to understand is that everything in our universe can be represented by recurring "fractal" like patterns. In computer programming, algorithms that utilize fractal geometry can be used to create software that faithfully reproduces the organic silhouette patterns that are found everywhere around us in nature.

And when we look at the mathematical concept of "fractals" in this way, it seems apparent that the naturally occurring fractals that it is based upon, are essentially just naturally occurring microscopic patterns.

Thus although they may appear to be singularly dissimilar, they are actually based on the ubiquitous presence of the broadly repetitive distal electron orbital energy patterns (i.e. "valances") that persistently appear in the formation of every elemental atom and organized molecule in our universe!

These persistently recurring atomic scale "outline" patterns, that allow the geometrically predictable fractal like surface shape of everything in our universe, are most likely grounded in the balanced periodicity of forces that define the elements that are listed in our "Periodic Table of the Elements".

Indeed, it could even be said that the breaking of the Primeval Singularities supersymmetry (in the "Big Bang"), that was caused by the (eventual) insertion of space into the dense chaotic compilation of the Primeval Singularities ever growing infinite probabilities, is what gave rise to the eventual separation of strings, membranes, and wavicles of quantum energy into repetitive arrays of atomic scale fractal like surface shapes.

In other words, the Primeval Singularity was never really static, but always dynamic, in that it (in potential) also contained the universal "Trifecta" of nascent Energy, Space and Time that "soon" (in the miniscule measurement of Planck Time) caused the FTL "Race of Cosmic Inflation" to begin!

This gave rise to countless numbers of "matter waves" that formed concurrently with an expanding field of force, called the "Higgs Field", which with its companion "Higgs Boson" began to transfer mass to the innumerable monopole strings of perturbation energy that were beginning to form. (How the Higgs Field and Bosons formed is still unknown, but is under relentless investigation at the Large Hadron Collider facility in Geneva Switzerland).

With the accumulation of "particle-like" mass, the matter waves began to push the initial supersymmetry of the Primeval Singularity apart as sub-quantum monopole particles began to spread out into dipole waves.

This is because at "apogee" these waves, or "strings" of energy appear to consolidate and take the form of "matter", which has mass and is effected by gravity. But at "perigee" they seem to lose almost all of their mass and gravity and spread out into waves that disappear into time, which is the opposite, or complimentary pole of space.

Hence quantum particles appear to be either a localized mass (i.e. particle), which occupies space, or a non-local energetic quantum wave that propagates through time by changing its apparent spatial position when it reemerges from within the universes quantum substrate. Which it does by briefly moving into an unpredictable, and therefore apparently "chaotic" quantum state.

This "insertion energy" when viewed relative to quantum space-time events would then superficially appear to act much like a "force of gravity". Hence the mysterious space-time bending "force of gravity" may actually simply be the more detectable portion of this process.

Moreover, it remains a clandestine process that on a deeper level acts like a "Strange Attractor" or "Dark Energy". Which recurrently draws the next most likely probability from within the apparently intervening quantum "chaos. But if this is so, then G-d does not "play dice", because G-d *is* the "Dice"!

Familiarity

LIKE THE PHENOMENON OF CONSCIOUSNESS, the concept of "familiarity" is quite revealing! Because the newness of things disappears as familiarity is assigned. This habit is quaintly termed the "habituation response".

But Zen, Ch'an, and Advaita Masters often persistently retain the heightened sense of awareness that is the hallmark of any new experience! This is no doubt allowed by gradually, but persistently, changing their point of mental perspective, and the brain-body functions that most certainly underlie it.

Because our brain's neural pathways automatically supply the information that they continually receive from our peripheral nervous system, through the *diencephalon* or *thalamic* region, to our *cerebral cortex*, any substantive neural change also implies an expanded neural network!

This is to say that any action, which is accomplished by an expanded cognition, can only be accomplished through the nucleus of axons that connect our brain's *diencephalon* to its *telencephelon*.

Hence, the remarkable plasticity of the human brain seems to almost magically emerge within our consciousness! (See "Cerebral Cortex" pgs. 505, and 522; "Diencephalon" pg. 524; and "Plasticity" pg. 565, for a deeper explanation of these anatomical terms.)

This fact provides an excellent "pointer", curiously as it were, beyond itself! Because the consciousness, which contains the activity of our brains and bodies within our minds, must present its purely conceptual information to an aware, but qualitatively independent observer.

Therefore, when the illusion of a conceptually dependent, but nonetheless impossibly independent objective "observer", our evident ego, is no longer identified with, whenever concepts show up, they are inevitably seen as fresh and new!

Indeed, the universe is seen with the fresh eyes of a child and mysteriously becomes ever new, or perhaps even "evergreen" again, as our brain's naturally occurring antidepressants, such as *dopamine*, *serotonin* and our own "feel-good" *endorphins*, are increasingly released.

But when our brain's pituitary "master gland" begins to incrementally release ever greater amounts of the powerful biopsychedelic "D.M.T." *(N, N-Dimethyltryptamine),* and our consciousness reaches its mind expanding "tipping point", we may again return to the spontaneous joyfulness of our youth! (See "Dopamine", pg. 525).

Hypothesis abound as to why and how this occurs. But one of my personal favorites proposes that this happens to help our brain grow a better "consciousness interface" with the subjective Source of its own intuitive sense of individual selfhood. Which then serves as the durable psychic locus for our ideational ego.

This is most likely accomplished by growing more connections between our brain cells dendrites and the microtubules which connect them "wirelessly" to the entire quantum universe! (see "Microtubules", on pg. 550).

Hence, whenever we look closely, we invariably see that whatever happens within our minds (including our purported "enlightenment") also transpires, in its own fashion, within our brains and bodies. And because they are composed of the same sort of underlying Universal-Consciousness, they are only separated by evolved degrees of transition, rather than of kind.

Like the gnarled old oak tree that has been transplanted to a Zen Master's garden, or the thousand-year-old redwoods that still stand, like sentinels from the primeval Jurassic past, around our campground. Regardless of their wild or domestic locations, they always remain just what they are; *trees!*

Indeed, this message is so profoundly simple that we must take care not to lose it entirely, while trying to clarify it in the ever multiplying complexities of our inquisitive dualistic minds.

The message is simply this; "We are 'That' by which our consciousness understands, and not simply the body that provides the material of our understanding!" In other words, our true identity is (for want of a better term or concept) the "Supreme". Which mysteriously gives rise to the Universal-Consciousness of which our entire universe is likely comprised.

But we will not be able to experience this understanding fully, until we are completely comfortable with having no particular identity at all![71] This is because that which is aware of consciousness, ironically must itself be "unconscious". What then might the nature of That, which is independently aware of our everyday subject-object consciousness, possibly be?

But by now, what with most all of my attention being focused upon our inquest, the previously luminous sky above us has grown too quickly dark, and with it, my brighter thoughts have finally grown dark too.

So, inching closer to our bright campfire for a "bissel more" warmth, I carefully pour myself another aphotic cup of hot coffee, to rouse my poor tired brain!

But unlike Diogenes curious daylight search (with a bright lamp) for truth in the company of injudicious men, we are compelled to do so in the quiet darkness of the forest night, in the steadfast company of wizened old trees (see "Diogenes", pg. 525).

(Dedicated to dear Frater Thomas Merton. A dedicated fellow monk, of a different "Order", who understood that all monks, regardless of their denomination, are seeking the "reawakening" of their essential oneness with the selfsame Supreme!).

The Point of Inception

In Regards to the indefinable "point of inception" from object to subject, and subject to object, almost no one ever thinks to examine not only the mechanism of our consciousness, but also the origin of the Consciousness within Life, by which all things are known.

Involvement

We Must Give Up the idea of somehow achieving any-sort-of "non-involvement" in our consciousness. The very idea is absurd, for our consciousness is involved automatically with every thought of which it is comprised!

The "Heavenly Family"

The Veritable "Heavenly Family" is the *Supremus* "Father", *Natura* "Mother Nature", and *Puer Sanctus* "Whole Mind". But in the superstitious fundamentalist Christian mind this basic biometaphysical truth often takes on a more purportedly "spiritual", yet nonetheless decidedly incorrect (if not childish!) anthropomorphic "Trinity" sort of meaning. (But be very careful who you say this to, lest you cruelly damage the uncomplicated faith of the innocent and simple minded!).

While in superstitious fundamentalist Jewish and Muslim minds, it often takes on an anthropomorphic (and by today's standards, a violent and demented) "Vengeful Deity" sort of meaning! And in the unimaginative and more rigidly intellectual "empirical mind", this basic biometaphysical truth often takes on a singularly narrow scientific and atheistic sort of meaning.

Nonetheless, all of the above are either secondary aspects of thought, or direct thoughts in themselves, that are subsequently arranged into conglomerations of relatable concepts. Nowhere can be found sufficient reason to think that our thoughts, which are contained in our dualistic consciousness, are not simply the phenomenal aspect of a single Self-sentient

Noumenon. Indeed, even our evidently invisible space and time are simply "objects of thought" too!

In fact, because everything that we experience is comprised of thoughts, the thoughts of which our (inner) universe is comprised do not always exactly match the supposedly "external" objects that they nonetheless attempt to accurately measure!

> *"Imagination is more important than knowledge. For knowledge is limited to all we now know and understand, while imagination embraces the entire world, and all there ever will be to know and understand."*
>
> Albert Einstein.

Pi

PI TIMES THE RADIUS (1/2 the diameter of a circle) curiously *almost* equals the circles surface area. But since π "Pi" is a non-repetitive, apparently endless sequence of numbers (i.e. 3.1412…), the total area of a two dimensional circle, or of a three dimensional sphere, is actually incalculable! Interestingly, this also means that Pi, so applied, could also contain all of the numbers that symbolically represent every possible three dimensional event transpiring in four-dimensional space-time.

Evidently "indefinable", as well as incalculable", is exactly what all things, including all numbers, actually are! Moreover, *we* are evidently the ever missing number(s); and not only in Pi, but for every psychosomatic based equation that altogether constitute our total experience of life! (*Hoc pro* אחי, Harold J.).

Not This Not That

THIS IS, BUT I AM NOT THIS. So what is "this" that I am not, if not the consciousness that contains both person and universe? But since I am not in Essence "that" either, all is my object and phenomenal to my noumenality.

The Endless Forest

THERE IS NO "PERSON", there is only knowledge. Within our brain knowledge arises in our consciousness as the concepts that represent our evident persons and universes. But conceptual "persons" have no more control over the appearance of the concepts - that contain every other apparent person - than they have over the evident appearance of their own (or other bodies) in an only evidently external world!

Knowledge accumulates automatically like the clouds that arise spontaneously from the surface of the sea, in the warming presence of the sun. Except that you are the essential "Sun", which illuminates your body-mind mechanism with the seminal concept that you seem to exist. (And by which not just suns and seas, but every other apparent thing is known to be!).

This is the knowledge that you are, which is the secret יסוד *yesōd* "foundation" upon which all of our conceptual worlds are built. The spontaneous appearance of this concept within the universe's (i.e. "our") body-mind contains the potential of endless other concepts. Just as a seed contains a veritable pantheon of endless forests, all patiently waiting - to someday Be!

Oh, That Again!

"REALITY", WHAT A CONCEPT! But whoever said that the universe is unreal? It is simply consciousness experiencing itself in the only way that it presently can, within the oneness of an only apparently "individual" universal diversity.

But its apparent subject is a complete mental fabrication, existing only to steer the temporal ship of our body-minds. But any vessel requires a captain, otherwise we may as well be a quanta of unthinking *eruv rav* "flotsam and jetsam".

Or perhaps we are simply a mass of mental perturbations drifting aimlessly around, in an insensate space-time sea.

But when we ask, "Who knows this?" Ironically, we're exactly right! It could only be some independent "One", who knows the contents of our consciousness. Good job, now we can all go home!

Fading Embers

As Last Night's Campfire Embers Slowly Fade, the dark sky above our humble little campground is just beginning to become lighter in the mountainous East. While in the shadowy forest, all around us, the woodland birds are starting to sing happily, anticipating the dawn of a new day and the promise of renewed warmth from the fresh golden light of the newly rising sun.

But I've been up all night, immersed in deep meditation, occasionally emerging into contemplative thought to carefully write down whatever's been intuited, or internally "seen".

Fatigued, and shivering a bit from the early morning chill, I must finally surrender to Mother Nature's insistent circadian rhythm.

So for now I must bid you, and the temporal universe of our shared subject-object forestland experiences, a heartfelt *gute nacht* "good night"! As I stumble groggily back to my tent, and finally collapse into the patiently waiting arms of dear, gentle, sleep.

Santa Cruz Mountains
Northern California, USA

Chapter 4
Arbutus "Madrone"

Mountain Men

While Hiking Along The "Creek Trail" Today, I cannot help but notice that the trunks of several large old Tan Oak trees are slowly bleeding red sap. This is an old woodsman's trick, to mark your path with an axe so you won't get lost in some unfamiliar part of the forest. Someone must have come along this way, no doubt happily playing at being a "Mountain Man" just as I did when I was young.

And although chopping bark with an axe is not good for the trees, they will likely soon heal, to endure the adventures of many more generations of intrepid young explorers. But children cannot know their paths in later life, some pleasant, and others not. Yet each, in turn, presents a different sort of journey, ultimately leading us all back home again. So understanding the truer nature of our various life paths and journeys seems important, lest we lose our way!

Perhaps the oldest and purest teachings about our apparent life journeys in the universe fall under the heading of Taoism, literally the "Way". Taoism holds that our evidently "personal" journeys are as much fiction as verity, because their fundamental basis lies in consciousness. In other words, all that we can ever know of the universe is contained within our minds.

Juxtaposed to this subtle psychic fact, which has been accepted for thousands of years in the East, stands the rigid intellectual perspectives and analytical traditions of the West. As a cumulative result, a tremendous body of knowledge and technology has emerged from our predominantly "Exoteric-Objectivist" views of the universe.

This is both a blessing, and a curse, for while our lives are certainly improved, we also seem to be ever teetering on the brink of self-destruction. And even our long cherished traditional exoteric religions have been inadvertently, if not disturbingly, debunked by modern science!

Our primitive fundamentalist dogmas regarding an anthropomorphic god induced creation, for instance, have been effectively overturned by modern cosmology and physics. While Darwin's theories of natural selection, paleontologists' numerous fossil findings, and the discovery and confirmation of DNA's biological purpose by Watson and Crik, have revolutionized our view of life itself.[(72)]

Yet many otherwise intelligent people still tenaciously cling to the comfort of their old and familiar (but unexamined) views of life, despite the fact that they must often surrender reason to do so! Absurd as this seems, evidently our early childhood experiences and training have a tremendous, and often lifelong impact on our psyches.

But we need look no further than "level-headed" (i.e. practical) tree trimmers for help with this issue. This is not to say that tree trimmers have some sort of inside track on the truth, since truth's provenance lies equally within us all. But because most tree trimmers are by necessity a practical people whose heads are generally not too muddied up with a lot of academic intellection, they soon come to understand from direct experience that it is much more difficult to unlearn a bad habit than it is to learn to do it right the first time.

But this in no way mitigates the necessity of learning to do it right, even if doing so is at first uncomfortable and difficult! And in this regard, the harmful ignorance of the average modern American, where the people, religions, and customs of the modern Mideast are concerned, serves as a disturbingly puissant example.

Not knowing the true history of our democracies greatest ally in the Mideast, the State of Israel, for example, renders the poorly informed mind vulnerable to the unjust and self-serving political schemes of the persistent radical modern revisionists. (Whom, we should not forget, are also our own countries sworn terrorist enemies!) Those who would blame the calamitous consequences of their own past aggression and continued explosive suicidal violence on those who were initially (and continue to be) their innocent victims.

But for an even closer example of the more widespread problem of unintended "disinformation", there are currently from 5 to 8 million Muslims in America, among whom almost 78% are U.S. Citizens. And although the rather large uncertain citizenship of nearly 3 million people certainly does point out an apparent deficiency in our current method of census taking, it also does not mean that they are all terrorists, or that they harbor terrorists in their midst!

Hence, rather than listening to the ignorant fear mongers, who currently want to banish all Muslims from America, we need to listen to our own better informed experts.

Who remind us that we are fortunate to have patriotic Muslims who are willing to endanger themselves and their families, both here and in the Middle East, in order to serve as translators and advisors (as well as soldiers) in the modern civilized worlds' ongoing struggle against today's radical fundamentalist "Islamic terrorists". Who are trying to legitimize their indiscriminate criminal actions by identifying themselves with an outdated interpretation of a major world religion that in today's world teaches peace.

Indeed, both Christianity and Judaism have also abandoned outdated practices that can nonetheless still be found, memorialized along with all of their suggested brutality, couched in the archaic Biblical Hebrew of the Torah (i.e. which Christians commonly refer to as the "Old Testament").

(Dedicated to the memory of an American Muslim hero. Army Captain Humayun Kahn, who gave his life in brave service to his men and country in Iraq in 2004.).

I

Ironically, the traditional religions of the Far East, and more pointedly the metaphysics and philosophies that they are based upon, often prematurely dismissed as mere "heathenism" and therefore incorrectly assumed to be simplistic and lacking in substance or verity, are found to be in agreement with many of the findings of modern science.

This is especially apparent in physics, which has always been considered foundational to the other branches of science. Such compelling evidence is widely emerging that we may live in a subtly sentient universe, that it is puzzling and somewhat difficult to understand the resistance of so many people to the acceptance of demonstrably sound new ideas!

This is doubtless another (rather potent!) way in which our illusory egos can enslave our minds. And indeed, it may even serve as a valuable lesson to the rest of us, who may believe that we possess a more open mind. For it does seem to follow that having and keeping an open mind might really mean that our consciousness is not being harmfully restricted in its focus by our pre-programmed illusory egos.

Our familiar ego chimera it seems, is especially prone to accepting the admonitions, emotions and fundamental world views of our parents and guardians, when our brains are young and still developing. A relevant question then is; "Why are some people able to overcome the psychological restrictions imposed by their conceptual egos, including their ever accumulating cognitive and emotional memories?"

In all likelihood, it has more to do with the presence of imagination, creativity and intuition in our psyches, than to a more than average amount of intelligence; those very aspects of consciousness which currently set us apart from the thinking machines that we have created.[(73)]

II

But affectivity, by its very nature, seems to be largely independent of the other more concrete aspects of our minds and is, therefore, less subject to the programmed constraints of our ego thought constructs. The more plastic aspects of our consciousness, along with affectivity, seem to reflect a different order, or quality, in the more recently emerged and personalized illusions of humanities individual consciousnesses.

Indeed, they seem to indicate the harmonious reflection of our universe's primeval unitary Awareness into, or from within, the tumultuous everyday consciousness of our dualistic minds.

Because of its probable foundational nature, this Awareness is likely not only necessary to the very existence of our universe, but is also clearly independent of our ideationally individualized minds. The essential *Nisarga* "True Nature" of our foundational Awareness is therefore less subject to the constraints of our autoprogrammed egos, even within the restrictive confines of our illusory dualistic consciousnesses!

The affectivity, which is our faculty of emotion, however, does not seem to enjoy the same independence, and can be quite unpredictable insofar as which master it may be convinced, or inclined to serve!

The volatility of human emotion and its capacity for such a wide range, from malice to love, would seem to indicate that it stands with one foot near our primitive instinct for survival, and the other near the reflection, in our experientially divided consciousness, of the universe's unitary primeval Awareness.

And while it is probably not wise to eschew (dismiss) our formidable intellect, we do tend to pride ourselves overmuch in it, and the encomium of pride is often less than a virtue!

Sadly, despite its many recent gifts to humanity, our powerful intellect far too often seems to generate quite strong, but nonetheless fictional, ties to our baser illusory egos. This is not only a problem in itself, but one that works at cross purposes with nature where the survival of Life itself on our planet is currently concerned!

III

Even if we believe that it somehow sets us apart from other creatures, it seems likely that our prodigious intellect only reflects a refinement of nature's basic survival instinct. Ironically, this refinement often expresses itself more violently, and with far greater cunning and cruelty, than it does in our fellow creatures that very often seem to possess it to a far lesser degree!

Indeed, the very presence of love, compassion, altruism and tolerance within our psyche is a probable reflection of our deeper Awareness. Which is always prior to our individual consciousness, and the refined intellect that it contains.

"Compassion is the greatest evidence of the existence of G-d." – An excerpt from *When Bad Things Happen to Good People,* by Rabbi Harold S. Kushner. This is another excellent "Life Guidebook", from a kind "warrior" of Truth.

Who, because he so courageously opened the doors of both his Synagogue, and his innermost heart to the world, I quite often think of with the same admiration that I reserve for my own insuperable little Irish, Free-Methodist, "Circuit Preacher" grandfather.

Who rode from town to town in a horse drawn buggy with my mom's big family of twelve, bringing hope and healing to the drought stricken farmers of rural, "Dustbowl Era" Nebraska.

Both men steadfastly served their congregations for many years, and if they had ever met, I'm quite sure, would have quickly become the best of friends! Because both selflessly took whatever they could finally "wrestle free" from the cruelly indifferent, cold embrace of heartbreak, loss, and sorrow.

Then they selflessly brought what they had learned about the transformative power of impersonal Divine Love, to the "Front Lines" of the broader world, *outside* of their Church's and Synagogues closed doors!

What could possibly be more powerful than the Force which brought our entire universe into being? The powerful "Love to Be", which is doubtlessly the Supreme Parabrahman's method of experiencing diversity, within its own essential "Elsewhere" unity, which was present before all else!

This is just one example of Nature's ongoing struggle to advance the "growth" of our humanity within this corner of the Divine's unfolding universe (see "Our Address in the Superverse", pg. 561). By "growing up" a still immature humanity, through the awakening of compassion within our consciously evolving human hearts (i.e. our hearts "evolve", as the Rabbis of old would say, through גמילות חסדים *gmēlut Chasidim* "acts of loving kindness".

And although it may not appear to be so, because of all of the "bad news" that's always being focused on, humanity is finally beginning to wake up! Because "good things" like these are starting to happen, with ever greater frequency, in ever growing places, all around the globe.

To the observant eye, especially when we take the "long-view", it is becoming increasingly evident that our world is rapidly changing. In fact, it is changing, "for-the-better", so fast now, that acts of brutality, that were once widely ignored because they were so "common place", stand out in all their horrific detail on the crystal clear screens of seventy billion wireless hand held audio visual communication devices!

And the world is almost instantly stricken, with both sympathy *and* even genuine sorrow, for complete strangers!

Moreover, while we human creatures tend to take great pride in both our intellects and our hearts, we also commonly observe these same attributes in our fellow creatures on a scale commensurate with their sophistication, or evolution, and concomitant (resulting) presence of consciousness.

Therefore, I'll expand our conversational parameters just a bit, and go out on a limb, and render a tree trimmer's simple observation, based upon a cherished childhood memory, and leave it up to you to parse out its deeper meaning for yourself.

Whenever we are not completely certain that we have finally learned *everything* that there is to know in this life, as my plain spoken, but nonetheless quite wise, country momma once said to me; "Son, life is not easy for any of us. And sometimes, as you know, it's downright hard! So you always need to be brave!"

"But no matter what you might have to do in your life to survive, and no matter how successful that you might someday become, never forget your 'country roots'. Don't forget to say "I love you!" to the ones you love. And always stay just like you are now son, never be afraid to be your *real* self, and remember to always be humble and kind."

I miss my momma, and I'll wager that many of you probably miss yours too! And evidently, because of all the Divine Love that's doubtlessly involved in creating and raising a human being, more than a few mom's probably rank quite high, in the hierarchy of superlative "heavenly" beings!

Now, if we can, almost every one of us, agree on this one extraordinary fact, how can anyone ever believe that humanity does not have the power to change the world through the beneficent power of Divine Love? Because brother, although it may not be obvious to many, the hand that fills and rocks the cradle, "rules an unruly world" with Divine Love already!

Evidently, a big part of our poor suffering world's problem, is that far too many of us loose our connection with our own deeper childhood roots in the difficult business of surviving and trying to thrive, in the tough experience of our shared dualistic World Dreams! Thus, we have to fight hard, just to hold onto an enduring remembrance of the bright childhood (guiding) "stars" of Divine Love, if we are fortunate enough to have a loving guardian to point them out to us!

Thus, our intellects, when liberated from our illusory ego's overwhelming influence, can become a worthy instrument in the service of our deeper, truer Self. Yet without the guidance of our subtle, but nonetheless ever present, inside compass of the "hearing heart", our intellect can easily loose its way in the constantly unfolding complexity and apparent struggle to survive in the tough virtual environment of our inner experiential universe(s)! This may even be the first real step in our journey toward Self-understanding, during our existential experience of "be"-ing in the natural universe.

Hence, achieving a clear understanding of our true-life situation, seems to be the necessary foundation for a deeper revelation that both we and our universe are in "Essence", exclusively this Awareness alone!

This is why I keep repeating the refrain, "Our *illusory* egos"! Therefore, while our ego is arguably necessary as an operative center for our psyche, can it be said too often that being only a collection of thoughts, concepts, emotions and such, it possesses no veracity of independent identity or existence? In fact, it can have no truly independent existence at all!

IV

Chinese Taoism is probably the clearest and most direct occult wisdom teaching, along with Ch'an, Zen, and Advaita. But unfortunately the term *Tao* is generally translated by linguistic scholars into English as [the] "Way".

In contradistinction to this *misprision* "inexactness", Tang Dynasty Taoist Master Hwang Po tells us, "There has never been anything, no past, no present and no future, they have no meaning. This must be understood fully, before you can enter the Way".

But, as earlier discussed, space-time is most likely the partial conceptual representation of a four dimensional reality that is, quite simply, beyond our sensorial range! The more contemporary Master Lóng Po also clearly states that, "Past, present and future have no meaning, as nothing has ever existed".

This is to say that nothing could exist, or not exist, without conceptual space-time to grant it the necessary dimension of space, and duration of time, to be perceived. Space-time, as we experience it, simply has no real existence, per se, independent of our perceiving consciousness.

The fabric of space-time is certainly present to be affected by what we perceive as mass and gravity, but they too are present only within our experience of them. And like the fuzzy "location vs speed" conundrum of a lepton, even the actual *in*-side and *out*-side of our universe remains rather uncertain!

V

Outside of our perception, form and force are merely implied. If anything, they must be the ineffable not "not-something", or not "not-nothing", of which our entire universe is apparently comprised. This is to say that the entire universe is, in Essence, neither something nor nothing!

However, no dialectic (reasoned argument) could ever truly describe or penetrate the consciousness of which it is comprised. Much less the pure foundational presence of the Awareness by (and of) which our entire universe is formed! So how could there possibly exist any "Way" to "enter", or to be understood fully?

A more probable explanation is that this is a colloquial translation of Tao, which has the more exact meaning of *Dharmakaya* "Buddha Mind" (see pg. 506). Therefore Master Lóng Po is most likely saying; *"You must fully understand that space-time and all that it contains has no real independent existence, and you must understand this before you can again realize, or actualize by direct understanding, the essential universal Awareness that you truly are"* (see "Hwang Po", Pg. 544).

Finally reaching the end of the trail, I can at last remove my hot and dusty short healed Tony Lama "walkin' boots" (since I don't have my horse with me today, and I even left my long rifle and short fiberglass fish'n rod at home!) and wriggle my poor tired toes in the cool clear waters of Pescadero Creek.

Oblivious to my hot aching feet, its crystalline waters splash happily past a nearby isthmus of taupe sandstone that juts out boldly from its mossy green banks. Soon invigorated, I wade across the chilly stream, take a comfortable seat on a big smooth sandstone rock, and ready for another adventure, slip my socks n' boots back on again.

Still Water

Looking About, I Spy A Rather Large Misshapen Stone rising above Pescadero Creek's cool clear waters like an inviting little island, and I carefully step up onto its uneven top. Now precariously poised on the unsteady stone, I stand motionless on one foot, like a Japanese Hokkaido Crane, just long enough to suddenly lose my balance.

Frantically flapping my arms I try to recover. But since I cannot actually fly into the air like a dancing Hokkaido Crane does, I must hop quickly away to another unsteady stone, or suffer an unceremonious fall into the ice cold water that surrounds my little rocky "island"!

This wobbly balancing act must be repeated time and again, to preclude a chilling tumble into the "icy" water that splashes merrily along below my unsteady feet. At first my balancing act is simply a curiously satisfying game.

But as I continue to play, I am incrementally transported backward, in the ever flowing stream of time's passing memories, to my earlier childhood days. And happily, the unmitigated joy is still there! In the simple pleasure of hopping from stone to stone, up a clear and chilly mountain stream.

But when I was young, the stream was almost always full of imaginary crocodiles and piranhas, and if you fell into its cold clear waters they just might make a quick meal of you! Nonetheless, their imaginary presence added some real spice to the "stone hopping" game!

But now, no doubt because I move just a bit slower due to the inexorable effects of time's ever passing years, I notice a brownish red crawdad (crayfish) lying motionless on the golden sandy bottom of the transparent pool at my feet.

Above him, a small school of silvery fingerling Rainbow Trout flashes by as I bend down to inspect the little stovepipe shaped, pebble studded homes of the Dobsonfly. For the larvae's protection, and for them to better catch any tasty morsels that might be floating by, the little "stovepipes" are generally cemented by the industrious little larvae onto the broad upper surfaces of stones that are immersed in the constantly rippling shallows of fast moving water.

When in my youth I fished these chilly mountain streams, the secretive little *hellgrammite* larvae proved to be the best bait to catch the wily larger Rainbow Trout. This is a timeless moment, and remarkably it seems that nothing has changed much in over fifty years, nature remains reliably, and comfortably, unchanged!

But, all good things must come to an end. And as the evening sun begins to dip below the towering tops of the surrounding redwood trees, I must begin the long and shadowy trek back to our distant camp.

A Campfire Discussion with Sifu Lóng

Finally, Back In Our Redwood Camp Again, my fingers have become so cold and stiff in the cool evening fog that I can scarcely grasp my pen! In the slowly fading light it has become too difficult to see what I'm trying to write, anyway! And in the distance, the now familiar "who-Who?" calling of a Great Horned Owl *Bubo virginianus*, seems to echo my own silent inquiry, "Who is it that hears, the 'hearing' of our listening?"

In the rapidly fading evening light, I fumble around with the old fashioned silver Bic® lighter that my quiet, country strong, "Oakie" daddy gave to me, before he finally left this *"Tired ol' world behind"* to *"Be back with your Grandma and Grandpa again!"* And then I try to kindle, then rekindle, the little kindling "tepee" that I had carefully erected earlier today to ignite tonight's stubborn campfire.

Sputtering at first in the increasingly foggy dampness, the little kindling tepee at last begins to catch fire! And soon a warm and cheery blaze begins to burn inside the small campfire ring of sooty, fire tempered stones.

Fingers of bright amber light soon begin to reach out from the smoldering wood to illuminate our campground's surrounding ring of massive redwood tree trunks, and tentatively stretch out into the darkness that silently nestles as quietly as a California Quail between them. And our cozy little redwood campground is now suddenly awash in our awakening campfire's happy golden glow!

As I settle gently back into my favorite camping chair with a heartfelt sigh, I can at last warm my poor fog chilled digits and water cooled toes by the bright light of tonight's cheery little campfire!

But now I hear a mysterious rustling sound emerging from the nearby Huckleberry bushes. And a ghostly shape begins to materialize from within the shadowy gray mist! Straining my eyes to see more clearly, a spooky, but nonetheless by now increasingly familiar shape appears to be slowly floating towards the bright circle of our campfire's golden light.

Ari: (Rising quickly and bowing) "Sifu! It has been too long!"

Sifu Lóng: "Yes, Ari (returning bow), the years have flown by like the wings of a hummingbird!"

Ari: "It is so good to see you again Master Lóng. Please, sit down and warm yourself by my humble campfire."

Sifu (Smiles) "Of course, but I am most curious about this 'humble' campfire of yours, it must be quite *unique*."

Ari: (Laughing) "I can see that your sense of humor has not abandoned you over the years Master Lóng! I hope that it is alright for me to record our talk?"

Sifu: "Of course, of course, no problem, but I wonder what I might have to say that is worthy of recording?" (Smiles.)

Ari: "I can only hope that my ears are worthy of hearing whatever wisdom my venerable Sifu might care to share?"

Sifu: "I see that you have not lost your ability to flatter a foolish old man (Laughs heartily.)

Ari: "If you are an old fool, then what am I?" (I join in the laughter, and after the merriment finally subsides, a few moments pass by in silence.)

Sifu: "What would you like to discuss?"

Ari: "Whatever comes to mind Master Lóng, perhaps a teaching tale?"

Sifu: "Hmm, I think that I may have just the right thing! A very old 'Shao-Lin' tale that was told to the young monks in olden times. It is called, The Tale of Robber Chi."

Ari: "Wonderful! And while you are telling it, if you don't mind, I will start brewing some Golden Lotus Tea."

Sifu: "Ah! A warm cup of tea would be perfect. The evening is getting chilly. Does the fog come in like this every evening?"

Ari: "I think it must have known that you were coming and floated in from the coast to greet you" – (smiles).

Sifu: "Well, in that case, I am honored to be chilled in making its acquaintance" (smiles and bows to the fog; see "Sifu" pg. 513, and "Shao-Lin" pg. 575).

Robber Chi

"Now, This Is The Tale of the notorious robber 'Chi', who lived long ago in old China. Robber Chi was ferocious; he preyed upon the wealthy, merchant and noble alike. Anyone that was brave and foolish enough to travel upon the dusty country roads carrying valuables was likely to become a victim of his heartless band of outlaws!

Robber Chi was arrogant and merciless, he would kill his victims with impunity, and he wore a necklace that he made from their severed ears. He became quite wealthy and powerful, so powerful and arrogant did he become that he overstepped his bounds one day and ambushed, robbed, and killed the Emperor's eldest son!

The Emperor sent a squad of his most highly trained martial artists and soldiers to hunt down and kill the upstart robber Chi. He ordered them to bring back Chi's head to the royal palace so he could impale it on an iron spike outside the palace walls for all to see. And so he could then spit in Chi's ugly face, just as Chi had done when he broke the Emperor's heart, by killing his beloved only son!

The Emperor's "hand-picked" men chased robber Chi's reprehensible band across all of China, until only he remained alive of all his gang of merciless cutthroats. Finally one day, they cornered him outside of a small village in the Northern provinces.

Robber Chi hid, shaking, under a wooden moon bridge as the Emperor's men closed in. Dust began to fall from between the boards as the horses moved overhead. Looking up, robber Chi saw a black and white spider lowering itself on a tiny glistening thread to escape harm.

Spying robber Chi, the soldiers let fly a deadly shower of arrows, hatchets and spears, and in the only act of compassion in his brutal life, robber Chi reached out and saved the little spider from sure destruction." (Master Lóng takes a long pause.)

"Every ten thousand years, the poor souls that are confined in hell have a chance of redemption. On this day, robber Chi looked up and noticed a small bright white light shining at a great distance from above.

After some time he spied a tiny dot moving closer at the end of a glistening thread. To his surprise, eventually robber Chi was greeted by the very spider that he had once saved! Grasping the spiders extended legs, Chi's tortured soul was slowly pulled up from hell, as the spider reeled the long silky thread slowly back into its tiny body."

Ari: "Thank you Sifu! What a wonderful story! I can see its value in teaching the young monks about the goodness that is always there in people, even if it gets buried deep inside them by ignorance and cruelty."

Sifu: "Yes, but perhaps its essential message is about what 'redemption' really means, and that heaven and hell are only the superstitious conceptualizations of the masses. Many people have the wrong impression about 'goodness' and 'badness', and how they affect the opening up of their "Buddha Mind."

Ari: "I understand, but please expand upon what you have been saying."

Sifu: "Many believe that they can 'good' their way into Enlightenment, or to 'Heaven'. When the simple fact is that there is absolutely nothing that anyone can do to become what they already are! There is no 'thing' and no 'one' to seek, because you are trying to become or find the one that is doing the doing, and seeking the Seeker. And there is no 'where' to go, because you are already there!"

Ari: "You have said that there is no heaven or hell, but you just mentioned 'heaven' along with Enlightenment, are you implying that they are the same?"

Sifu: "Exactly! In fact, I am saying that you are already in heaven. Heaven is your natural state of being, and awareness of it while you are still having the experience of being in a body, is what constitutes "Awakening", which is even prior to Enlightenment! But I must also add that no one understands or can even accurately describe what this might be.

Ari: "But there must be something to do or to understand, otherwise aren't the monks and nuns all just wasting their time?"

Sifu: "Very good Ari! Yes, but all that you can do is to correct all of the misunderstandings that make you think that you are not already Enlightened. This includes every concept that is grounded in duality, such as the concept of good and evil that the Tale of Robber Chi deals with.

The actual disappearance of the illusions of perspective, regarding the objects of our consciousness, can only take place after this is done. The main point is that our ephemeral ego identity, which is just a collection of concepts, is totally unreal, including its functions as a necessary operative center within our mind"!

Ari: "But, if our ego is no longer in charge, even though it is not really who we are, how can our mind and body possibly survive?"

Sifu: "This function is quite capable of taking care of itself without the misconception that you are the ego in charge of it, just as the body carries on all of its other functions without any interference or unnecessary help.

But as long as you believe that you are only the ego, the body, the mind, or any combination of them, even just the tiniest bit, you will be trying to do something to save your illusory self. Even if you don't consciously realize it, you will be constantly trying to buy, bully, bargain, or steal your way into heaven!"

Ari: "Is there any advice, Sifu, that you can give us to overcome this problem?"

Sifu: "Techniques are useless! There is nothing that you can do other than totally accept that your body, all of your acquired skills, memories, thoughts, and even the person that you imagine yourself to be, will sooner or later 'pass away'. Only when you fully believe and understand that you are not the ego, the mind, or the body, and every doubt is removed, will the unreal 'you' begin to disappear."

Ari: "And is 'disappearing' a good thing?"

Sifu: (Smiling) "Neither good nor bad, but necessary. When you have disappeared enough, the Tao begins to reveal itself, because you were standing in its way. Eventually, you will become so small that one day a tiny puff of wind will blow the last illusion of 'you' completely away."

Ari: "Then what remains?"

Sifu: "Only the Tao, only the Buddha Mind remains, because it is what is there all along. Everything else is just a fleeting mirage. "

Ari: (Bows) "Thank you Sifu, you have come a long way, and I am most grateful for your teaching. Temple Lin Quan Yuan is long ago, and far away, but also, never truly forgotten (See "Lin Quan Yuan", pg. 550).

Sifu: (Sighs and returns bow) "Have you forgotten what I taught you? There is no 'space' Ari, and there is no 'time'. There are also no 'objects' and more importantly there are no 'subjects' either. And from the apparent beginning of the entire universe, nothing has really happened at all!"

Ari: "Yes, Sifu Lóng Guì Xìng, I have not forgotten; I was only speaking metaphorically."

Sifu: "Um hum…then let's see how well you remember your Kung-fu lessons." (The sound of fast movements, rustling garments and scuffling, punctuated with an occasional grunt or short burst of mirthful laughter. A few minutes pass…).

Ari: "Enough, enough, I relent!" (Breathing heavily). "I see that the years have not slowed you down Master Lóng (laughing).

Sifu Lóng: (Chuckling good naturedly). "And I'm happy to see that you have not forgotten your lessons, but you do seem…" (We hear a few indistinct words, some static noise, and the recording abruptly ends).

Sifu pointed out, before he had to leave, that I seemed "a little bit out of shape", and had "slowed down some over the years". But he then added that I had become more "cagey" with age, and that this is "better than youth and speed anyway"; which did make me feel a little better!

He also reminded me that the real goal of Shao-Lin Kung Fu is learning to direct and concentrate the flow of Chi (Life Force), which can only fully happen when we become one with the Tao. These are his final words:

"Your practice now is to meditate upon the meditator, and constantly to watch yourself from outside the mind. In this way you will slowly break the habit of identifying with your senses, and the thoughts that go on unceasingly in your mind. When this happens, you will be able to effortlessly direct your Chi like the venerable Masters of Temple Lin Quan Yuan".

This seems like very good advice, which I am happy to follow and share with my fellow *Rōmi* "Travelers" in the universe. For indeed, are we not all in our *Gypsy* "heart of hearts" just that?

(Dedicated with *dai mahalo* "great thanks" to my inimitable Goju Kai Sensei, Al Monar, and to my wise Nikko Jujitsu Sensei, Professor Sig Kufferath. Whose long-ago lessons have on more than one occasion saved not only my life, but a few others as well! And whose more sagacious words have thankfully served as a reliable "guiding light", through the difficult "darkness" of some of my life's more challenging days; *Arigatōgozaimashita* and *mahalō eia kekahi* "thank you"!).

A Grand Illusion

Perhaps No One Should Attempt to fully spell out the truth of life, otherwise only more concepts will be created, to further cloud our already beleaguered intuitions. Certainly any understanding that goes against our true nature is unsound, but living in ignorance does no one any good either! So, do not interfere with your busy mind, simply let it do its work, and live your life as best you can.

Only understand that we are its impersonal Witness, which certainly includes witnessing not only our illusory ego entities, but also the body-mind vehicles by and through which our every phenomenal experience arises.

This true "You" (the subjective Awareness, by which you become an "I") is prior to every experience, and therefore even to our consciousness and conceptual individuality. Lacking a true subject, the subject/object (i.e. ego) of our everyday experience, a faux entity, is merely a virtual center of perception.

Our mind is illuminated in the reflection of the impersonal witnessing Awareness, which is the true "I" of you and me. This reflected illumination is what we experience as our everyday consciousness. And in the activity of the experiential machinery of our body-minds, it entertains the conceptualizations of our individuality, and relative realities.

The entire universe is thus contained as the apparent individual objects of consciousness within our "binary" (i.e. subject-object) minds, which make their transitory symbolic appearances in the complex electrochemical activities of our brains.

Thus, our sentience automatically arises, but also abruptly ends with the death and dissolution of our bodies and minds. While the impersonal cosmic infrastructure of Universal-Consciousness, which makes everything possible, remains completely unaffected.

Of course this does mean that any real introspective success can only be gauged by fully understanding who and what we truly are. But in any case, no one finally remains to gloat over whatever could be finally fully understood, when it finally is!

Such is the quandary of our implied existence, and the truth of our being, which ever underlies whatever little consciousness that we might temporarily enjoy. Yet another thorny issue, rarely addressed in metaphysics, remains. For even if this esoteric understanding seems well and good enough for us, what about the evidently "illusory" people, places and things that we invariably come to love?

Terrena Familias

Amongst Humans, families are nature's best current method of species continuation. Psychologically, families also serve to preserve the emotional security of our latter years. Indeed, outliving everyone that we love, to be finally left alone and perhaps even helpless, is a threat to both our psyches' and our somas!

If we die first, then we are out of it. But when we outlive the people and creatures that we love, our psychological suffering only deepens with each and every heartfelt loss!

Their death becomes our bereavement, and their loss is felt for as long as we continue to live. In this way, even the brave who may have little fear of pain and death, suffer just like everyone else. When this understanding finally comes to us through the experience of living, we begin to question our existence and may even wish for some sort of supernatural "Grace" to protect or save us.

But we need look no further for Grace, for it has already been given. For Grace could as well be called the experience of life itself! Some call this a "miracle", but it is not, because everything that exists does so only as an act and percept of Universal-Consciousness. Nature simply has no need for dubious (unlikely) miracles in order to accomplish the seemingly miraculous!

Because life is a dualistic, cognitive experience, first our soma is born and only later is our psyche "born". Moreover, we do not choose to come into our universe, but are here because of love, nature's love of "Be"-ing.

Our bodies are products of the ecstasy and love of our parents, and of theirs before, stretching back to the first arduous emergence of life from within insentient matter. Everything in our astonishing universe, from "alpha to omega", is thus the artwork of unfolding Universal-Consciousness!

All living things are members of the same great family. The ever evolving family of Life, which works both for and against itself, through apparent loss and gain, by constantly dying and taking on different forms in order to preserve and advance itself.

Thus life "feeds" upon life, apparently indifferent to cruelty and blind to suffering, until we begin to somewhat understand by glimpsing the inner connectivity of our impersonal Universal-Consciousness. Which is actually the indefinable fundamental "Essence" of our undifferentiated primeval universe.

Ergo, we need look no further for answers than Life itself, and the conscious universe that appears as if "before" us. Because there is nowhere else to look, and no real "person" is ever really present, to do any real sort of looking anyway!

Do It Right!

One Day We Hired A Man to paint the inside of our house, and Ray showed up at our front door. But he had to duck to come inside, because Ray was seven feet tall! As a painter this gave him a distinct advantage, it was so easy to reach the high places that he didn't even need a ladder!

Ray's assistant moved the furniture and removed the things hanging on the walls, then he taped or tarped everything else to protect it, and painted the walls below Ray's waist.

They did our entire house in a single day, and when his helper inadvertently glossed over a few spots, Ray came back and repainted the whole wall, for free! Over the years he came back two more times to paint our house. Each time he brought a different assistant, and each time Ray was a little grayer.

When we called him a fourth time his wife said, "I'm so sorry, but Ray has passed away". As I recall, every time that Ray, a Master Painter, gave us a bid he would say; "Want to do it? Want to do it right?" Now who can argue with that!

Simply put, direct doing only follows upon direct seeing. Masters it seems - like chameleons - may appear unexpectedly, assuming all sorts of shapes and sizes! But then, are we not all essentially just camouflaged "Masters", naïvely existing unawares, until we finally awaken to the fact?

Empty Walls

THE ROYAL EGYPTIAN MAU, Sacred Burmese, and Buddhist Birman cats are revered not only because of time honored superstition, but because a Master soon realizes that his pet is quite easily enlightened!

No matter the Master's level, his cat soon faces him with full awareness (monk's chance encounters with suspicious dogs, on the other hand, are the grist of many humorous canine Zen "tails").

But seriously! Everything (and everyone) is intrinsically meaningful, even if it (or they) may at first seem of no great consequence in a cursory estimation of our life's causality. This is a great mystery, but true, because everything takes place only within our consciousness.

Only Awareness is greater, because it makes possible our cognition of every experience. And moreover, evidently even the smallest and apparently most insignificant of things is nonetheless quite necessary to the existence of everything else in our incalculably large Superverse.

Life is thus full of wonder, just waiting to be directly seen! Have you ever observed a cat contemplating an empty wall? Experiencing is itself the wonder, not the object of consciousness which is being experienced.

An object is essentially an ideation possessing no demonstrable concreteness. It is merely a concept, and therefore no different than an appearance in a dream. The experiencer decides what is real. Perhaps this is the only real "choice" that we ever really have!

But our usual experience of "reality" is also just a concept, and thus cannot embrace the actuality that unites an experiencer with their experience.

The master and his cat, for instance, simply see directly that everything is the play of Universal-Consciousness. The imaginary line that we place between subject and object evaporates from this deeper perspective.

Hence, moving in accord with direct seeing gives both cats and masters a powerful advantage! But whatever other advantage that this event may generate eludes description, because the only thing that may have changed is our point of perspective.

The universe still is what it is, pure Consciousness, powerfully unfolding as it will, except that this is now, more directly seen! *"I have lived with several Zen Masters – all of them cats"* – Eckhart Tolle.

(Dedicated to Master Gogen Yamaguchi "The Cat", founder of Goju Kai Karate Do).

Power

Contrary To The Rather Cynical Opinion of my clearly disillusioned fellow Jew, Karl Marx, the source of true power is not always "the barrel of a gun!" Many events are extremely powerful, yet they do not really "possess" power.

Power emerged when sentience within life evolved past a certain threshold. And it made its debut in earnest when the illusory human ego finally appeared, and gained control of our species' rapidly "evolving" mind.

Before life and mind eventually appeared in our universe, there were only powerfully "fuzzy" ephemeral events that were transpiring within the impersonal Force of the Universal-Consciousness that brings them forth. To be sure, these events may go on, but they do so unnoticed until consciousness appears within a sufficiently evolved mind to quantumly experience them into "existence".

With the appearance of our subject-object consciousness, apparently concrete phenomena are born, and the objective universe comes into focus within our minds. Because all phenomena are brought into existence by objectivizing them within our dualistic individual consciousness, they are "conditioned" into existence. In other words, they are conditioned into phenomenality by the cognitive vehicles of intermediary space, and sequential time.

Thus all phenomena are subjected to the so-called "chain of causation". But ours is not an independent causation. And because we are essentially neither subject nor object, both of which are objectivized into existence, we are also not subject to any reputed "chain of causation"!

However, both our subjects and objects necessarily have a causal, phenomenal relationship, because both are bound together by our experience of an apparent space-time. Our super-subjective "I" exist(s) beyond, or prior to both, but our "sense" of I enters into causality when our super-subjective I mistakenly identifies with our body.

Then our super-subjective I, becomes a prosaic phenomenal entity. An "entity", or psychosomatic individual, is comprised of a body, created out of the elements of the universe, and the Life-Force, which together gives rise to our familiar individual consciousness.

Hence, our illusory phenomenal individual-self is a "tripartite package", and if any part is removed, our phenomenal individual-self ceases to exist! Yet, because we cannot directly see either our Life-Force, or the unconditioned all seeing "I" by which we know that we are, we naturally want to identify with the body that we see directly before us!

Moreover, we have no choice but to identify with our individual consciousness when it arises, because it forms a unity with our body and the Force of Life that brings it forth.

Our objectification of the universe begins automatically, as our whole-mind splits into subject and object, and we enter into the phenomenal, or apparent, "stream of causation". When the subjective portion of our split-mind finally elaborates into our apparent ego, it becomes our mind's "C.E.O." and begins to make decisions that generate apparently causal consequences.

But any apparent causation is bounded by what we conceive as space and time, and therefore our volition and causation may be regarded as one.

In other words, all phenomenal volition is itself the result of an ideational causation, and our involuntary intermediary "volition" merely continues the chain of apparent causation. The individual is thus nonexistent, being the result of a confluence of ambiguous potential quantum events, which in turn give rise to the causal chain of phenomenal events that automatically transpires within our conditioned individual consciousness; all of which is seemingly guided by our illusory egos!

But, our ego's seeming power actually accrues from the impersonal Universal-Consciousness that gives rise to our entire universe, and thus it contains, but does not "possess", both Universal-Consciousness and the Force of Life. While the independent Awareness, prior to both Universal-Consciousness and our apparent individual consciousness, is "That" by which our mind knows that it is conscious.

Whatever goes on in our individual consciousness, therefore, is subject to both apparent temporality and spatially bounded causation.

But we are essentially intemporal, thus spaceless, and therefore not bound by the rules of any causation. Our automatic "I" experience however gives rise to the "I am" concept and thus appears to become bound up within the chain of phenomenal causation, which automatically arises.

Thus, whatever arises in temporality is actually based upon *in*-temporality. These two are the opposite poles, of our "Noumenally phenomenal" universe.

Ultimately, both result from the appearance of Universal-Consciousness from within the Supreme. So the Supreme is intrinsically neither, yet both, and is therefore the penultimate Power, while the illusory "power" of our conceptual egos amounts to mere vanity.

And as long as we continue to identify ourselves with our conceptual egos, we too remain a passing vanity, arising ideationally from within a phenomenally recurrent dream. A grand dream indeed, but a dream that is nonetheless not our own!

Bad Breath

Like The Amusing Anecdote, "Bad breath, is no breath at all", bad self-doubt, is Self-denial. Because "bad breath" and "no breath", like self-doubt and Self-doubt, are also quite different! Self-doubt (with a small "s") is simply a lack of confidence in the brain's executive faculty.

While Self-doubt condemns the doubter to identification with their body-mind and the continued suffering of worldly existence. Therefore what is actually "bad", is *no* understanding!

Yet a good many physicists, and even more than a few strict metaphysicists, see no practical use for faith. But make no mistake about it, faith is an essential step as we finally begin to settle into awakening from our habitual identification with our persistent waking dreams! (Just as it is when we settle into our seats at the beginning of any new semester of school). But true faith cannot arise out of ignorance.

If this understanding is missed, the next step of psychosomatic disidentification cannot be reached, and Nature's entire evolutionary ladder of sentience becomes our inadvertent trap. And then the chutzpah of our phenomenal mischief within Mother Nature's persistent waking dream begins!

Mischief

Of Course We Should take responsibility for whatever we do! It is only when we try to take responsibility for doing whatever we think that we alone are doing, that we run into problems. From this simple habit, our ego is born. This "doing" is the fatuous (foolish) illusion of our purported "free will".

But, since ultimately, and fundamentally, all things are interdependent with everything else, in the end we must do whatever we can't *not* do! Which, of course, is everything that the universe has already done, and will ever do, since the "Big Bang" set everything into motion!

Moreover, below the level of subatomic strings, our apparently substantial universe disappears back into the basic principle of a preconceptual, "fuzzy" potential identity. This nonetheless still requires an experiencer, in order to enter into reality, because experiential reality is always contingent upon the functions of our dualistic consciousnesses. Otherwise, all of the numberless universes would exist only in potential!

Extending the Force of Life, the essential Universal-Consciousness creates a body to dwell in. Then it "ignites" the potential of universal existence by illuminating it within our apparently individual consciousnesses (a process known as the *"Kav"* in the Jewish Kabbalah, and the "Quickening" in ancient Rosicrucian Mysticism).

But we have no willful causal impact whatsoever on the ceaseless arrangements of potential energy and matter that underlie the presence and change that we experience within our individual consciousness. Something is most definitely "out there", beyond the reach of our eyes, ears and nose. But it is self-existent within an impersonal Consciousness that is not only universal, but apparently fully understandable only to *it*-Self!

When the Life Force eventually flees our body, our mind and the entire universe disappear back into the realm of unmanifest potential. Indeed, the "we" that we think we are, is only a passing figment, that is created by our brain's faculty of imagination, along with everything else that is stored away within our "treasure house" of ephemeral memories. But without these thought concepts we could not be conscious, and would be unable to function as we do, if we somehow were!

Our experiences of mind, body and universe are thus all contained within our phenomenal consciousness. Hence our purported "Enlightenment", is merely the initial arousal of Universal-Consciousness from its long sleep within insentient matter. Indeed, we are already fully "Awakened", because it is our natural condition!

Thus the imbalanced and worrisome individual consciousness that is associated with our illusory ego, is only a disquieting and intransigent delusion. And when responsibility is finally given back to our universe's veritable infrastructure of Universal-Consciousness, the illusion of our being a body-mind begins to disappear. Just as a mirage-body of water seems to miraculously evaporate in the desert heat, by the time that we finally reach it.

But of course our illusory ego center fears and resists its own apparent disillusionment. And mischief abounds, until our false ego self eventually begins to fade away and lose its hypnotic power over our minds.

When our illusory ego entity finally relinquishes control, including its search for the ultimate control of its own mortality, our ego's control functions remain, but our ego is no longer identified with. It is evidently only then that we begin to see more directly that our entire universe is firmly rooted in our brain's own sub-quantum microtubule "doorway" to Mother Nature's Consciousness Divine! (See "Microtubules", pg. 550).

(In traditional Kabbalistic symbology this "doorway" is represented by the very first letter of the Torah, ב *beyt* of Bereshēt ("Genesis"). Which literally means "In"; as symbolized by its perpetually open "doorway").

As the Prophets of old would say, this sort-of-thing is بصدق *bisidq* "truthfully", and באמת *be-emet* "clearly", قسمت *kismet* "fate"! This is to say that "Man proposes, but the Divine disposes".

And evidently our expanding Cosmos has already done completely so, as it continues to endlessly inflate into the great *Mouna* "Elsewhere", on its endless journey into the trackless void of the great "Neverevermore". (See pg. 268).

In other words (for example), young Neil DeGrass Tyson's "chance" meeting with his childhood hero Carl Sagan, which eventually lead to Tyson's becoming a world class Astrophysicist, was actually in all likelihood no *chance* event!

(Dedicated to the intuitionally gifted Rosicrucian Professor, Irwin Watermier ז"ל, who introduced me to הפתח *ha-Petach* "the Doorway" of the Kabbalah, many memorable years ago.).

An Eternal Flame

HAPPINESS IS BALANCED BY SADNESS, wealth by poverty, and life by death. No one escapes the "taxes" of our existence! So today I bid a final farewell to "Yoda Bear", my faithful old Birman cat.

He lived for eighteen years, four years longer than the Birman average. But now that he is gone, I miss my dear old friend, and no amount of metaphysics, theology or philosophy can possibly replace him!

Yet it does no good to either deny our losses, or to repress our sadness when we inevitably lose the ones that we love in life, because hardening or denying our poor wounded hearts, only makes such unsettling matters even worse.

Even saints and sages mourn! And anyone who does not "shed a tear" for the suffering of those once held dear, and for our inevitable losses of life and love, tragically lacks compassion, if not for others, then minimally for their own, now poorer creature selves.

Regardless of what they or others might otherwise say, lacking in sympathy, as well as empathy, how could they possibly be truly enlightened? For our emotional heart is evidently just as necessary to awaken a compassionate Self-understanding as our rational mind is!

But what is this *Heart* that can "hear" without ears, "touch" without hands, and even understand what the greatest of minds may never know? The inscrutable "answer", given by those who apparently do know, is generally a simple but profound silence, and unpredictably sometimes even the *Kwatz!* of a Zen Roshi's frightening "Lion's roar" (pg. 508).

So let's take a silent pause (and if you're Jewish say a *Kaddish* "memorial prayer") in memory of those that we have lost, but will always love; person and creature alike!

For this unselfish love knows no bounds, it is the living flame of the Divine. But it is also, paradoxically, the "Eternal Flame" (of Life) that eventually consumes every living person and creature that we come to love in our brief experience of life!

Because the simple act of becoming an organizationally alive conglomeration of a trillion-or-so individual cells (despite our living bodies constant efforts at maintaining its own homeostasis by negative entropy), evidently creates a constant, and slowly accumulating occult "entropic debt" in the universe, which is eventually paid back fully by our illusory deaths!

Yet, what we experience as an inwardly abiding *Schenah* "percipient Divine", is actually the mirrored presence of G-d. Both within our Hridaya "Heart of hearts", and within our astounding (and sometimes quite confounding!) inside/outside universe. Thus, inseparable from our truer, deeper Natures, the essence of our inwardly abiding *Schenah* is *Agape* "Impersonal Divine Love".

Moreover, because it is not only eternally aware, but also temporally conscious as the sentient subject within our minds, it also suffers as we do within our lives. But it is also infinitely stronger than any sort of suppositious agony or death!

Nonetheless, as the years relentlessly pass us by, life eventually misplaces our loved ones, like so many lost keys. But sadly, no matter how hard that we might search, they are now in secret hidden "places", where they can no longer be found.

Then our mutually shared experiences are cruelly swept away by the ever flowing river of steadily passing time. And quite soon, they become locked permanently away behind the slowly shutting doors of our forgetfulness. Which sadly, can then no longer be opened.

And since they can no longer be refreshed and shared, all that remains are the slowly fading memories of people and events that will no longer pass our way again. And the joy in our lives becomes tempered with sadness, even if we understand very clearly that such is the fleeting nature of our shared temporal existences.

But because our lives may now seem to be replete with sorrow, we may feel compelled to withdraw not only from forming bonds that we know will eventually be broken, but even from our own sacred inner stillness, and life's precious quieter moments.

We may even unconsciously fear that deepening our investment in life might only add to our burden, by exposing our sadness and our pain. Conversely, if we try to escape our pain by chasing after "happiness", we risk becoming trapped in an endless cycle of fleeting happiness, repetitious disappointment, and the inevitable continuation of our losses and suffering.

But eventually we learn that this *Janus* "two faced" sort of "dualistic happiness" does not endure, so we begin to seek a more lasting peace. Yet the only way to find this, שלום עמוק *Shalōm Amōk* "Deeper Peace", is to *not* pursue it. Because we already are, in "Essence", what we are pursuing!

Just as the River Jordan appears to spring up spontaneously from the ground near Tel Dan at the foot of lovely Mount Hermon, near the *Etzbō ha-Galil* (the "Toe of the Galilei"), all things (conceived and unconceived) gently flow from the great Universal-Consciousness that lies deep within us all. And its very nature is *Ananda* "abiding Peace" (Videre licet; *Shalōm Amōk* "Deep Peace", *"Agape"*, etc.).

But only the fearless and selfless are able to embark upon the inner journey that is the transformation of our ephemeral individual identity, back into the all-inclusive root of impersonal Universal-Consciousness (see "Ananda", pg. 504).

Because the truly brave do not submit to loneliness, fear, or pain, they can continue to nurture themselves and care for others. However this does not mean that they have no feelings!

Yet even bravery and hard-earned knowledge in life is not enough. It is making a start toward selflessness, but only if there is no mistaken intention to act compassionately, practice self-inquiry, or even to sit in quiet meditation.

Deepening will only come in its own time, and of its own accord. Yet if we continue to do nothing, only *nothing* will transpire! But if we act compassionately and without intention, "nothing" will happen, yet our entire experiential universe is thereby wordlessly transformed, by the endlessly creative power of Divine Love! (*Hoc pro* אחותי, Sharon J.)

Experience and Existence

Our Experience Of The Universe is infinitely varied, but ultimately derives from a subtle admixture of basic, opposing primeval forces. Hence allegorical Adam *(Purusha)* and Eve *(Prakriti)* appear sequentially in time and occupy different spaces, with the first appearance of the thought *Ehyeh* "I am".

Almost simultaneously the birth of identity with our body-mind happens. First the principle of our identity occurs and there is a singularity of consciousness, then duality bursts forth, and the entire illusory Universal Manifestation unfurls in its wake!

But there is no real dividing line in totally interdependent opposites. Therefore there is no "inner spirit" and no "outer body", and no "inner space" or "outer space", and if the universe is essentially spaceless, it is also effectively timeless! *"He is the First and the Last, and the Outward and the Inward, and He is the Knower of all..."* The Holy Koran; Surah al-Hadid, 57:03.

These two aspects are known as the esoteric עמודים *Amudim*, the Kabalistic "Pillars of the Absolute", which support impersonal Universal-Consciousness' entrance into our apparent individual consciousness. This is because the two are actually only opposing aspects of the same Universal-Consciousness, which simultaneously unites and holds them apart!

Front and Back

UNLESS A HOUSE OF PRAYER IS *KADOSH* "HOLY", which is to say sanctified by the deeper presence of Divine Love, it is likely not worth striving for! But if the "Shalōm" of impersonal loving Peace is present, *Hamakom* the "Kingdom of the Divine" may be realized anywhere on earth, "as it is in Heaven".

This is because *nirvana* "noumenal", and *samsara* "phenomenal" actually appear at the same moment, and into the unremitting flow of our consciousness, endless universes emerge! Thus the Buddha said, *"I see universes without end, because I am universes without end"*... (See "Nirvana" pg. 510, and "Samsara" pg. 513).

We consist solely of this Universal-Consciousness. It is our subtle so-called "Buddha Nature", which is quite content to experience the turmoil of life, as well as its inevitable gains and losses. These it can easily bear, because it does not suffer from our merciless habitual involvement with an endless array of polarized ideational experiences, in an only consciously apparent universe!

Thus phenomenal changelessness and change, subject and object, front and back, left and right, and up and down automatically follow each other as they dance endlessly about in our consciousness.

Apparently, the only way for us to escape our habitual involvement is to "wake up" from the entrancement that holds our attention, and overpowers our minds. But in the resulting quiescence, there appears no substantial alter identity, to reassure us of our own presence!

Our illusory egos are merely a function of our brains, because of the presence of an executive center in its frontal lobes. But, our ideational ego center is also necessary for the experience of our independent identity to open up cognitively within our minds and bodies. It is essentially, the "Executive Software" of the biological Metacognitive Processor that operates our brains'.

Without an ego center, we might as well be a stick lying on a lonely forest path, or an insentient lump of rock languishing in a chilly mountain stream! But thoughtlessly identifying with our ego instrument of consciousness, as it mechanically creates the reality in which it seems to exist, is strictly an act of long habit.

Ironically, our bodies and minds function best when they are not identified with, because they are then in accord with their own true Nature (i.e. Which the ancient Chinese Masters called the "Tao"). This is to be in harmony with our Universal-Consciousness' essential love of "Be"-ing, rather than living in egoistic opposition to our own essential Natures.

Otherwise, what are we? We are apparently a curious conglomeration of body, mind and ego that continually weaves together the contrasting dualistic aspects of our emerging consciousness, and is thus indispensable to it.

But finally, what is it that links all three? It is most likely the very same Universal-Consciousness that unites the entire universe of our experience, in its "Love" of being! But if we have in essence no control, except perhaps in Self-understanding, where then does our most essential responsibility in life then lie. And moreover, must we not also then ask, "If so, then why"?

Tzedek

BECAUSE OUR SHARED DUALISTIC WORLD is largely in the grasp of unaware illusory egos, when injustice exists, it can be allowed no quarter! There is, and should be, for instance, no just statute of limitations for homicide, and the same must certainly be true for even greater crimes against humanity!

Without question, brutality and murder are to be summarily confronted, overcome, and those responsible for them brought to justice. Some truths are so powerfully salient, that they are simply self-evident!

Hence, the exercise and defense of justice is often on the front lines in Nature's ongoing evolutionary battle for the ascent of a more aware "whole-mind" sort of consciousness within our hominid species, ultimately no doubt, to ensure the continuation of life upon our planet.

Nonetheless, our entire objective universe is contained within the consciousness that is constantly emerging inside our brains.

Consequently, when people believe that they are personally in charge of whatever happens, just about anything might manifest out of their dualistic mind's endless potential, including the unthinkable! Indeed, history bears witness to this ongoing ruse, and disturbingly at times, also to its devastating consequences!

Justice, but not vengeance, is the natural response of awakening whole-mind consciousness to brutality. Therefore the religious instinct of humankind carries the deeply important message, *Tzedek, tzedek sheli, omer Elohim!* "Justice, justice is mine, says the Divine!"

Thus deep within our emerging consciousness there appears the archetypal image of the unwavering archangel warrior, waging an endless battle against the heartless demons of hell.

Perhaps even more importantly, this is why little cowboy's and Talmud students mothers, teach them that some things in life are worth more than gold, and that it's always best to be humble and kind. And it's also why grown up cowboys and Rabbis don't judge a person by the contents of their pockets or the color of their skin!

Because not just *most* American Cowboys, but *all* Israeli Citizens serve their country when they are called, and a country takes all kinds of people to make it go. And you quickly learn to care about any one of them who is brave enough to be fighting for your countries freedom by your side!

As a result, all of the petty considerations and hurtful opinions about politics, sex and religion, that you later hear going on all around you when you come back home again, seem quite divisive, mean spirited, and just plain *meaningless* when you see people die, fighting to keep not only themselves, but you alive!

And you come to realize that, just like you, they are doing it as much out of "brotherly love", as from a patriotic sense of duty.

So, as strange as it may at first seem, I have learned from personal experience that if you take the "cows" out of the picture, Cowboys and Israelis actually have a lot in common!

But there is also something more that I know we all share, Americans and Israelis alike, and that is a personal understanding of the terrible price that's paid for the wonderful egalitarian freedoms that we enjoy. And because of it, more than anything, we are fighting to reestablish the peace, and bring an end to the harmful social instability and terrible violence of terrorism and war!

(Please pardon this little "Intermission" in our mountain story, but if you have never had the pleasure of hearing a good "Modern Country Song", just listen to this one, and you will not be disappointed! Because it speaks directly to the "heart of goodness" that is the true source of the insuperable strength of our nation.

And, it also reminds us that Love still has the power to heal our suffering world! Because, after all, it was the "Love to Be" that brought our entire universe into existence! (I.e. *Damn Country Music;* "Humble and Kind", by Tim McGraw.)

Nevertheless, the appearance of violence on the "world stage" is almost invariably a ragtag collection of unaware egos, playing out one part or another. Therefore, the battle raging between apparent "good and evil" appears to be interminable!

Good and evil most certainly exist within the cognitive field of our dualistic minds, along with everything else that we can be cognizant of, but there is really no inner and outer, no self, and no other. There are only the perceptions of our consciousness, which contain every experiential thing in our apparent universe (see "Elohim", pg. 506).

This is not to say that evil should not be confronted summarily, yet sadly everyone generally gets hurt in the ensuing drama! In the Hindu epic poem the "Bhagavad Gita", for instance, the hero warrior Arjuna is counseled by the Divine to fight bravely, but deal justly with his ruthless enemy.

And because this symbolic confrontation actually wages endlessly on the front lines of emerging consciousness everywhere, "heroes" come in all shapes and sizes. And the most meaningful of our recalcitrant personal battles are often also bravely fought, with a kind of "quiet desperation", deep inside ourselves, where almost no one sees. But nonetheless, this is a daily existential struggle that almost everyone understands!

Hence the most meaningful battles of emerging mind are also being fought, through the beneficent power of love that lies within the selfless deeds and decent acts of common, everyday people! And no doubt our endless physical wars will not decrease, or perhaps even one day mercifully cease, until our deeper center of Divine Consciousness emerges more fully within our evolving human species.

Therefore, labors toward this end, while defending justice and freedom with due mercy as the need arises, seems the best use of whatever "power" that we may possess during the short span of our lives.

Because, allowing the Shalōm of Mother Nature's great Tao to manifest harmoniously in the actions of our mind, is evidently a rather potent form of faithful natural prayer, with the power to reestablish the superlative accordance of our mind with Life's own munificent power! (Piae memoriae, Chanah Senesh, see pg. 540).

Accordance

SCIENCE, RELIGION, AND PHILOSOPHY all take different approaches, yet all three miss truth's innermost mark if there is no recognition of the essential mystery of the underlying Universal-Consciousness, which is the true יסוד *Yesod* or "Foundation" of all of our cognitively shared experiences in the universe.

But if the essential *mouna* "mystery" of the universe's inevitable emergence from its superlative Source is acknowledged, they are then again in accord. They then merely represent different approaches in our attempts to more fully understand the endless mystery of our universe, and of our mind's rightful place within it (see "Mouna", pg. 509).

Indifferent to the ignorant and often self-serving manipulations of man, the deeper truths and rules of Nature stand alone and rely upon neither race nor creed to prove their point!

Thus, these truths contain a certain sacred quality, that of קדושים kadōshim "holiness", which stubbornly resists the harmful manipulations of disingenuous men and greedy corporations.

Hence, veridical "Truth" unlike our current world's many half-truths, not only unites us through its intrinsic inclusiveness, but it also unreservedly serves the welfare of all humanity!

Therefore in a very real fashion, this sort of אמת עמוקה *Emet Amōkah* "Deeper Truth" is indeed "sacred"! Even if its sacredness is not always apparent at first glance, to the casual eye.

Consequently respect, and even assistance, should always be given to genuine seekers of, and fighters for, veridical Truth. Whenever and wherever it is due! And when it is, history shows that remarkably beneficial events can sometimes transpire!

"Science is a cooperative effort, a community of minds spanning the generations." – Neil DeGrasse Tyson.

Margōlit of Great Price

TRADESMEN TRADE IN PRACTICAL SKILLS, artists in creativity, businessmen in coin, scientists in discovery, soldiers in war, lawyers in law, police in protection, priests in prayer, intellectuals in ideas, and doctors in healing.

We all serve each other in trade, and when we forget this simple fact a lack of empathy ensues and trouble is surely not far behind!

But asking nothing for themselves, the quietly Awakened heal ignorance by trading in absolute truth, and thus secretly serve the welfare of all.

Yet, largely unrecognized as being *margōlit* "pearls" of great price, nonetheless, the quietly Awakened are among Mother Nature's newest, and perhaps even greatest, current treasures of them all!

Zim-Zum!

MIND IS THE REFLECTION OF SPIRIT, which sprouts forth quite suddenly from within the pre-universal void, giving birth to a multitude of energy "strings" and quantum "membranes" that subsequently become the building blocks of our known universe.

But when the Divine is no longer quiescent, by expressing its endless potential, it appears no longer to be One. Hence, the evolutionary process of *teshuva* or "return" to wholeness begins automatically at the moment that our universe's energy strings begin to vibrate into being, *Zim-zum!*

Yet in order for the quantum domain of consciousness energy strings that comprise our universe to appear to exist as separate entities, space and time must simultaneously emerge, because space, time, and energy, are actually only interdependent aspects of our universe's fundamental principle of identity, or Oneness.

They are wholly contingent upon one another, existing independently within our species' (currently) dimensionally limited consciousness of the universe, only as a matter of perspective. Nonetheless, apparent "matter" emerges as the result of their interactions (see "String Theory", pg. 581).

Thus, the energy contained within matter appears to be a factor of the upper limit of its ability to move within its companion aspect of space-time. Therefore in our four dimensional experience of a three dimensional universe, $E=MC^2$. This is to say that the amount of energy that can be contained in the interaction of the primary constituents that comprise matter, will always be equal to the sum of the amount that they can bear!

And this equates quite nicely when we use the consistent speed of essentially "massless" unhindered light photons, a type of "lepton". As a particle *space* is necessary, and as a moving wave *time* is necessary, hence "space-time", but in not exactly being either, a photon also displays no appreciable mass!

Moreover, all three are aspects of perspective, necessary to the appearance of the other. Thus the subtlest form is able to move the fastest, because the aspects of space and time, which invariably must accompany any-sort-of "dimensionally expressed" energy (i.e. a complex conglomeration of structured quantum energy/strings with a definite organized form), will exponentially hinder any greater amount of its dimensional presence.

In other words, as energy approaches the speed of light, it begins to "collapse" under its own weight. And it does so in proportion to the rate at which its potential particle mass is being expressed. Thus, as the energy of a wave becomes increasingly present, so does its particle mass and subsequent gravity, and the speed of light eventually cannot be exceeded.

By simply going "very fast", if we try never to slow back down, curiously, eventually the equivalent of a Black Hole singularity is created!

Conversely, if relative mass, such as that of an imploding star, increases sufficiently, the rate of its total particle/wave motion, which is actually a product of its cumulative underlying strings quantum microwave vibrations, slows down almost to a halt.

Consequently, energy, space and time begin to remerge due to the accumulating force of runaway gravity, until only the fundamental principle of its "super-crushed" constituent particles identities remains. And because even our universe's highly energetic photons of light can no longer move around in this über-dense environment, it is euphemistically called a "Black Hole".

And so by energies essentially not moving at all, a Black Hole singularity comes to exist! Ergo, gravity can also be thought of as a measurement of the deformation of space-time between conglomerations of identical string identities that are being stretched and vibrated into different forms, by the expansive force of the Big Bang.

The record of the sudden FTL inflationary initiation of this process is present in the ubiquitous cosmic background microwave radiation that bathes our universe, and in the subsequent impossibility of "Faster-Than-Light" motion by post quantum scale particles (and most recently with the discovery of the propagating "gravity waves", which were predicted by Albert Einstein!).

But cosmic potential being infinite, or at least "googolplexes upon googolplexes" beyond our ability to calculate, there are effectively an infinite number of universes. Each one likely separated by the smallest possible quantum of space-time and energy identity.

Yet even infinity, evidently, is not necessarily *in*-finite, if it can be represented by a nearly incalculably large number (viz. a "googolplex"). A number so grand that several of our universes might end before our fastest computers could fully calculate it! (See "Googolplex", "Graham Number Sequence", pg. 537).

Every universe, in this paradigm of our multiverse, is thus comprised of a "googol" of interactive energy strings (or perhaps mem- "branes"). Moreover, each universe also likely extends in a dimensional "right angle" to its nearest neighbor. From every sub-quantum string within it. In a googolplex of energetic lattices that differ by only one quantum dimension in their fullest expression.

At which point there appears the emergence of a new Primeval Singularity, another Big Bang, and a new parallel "pocket" universe is born![74]

But what is being continually "stretched" is actually the consciousness energy identity of the perfect initial Primeval Singularity. Which made its eventual appearance in a multiplicity of dimensionally separated universes.

But always at its basis are *expansion, attraction* and *stasis*. Our vast multiverse literally "evolves" between these guiding expressions. It is impelled to do so by the impossibility of exhausting its own infinite potential, which emerged together with its phenomenal expression within space and time.

But if the space-time balance in our universe is maintained by an energy/mass "seesaw", naturally the question then arises; *"What happens when an evidently unstoppable 'Consciousness Force' meets with an immovable mass of infinite potential 'Identity', in an apparently spacetimeless void?".*

To which the correct answer must be; *"The unbreakable supersymmetry of the Primeval Singularity is, 'in potential', broken".* And consequently, an endless Superverse is 'potentially' born.

Evidently this is just the sort of thing that happens when cosmically deconstructive extremes are approached in a universe essentially comprised of, and bounded by, an über Energy-Consciousness!

But eventually there is an overall universal resolution, when the universe finally fully awakens and takes control of itself on the quantum level. (An event that is becoming increasingly apparent in the currently emerging Singularity of cybernetically assisted Consciousness that is just beginning to unfold all around us!).

Or in the interim, an individual noumenal and phenomenal reintegration of the underlying principles of space-time manifestation with the essential spacetimelessness from which they emerge (i.e. our re-"Awakening").

In the grander scheme, this signifies the eventual cessation of space and time, and the dissolution, or "reintegration", of our entire Superverse. And incredibly, within the endlessness of Cosmic potential, this may already have "spacetimelessly" happened!

Or more precisely, it has neither "happened" nor "not-happened", within the infinite potential of spacetimelessness. Which endlessly expresses itself in the forever "Now" moment of our waking consciousness!

In the meantime, because our consciousness is even more subtle than light, it is evidently also possible for the Superconscious foundation of the universe, in which we experientially find ourselves, to achieve this balance and reawaken to its authentic Self "within" us.

This is allowed because our consciousness becomes ever more subtle as it approaches, in reflective stillness, the Source of its own presence within the evolving chemical soup of our human brains.

In other words, our universe's "Mind" stretches too! But what we imagine is our ultimate Awakening, is perhaps only another endless beginning. As Wu Roshi once confided to me when I was a young probationary Zen monk; *"The river of life is [impersonally] One, cascading over the precipice of manifestation into apparent temporality. Each droplet in its fall has an ephemeral [personal] identity, but is soon gathered back into the primeval One."*

A Bitter Harvest

THE FOREST CREATURES around our redwood campsite neither sow nor reap, and because they simply accept whatever nature provides, they are content. They do not ruminate about the past, or make plans for their future.

Following their inner knowledge, with no effort of thought, they entertain no worries. Clothed in "garments" freely given, they have no need to sew or dress for the changing seasons.

The forest is their home, and its bounty their possession, so amongst them, there are no rich or poor. Giving no thought to either wealth or station, they do not compare themselves to one another, and do not believe themselves to be either better or worse than their neighbors.

Unconcerned, they live in harmony with the moment, and accept whatever comes their way. With them, birth, life, sickness, old age and death come and go naturally, like the rising and setting of the sun.

They do not delude themselves with thoughts, and therefore give no thought to being, or of not being, in control of their own destiny. In times long past we were as they, but have forgotten the inner wisdom of our own true nature.

Perceiving ourselves as separate from nature, we burden ourselves with the yoke of an assumed causality. Incorrectly believing that we are somehow separate entities we struggle to survive, and strive to pursue life, liberty and happiness; things already freely given.

Amassing wealth and power, we soon become burdened by our perceived possessions and strive to protect them, and eventually violence and even premature death may ensue!

We make our plans and sow our seeds, only to lose whatever advantage that we think we have gained, and sadly, bitterness may become our final harvest. How sad that it almost never occurs to us, that abiding peace is only found in what has already been freely given.

For our greatest possession is the abiding שלום *Shalom* "Peace" which cannot be attained, and cannot be taken away, by anything but the dark veil of ignorance; because it is always, already ours!

Strolling in the Wilderness

After We First Realize that the inner flow of our everyday consciousness might be a trap, our experiences of person and universe may start to become unpleasant. This is the spiritual seekers infamous *Golgotha* "dark night of the soul".

But when we eventually learn how to pop out of the phenomenal stream of our experiences, and to keep our identity out of the quicksand of our illusory individual consciousness, our apparent persons and universe(s) may start to look beautiful again. This is a wholly natural occurrence, emotionally felt, and like everything else, it is already written into the very (non-supernatural) fabric of our shared unfolding universe!

Or it might also be metaphorically said that: "Eventually the seed of creation flowers, and subsequently its fruit ripens, into a fuller manifestation of our universe's essential Awareness."

But it is a heavenly immortal "fruit", growing upon our universal *Etz Chayim* "Tree of Life". And by consuming it our faux "we", which is occasioned by our wrongful identification with our illusory egos, will most certainly "die".

Nonetheless, the impersonal divine Subject of our ego's "we(s)", our deeper, truer divine Selves, also simultaneously reawakens! And then, evidently quite suddenly, we simply somehow realize that the *Gan Eden* "Heavenly Garden" has always been within us, and we within it! And thus, some have said, we become "Self-Realized"!

This is the inevitable עולה *oleh* "rising up" (syn. Aliyah) of our essential Awareness from our symbolic fall, into the Kabbalist's "pre-Malkuth" Purgatory of "Adamic Worldly Illusion". And the purported subsequent *Ananda* "Bliss", of our spontaneous reawakening in الله *Allah's* (syn's. יהוה *Yahweh's*, G-d's, etc.) intended "Kingdom of Malkuth".

But, it is even more exactly Universal-Consciousness' eventual reintegration of Its own ideationally separated identity and function. In other words, its "form" and "force", or *nama* and *rupa.* Which permits its expression as (and within) our individual-consciousnesses.

This constitutes the only veritable *Shalōm,* or "Abidling Peace" that's available to us, for as long as we seem to occupy our bodies. But of course, so long as our identification with our Adamic "nama and rupa" prevails, our dipole perspectives are endless! As are the resulting erudite dialectics of scientists, philosophers and scholars.

And almost no one can reach a lasting agreement! To the clear detriment of anyone, and everyone, that depends upon them to accurately understand and clearly interpret arcane matters in such rarefied fields!

Thus, the presumptive strife of intellectual divisiveness never ends. And evidently only the tincture of time, and the sometimes bitter medicine of experience, and soothing ointment of calm reflection, have any hope of providing the requisite cure of humility!

What to Do?

APPARENTLY ALL THAT WE CAN DO, is to live our lives in accord with what is "natural". In other words, live our lives as usual, while directly perceiving the unfolding of our life as it truly is. Because the universe is always in accord and requires no assistance!

It is only the rapidly spinning *merkaba chakras* "chariot wheels" of our body-mind's somnambulistic subject-object consciousness that have temporally drifted away from the truer, direct thoroughfare (freeway) of our naturally unfolding lives.

But it merely requires a shift of our errant perspectives, to get them out of the ditch of delusion, and back onto the thoroughfare of a more direct verity.

This shift reveals the ongoing *prajnaic* unfolding of events within our consciousness, according to their deeper, truer nature. Only this accordance endures, while all else sooner or later simply fades away!

Until this is understood clearly, prajna teachings (*pra* "prior" + *jna* "knowing") are only the equivalent of so much divisive gibberish. To paraphrase the acerbic wit of the perspicacious American Transcendentalist Ralph Waldo Emerson; *"...it is sometimes better not to have opened a book at all, than to be drawn out of our own self-understanding by the persuasions of a Self ignorant man!"*

The Freeway

LIFE IS LIKE A BUSY ROAD WITH TOO MANY TOLL BOOTHS! But meditation is the open country "Freeway" (i.e. the "free" way) to our reawakening. Because it takes us out of our body-mind's constant subject-object, self-other objectifications! It costs nothing, and regular practice provides us with a rapid method to become reacquainted with our truer "impersonal" nature.

Eventually, through the persistent כונה *kavanah* "attunement" of diligent meditation, we begin to wake up, and our involvement in the delusions of our discrete separateness ends. Albeit, our observed universe never ceases to be! And why should it, no one ever said that it is unreal! Only our perception of it is in error.

But beware, because the road to "Quietism" is a misleading solipsistic "turn off" that only leads to a harmful and empty "dead end". Yet, ironically, by persisting we may eventually come to discover that meditation gives us everything that we need, except for the unreal "needs" that we so often believe that we want! (See "Solipsism", pg. 514).

This Too

"This Too Shall Pass" is a well-known, but often misconstrued phrase, which really advises us to simply, "Be patient, and try to endure the unacceptable." Or it may be more precisely construed as a reminder that even our good fortune will someday end.

It demonstrates the lesson of impermanence. If you are happy with this definition, strive to treat others fairly and with kindness, and can accept with equanimity whatever comes your way, then this may be good enough. As far as it goes!

But when we clearly understand that whatever actually is, always was, and shall forever "Be", then our Self-ignorance, delusion and suffering may finally come to an end, before everything passes again into darkness, through death's "one way" door.

At the very least, we need to clearly understand that even our own purported "death" may not be forever. Know only that either way, this little "you" cannot endure beyond the grave!

Inside Outside Upside Down

At First Consciousness Arises in our body-mind, and it seems separate from all that it surveys. Next, a certain understanding is gained that there is no real separation between our experiences of inner and outer, or past, present and future. (Or even between our ideations of left and right, up and down, or self and other!).

Finally all is seen to be Universal-Consciousness. And eventually even this disappears back into the universal substrate of the Supreme. Therefore, with all due respect to the scribes recorded statements of the historical Buddha, we are compelled to ask; *"Who then is, was, or ever will be, 'present' to suffer?"*

But, then again, even well-meaning scribes have certainly been known to make a few inadvertent mistakes!

Agnos Agnosia

Agnos "Without Understanding" + *Agnosia* "Not Understanding" Means; *"Not even knowing that you do not know!"* This is not only a rare medical condition of mental impairment, but ironically, it is also the most common everyday misunderstanding of our everyday lives!

While on the other hand, intuitive *prajna* is; *"Knowing the 'Nothing' that knows nothing about not knowing."*

This is not only a rare healthy condition of the human mind, but it is also the uncommon understanding sometimes called "Enlightenment". And, no doubt, it is most likely the seminal meaning of Plato's *anamnesis* "lack of forgetfulness" as well!

Dai Shen-Zen

If You Grow Up In The Forest, or spend enough solitary time there, you will probably eventually discover that it is filled with what the ancient secret society of 14th century Japanese Ninja Assassins (who helped to unify Japan and eventually became the Tokugawa Shogunate's equivalent of our modern day Special Forces) once called *Kami* (Japanese for "Nature Spirits").

Depending upon your point of view, "Kami" are either elemental spirits, guardian spirits of the trees and animals, or perhaps something not really supernatural but nonetheless indefinable, and likely even essential in nature. It may even be that they are merely a subconscious awareness that whatever we see, smell, taste, hear or feel is really just our own mind!

But whatever their true nature might be, if you are alone in the wilderness for long enough, you may come to sense that it is full of "nature spirits", "ghosts" and (maybe) even gnomes, faeries, sylphs, jins, minihunes, or even some sort of "mountain demons! *Perhaps* they are objectively present, but they are most certainly present in our minds, as is our experience of the entire universe.

When the wind howls like a Banshee in the night, raising the hairs on the nape of our neck. The wind, the "Banshee", and even our "goose bumps" only exist as perceptible objects within our consciousness. But they existed in nature long before they came to inhabit our stories, books and art, or even our imaginations!

They existed as potential in the Consciousness Force of the universe long before we began to name the objects of our experience. But it is we who transform our own "ghosts" and "demons" into the presence of Kami (and various sorts of spirits and "Little People"), in the subconscious wilderness of our own primitive minds. And in the quietness of nature, they sometimes make themselves known.

But in point of fact, the lines between our inner and outer universe, and between our smaller personal selves and greater impersonal Natures, are so indefinite as to have no substantial verity. Because they are like the imaginary lines on a map which have no purchase in reality, and cannot be found anywhere upon the whole face of the earth! This is to say, that our "greater nature", is actually (all of) Nature itself!

In the solitude of *Dai Shen-Zen* "Great Nature", we come face to face with the deeper primitive nature of our own existence. Which the talented British author, Joseph Conrad, appears to somewhat pessimistically opine (in his classic 18th Century novel, the "Heart of Darkness") is humanities ever-untamed, and therefore *always* dangerous, "animal nature". Which reflects the generally confrontational nature of humanities relationship with our own Mother Nature at that time!

An attitude that was likely based upon the (then widely believed) allegorical tale of Adam and Eve, where G-d says to Adam; *"Be fruitful and increase in number; fill the earth and subdue it. Rule over the fish in the sea and the birds in the sky and over every living creature that moves on the ground."* – "Bereshit (Genesis) 1:28.

But since Joseph Conrad could just as well be saying that isolation can be dangerous to an imbalanced mind, no one can (reliably) decide upon which one is actually right!

These are common literary theories, and often a point of heated classroom debate! But it could also be just as likely, that spending time alone (in the midst of the majestic natural solitude of a redwood rainforest, for instance), can provide us with the opportunity to discriminate our present fantasies from a far truer reality.

And moreover, to quietly address the "demons" borne from our own self-ignorance, and the "ghosts" of old memories that we've dreamed up, from within our imaginary pasts.

Yet this can only happen when we closely inspect the processes of our own consciousness. But as my Shamanic Cherokee uncle Leonard would say, "Solitude is strong medicine"! (Thus until we gain sufficient strength and clarity of understanding, it can be dangerous, and must be expertly administered, in carefully measured doses.)

Nevertheless, with diligent and persistent meditation, we may even eventually come to discover that deep within us there exists an ever present still center. An inner "Sanctuary" in the midst of our ever busy so-called "body-minds".

But until we understand what is really going on in our own consciousness, we have no choice but to rise and fall on the unending waves of our own perspective dualisms, continually experiencing a world of intermittent happiness and sorrow.

I

In happiness there is no urge to seek a resolution. But when we experience suffering, we naturally want to stop the pain. And perhaps unknowingly, we stand upon the threshold of a great opportunity. Otherwise our suffering offers no supernatural redemption, and is simply an unfortunate occurrence that is best avoided if we can!

The question naturally then arises; *"In this seeming world of constant change, how can we possibly avoid suffering?"*

"Split" or dualistic subject-object consciousness, which is the predominant human condition, much too often engenders (produces) unsettling, unpleasant, and sometimes downright painful states of mind!

One such condition is the result of the strong bond that we often form with the apparent "others", that appear within our consciousnesses, and with whom we share our life journeys.

But it is important to understand that they are having the same general experiences as we are. This is to say, that they are also quantumly connected (in impersonal Universal Consciousness) as the collective mind of "us", you and I, which always includes them as well! Indeed, when we lack this knowledge, at best we can claim only a tenuous grasp on our own sanity!

This bond is often occasioned with an almost overwhelming feeling of personal loss when it becomes damaged, or otherwise broken for some reason. Unfortunately, and often with little reflection, we accept vivid descriptors like "bereavement", or "divorce". Which then acquire a grandiose, if not terrifying status, to match their emotionally conflated conceptualizations. Which are generally based upon our painful past experiences!

And to make such grave matters even worse, strong negative emotional experiences often generate harmful secondary emotions. Like the aftershocks of a devastating earthquake. These potent experiences can erode our mental and physical health. And eventually they can even undermine the very quality of our lives!

But what are our emotions, other than thoughts conceived from percepts? Emotional thoughts that are automatically generated within our own consciousness. (By a reaction to other thoughts and percepts). Which causes a release of various glandular chemicals ("hormones") into our bloodstreams, nervous systems, brains and organs, in response to our (now psychosomatically "felt") emotional thoughts.

Without a doubt, they are linked to the body and brain that support our minds. But if we trace everything back far enough, our birth actually begins with the very first thought that we exist. Which indelibly marks and anchors the remarkable appearance of sentient awareness within our apparently individual consciousness.

Our first notion that we exist, then becomes the epicenter of our (rather unstable) sense of constant beingness, within an ever shifting experiential universe.

In the absence of our subject-object consciousness, and the sentient presence of an independent Awareness to receive it, our body experiences "neurological events". But they are not yet "pleasure" or "pain".

Infants, for example, are certainly both aware and conscious. But the awareness that they seem to exist as an "entity" does not appear in their consciousness, until their brains are sufficiently formed to support it.

In fact, it is the elaboration of consciousness within our emerging minds, which allows our essential Awareness to be reflected therein. When this transpires, the organism, be it a worm or a human being, is only then *sentient*, or "awake to the experience" that it exists!

II

Our mind is thus comprised of various kinds of concepts, in various states of stability and flux, all contained within our evident consciousness. Therefore what are *concepts*, if not simply forms of "imagined" information? Because they are, "so-to-speak", essentially (mentally) imaged "in" (thought) "formations".

Emotions, for example, are just another kind of mental information. But they also provide a very important link between our awareness of being, and the other ever changing concepts that comprise our experiential universe.

But because our mind's "eye" also needs to "see" a concept in order to register its existence, our essential Universal-Consciousness has also evolved the clever concept of "space"! Which allows the expansion of thoughts in its subjective object's (i.e. "our" evident ego's) mental "eye".

But because our mind's subject/object eye is only capable of focusing upon a single thought at a time, which monopolizes the space that it occupies, the unique concept of experiential time has been evolved within our consciousness, in order to set it free!

The "emptiness" concept of space is therefore necessary, for the "fullness" concept of dimension (size) to appear. Likewise, a "stillness" concept is necessary for the unidirectional concept of a fluid time to become cognitively manifest in our dualistic minds. As a result, our sensation of space-time can also appear to *dilate* "expand and contract".

This allows the "objects of consciousness", that are stored away in our memories, to be compared with those currently appearing before our mind's eye. (And also to those which our imaginations anticipate might yet come.)

But time and space, just like our imaginations and memories, are purely mental constructs. Like everything else that appears in our consciousness, they have in reality almost nothing to do with either an actual outside universe, as we commonly experience it, or an inside observer, as we so strongly seem to be!

This is true, whether what we observe appears to be positive, negative, or something in between. And this is not so difficult to understand, if you just consider that we simply cannot be anything that appears to be "before" (i.e. in-front-of) us! Therefore, we are none of the processes of conceptualization. Which includes all of our apparent objects, as well as all of the apparent subject/object "egos", of our individual consciousnesses.

It is important to realize that our minds will always contain the fundamental conglomeration of concepts, that will eventually become our illusory egos. As well as our ego's own *persona* "personality", which we commonly present to other illusory egos.

Hence, in summation, we are not only always not our ego, but also not even the mind that contains it! This is evidenced by the fact that our seemingly central illusory ego entity always appears temporally and spatially before our necessarily contiguous (i.e. "percipient") Awareness' awareness of it. Simply put, we cannot be whatever appears before us, even if we strongly believe that it must be who we are!

III

The important question therefore is not "who" we are, but *what* we are. Because what we are must be "That", by which we know that we are! (Rather than whomever, or whatever else that we might erroneously believe ourselves to be.).

But even if we have established this, once and for all in our understanding, the consequences of our consciousnesses operating in the manner that nature has currently achieved through us, must still be contended with.

In other words, even if our position is clearly understood, a dis-ease in our brain or body is still experienced as pain. Yet "suffering" the *experience* of pain and pleasure is an entirely different matter!

Nonetheless, where our body is concerned, it is most likely created by the activity of the Life-Force operating within a not quite "insentient" energy/matter. And as a likely point of fact, the entire universe is arguably comprised of a fundamental sub-quantum Consciousness "stuff" that is universal, and it is therefore, necessarily, also impersonal.

But, let's suppose that we have also "once and for all" established that we cannot be any sort of an individual spirit or soul (see *Atman;* under "Tantra", pg. 586). We may even understand that it is as fruitless to ask why a personal god would either want or allow our suffering, as it is to ask why an impersonal god would either want, or allow itself to experience suffering!

But these are actually only questions asked in the wrong way, questions that are not really questions at all, because they are based upon a fundamental error in our understanding. They are merely "tautological", that is, equivalent to statements that contain no real information. Like the explosive sound that escapes our lips, whenever we strike our finger with a hammer!

In other words, "meaning" is not intrinsic, it is assigned. And until we awaken to the fact that we are not merely the body, brain, or even the mind which thus assigns, we will continue to suffer certain unpleasant consequences.

These conceptual consequences occur as a result of entertaining the belief that the activities of our consciousness, which reflect the ongoing physical processes of our body and brain, are objectively real. Even if, for the sake of argument, we were to eliminate the somewhat controversial idea of a כוח החיים *Koach ha-Chayim* "Force of Life", this still holds true!

The first consequence suffered is belief in the concept that we are the psychophysical subject of all of the objectified concepts that appear, as if, "before" us. Next, we begin to "suffer" the receiving of various ever changing experiences, as a result of the unremitting flow of concepts that automatically transpires within our consciousness. (As if they were somehow real separate external objects that are separated in a discreetly external space and time!).

When our involuntary concepts of emotion, brain, and body begin to function, we then judge whatever we experience to be either "pleasant" or "unpleasant". And when our investment deepens with our ever growing memory of these compelling and apparently personal experiences, the belief that we are a body existing in an external universe soon becomes our everyday reality.

Then the concurrently emerging concepts of emotion begin to cause the release of a cascade of psychoactive hormonal chemicals that give us constant visceral feedback information on our resulting subliminal state.

Emotional information tends to generate more emotional '*in*' formations, which then deepens our investment in the automatic ongoing subliminal processes of our illusory subject's objectification of the universe; until we become almost hopelessly involved!

The secondary effects of all of this ongoing deep-level emotional activity are no doubt registered causally in our brain's shared higher level cognitive, as well as its intermediary motive aspects, via its numerous vertically organized connective neural nuclei.

IV

One such negative ancillary (additional) consequence, for example, is the destructive conceptual construct of "guilt". And ironically, guilt seems to be a malady that most often plagues those within our midst who have a refined sense of ethics and morality.

Hence, guilt is widely recognized amongst both Jews and non-Jews, as being a very "Jewish" state of mind! Therefore, understanding the phenomenon of guilt seems an important thing for anyone to do, and even more so, if (like me) you too should happen to be a Jew! (So, let's "get down to business!").

The *rostral cerebral cortex* and *diencephalon* are both intimately involved in guilt's eventual psychosomatic construction, which is essentially a multilayered, phenomenally linked combination of cognitive and emotional, conceptual in-formations.

Guilt, for example, like loneliness and other emotionally based ideations, is a conceptual conglomeration of information that is linked by conscious and subconscious memories with other neural events that are only apparently current.

These are linked with other anticipated conceptual events, which are in turn connected to long forgotten past experiences and admonitions. All of which are both physical neural events as well as mental conceptual experiences!

To accomplish this rather enigmatic feat, an elaborate neural network adjacent to our language center, in the temporal lobe of our *telencephelon,* includes these extended connections in our brain's network of interconnecting *arboreal dendrites* that comprise our *limbic* system (pg. 549).

In other words, our brain's neural network can only change in this way by growing new connections within its limbic system. This is achieved by forming new neural connections and pathways between our *prefrontal cortex* and the diencephalic pathways that connect it with our brain's globally linked limbic system.

(For example, with the *hippocampus*, the only structure in our brain that forms new neurons; see pg. 508, and 543).

Indeed, quite recently we have discovered that every time we recall a memory, we also must grow new neural connections. As a result, we actually change the memory every time that we recall it! (Which also illuminates a serious "ever present" potential problem within our adversarial legal system, whenever this fact is not properly taken into account).

Our limbic system is also connected caudally (towards the rear), by way of our brain's *pons* to our *cerebellum*, and through our spinal cord to our peripheral nervous systems. This connects it to the rest of our body, and sensorially to the rest of our Universal-Consciousness (i.e. consciousness is our sixth sense).

Which expresses *it*-Self energetically, as an apparently external universe, within the cerebral mechanism of Its evolved biological machine of space-time experience (i.e. "us"). This essentially "reactionary" impersonal universal mechanism is what we so often mistakenly believe is our very own, strictly personal, and physically discrete, body-mind!

Moreover, because of our "body-mind's" necessary subliminal psychophysical and universal quantum interconnectedness, complex deleterious *cortisol* generating conceptual constructs, such as guilt and loneliness, are not only shared with others, but can also manifest as stress disorders in specific neural sites in our bodies.

This can lead to physical as well as mental disease (cortisol is a systemically harmful biochemical byproduct of stress).

Unfortunately, continued episodes of these and other stress related disorders, such as obesity, can even cause "epigenetic" (i.e. heritable) diseases. In short, whatever happens to us personally, by varying degrees of heritable extension and genetic separation, may eventually happen to us all!

Indeed, the ability of our hominid mind's refined sentience to generate (and even "engineer") rapid genetic changes is apparently remarkably unique in the animal kingdom!

Of course the reverse is also true, the experience of positive constructs can lead to our physical and mental wellbeing. Not only for us presently, but perhaps even for future generations!

For instance, the language, moral and ethical areas of our brains develop (for the most part) automatically. This is because, as we continue to gain new experiences, we are also growing new neural connections.

This recording process is occasioned by the enduring presence of our inherited cerebral "hard wiring". But in less evolved animals, we generally call this "instinct". Nonetheless, both reduce down to the double helix design that contains the biological "coding" of our complex hominid genetic "algorithms". Which are written using only four organic chemicals!

But, just as another language can be learned by diligence and persistence, even our deep seated moral and ethical compass remains amenable to correction and fine tuning. It is important to realize that our every thought, feeling and action, indeed our every concept and experience, is connected to some underlying function of our brain, and by extension to the entire multiverse!

No action of our body or brain, however, fully explains the presence of the Awareness within our consciousness that knows the contents of our minds.

Indeed, this is actually the wrong perspective, since our universe's fundamentally necessary identity principle of Presence, is necessarily synonymous with our Awareness of it!

Thus, it is the *prima causa* "first cause" which precipitates *de rigueur* "necessarily" the primary and independent thoughts or concepts of all of our subsequent experiences, which are also conceptual in nature.

Note: The nine *italicized* anatomical terms on the previous two pages are further clarified in the Glossary and in the Appendix at the back of this book.

V

As such, aware Presence, which can also be said to be our mind's foundational concept of "I Am", spans the qualitative gap that stands between our cognition and our emotions. And indeed, between all manifest phenomena and their ever present, but unmanifest veridical Witness.

This is because our body-mind consciousness is the "mirror" in which the sentient light of the transcendent Supreme is being continually reflected. Thus, the reflected, but "unaware" in the common sense, "Essence" of the eternal Supreme becomes commonly aware within our everyday consciousness, as its witnessing Presence. This involuntary occult process grants to our binary consciousness its subjective sense of being.

This is the primeval "I Am" concept, by which our living organism becomes temporarily sentient within our conditional temporality. Our illusory sense of conceptual "beingness" therefore ceases with the death of our body. We are in fact always alone in, and as, our ever contradictory experiences of dualism.

But we are never truly isolated from our essential oneness with the principal Awareness of the Supreme. Because, our every experience transpires through an intelligent biological mechanism. Which is foundationally comprised (by the presence) of our subtle evolutionarily elaborated Universal-Consciousness.

And they are granted therein by the subtle, yet nonetheless indispensable, "Presence" of the Universes ever-witnessing foundational Awareness. Which is ever the first and last cause of our discreet dualistic consciousness!

Evidently we have no real choice, in the limited dualistic experience of our grander universal existence. Therefore, not only pain and suffering, but pleasure and happiness as well, will inevitably appear within our cognitive experiences of life.

But now an important question is revealed, "Who then is it that receives and interprets the constant flood of our ongoing dualistic experiences?"

Because if we are not the one who receives, but instead are the one who interprets them as being real or not, then we can only be the unmanifest Awareness by which both our subjects and objects are known. And we can now ask, "Who is it then that suffers either pain or pleasure?"

Or perhaps a better question is, "Since the one apparently experiencing, is actually only an insubstantial experience that also appears to an independent Awareness, is there any real difference between the evident experience and its purported experiencer?"

Are not both bereft of any but an implied reality? The seeds of suffering may be planted, but who is there essentially to experience their bitter harvest?

Thus our life is actually a matter comprised of constantly shifting perspectives. Perhaps the most important question then is, "From where are we looking?" And apparently our point of perspective, makes all the difference in whatever is seen by whom!

Beyond Enlightenment

We Are "That", by which we experience enlightenment. Therefore enlightenment must be occurring within our dualistic consciousness. But since neither exists without the other, subjects and objects are actually one, being experienced as such in our state of ostensive enlightenment.

But, of course, in the end even our supposed "enlightenment" is subsumed into the unknowable *Tat* "That", by which it is known.

This ineffable "That" then, is all that there ever is, was, or most likely could ever in consciousness *be!* And consequently it includes not only the truer "One" in every one of "us", but also every ephemeral iteration of "we"! And whether or not we even understand our true-life situation, is therefore of no real consequence at all!

Evidently there is only our "experience", and even it has no intrinsic meaning. But delusion results when we believe that an illusion is real. Consequently, in this regard, most of us are at best lost. And we only "exist" substantially as concepts. Which are neither "dead" nor "living". (Nor do we exist as any sort of conception that could ever lie "in between"!)

So what then, other than "nothing much" at all, could we possibly *Be?* And if simply understanding this is all that there really is to our supposed "Enlightenment", after all of the work that we must do in order to acquire it, it doesn't seem like such a "bright and shiny" prize to me!

How Now, Brown Cow?

How Can "Now" Be Both Present And Changing? Unless "change" is in some cosmic cowboy's, "How now, brown cow?" joking sort of fashion, also "Now's" true occult state? But if change represents a bona fide "force", rather than merely a "state", then change must also be able to somehow spatially change itself, in order to temporally change its place in the ever present moment of our Universe's "Now"!

Conversely, when we try to use our everyday mind to understand how this could happen, then we might as well be asking the Cosmos a cowboy's open ended "How now, brown cow?" sort-of "cosmic" question!

Indeed, when we (perhaps like Nietzsche, in utter frustration) ask a brown (or Zarathustrian "speckled") cow, that's peacefully grazing in some green mountain meadow, if it knows what the secret meaning of life might be. The brown or speckled cow almost always wants to answer back (you can see it in their eyes!).

But when it tries to speak, it almost instantly forgets how! And then it even forgets, what it once thought that it knew, and then it always goes back to its more usual bovine task, of munching soft green grass once more.

Because the Cosmos doesn't ever "speak" to us with words. And our affable green mountain meadow companion, was really always just a sleepy brown (or even perhaps a philosophical Deutschland "speckled") cow!

Otherwise both "now" and "change", could not exist, exchange, or even change! Therefore both the concepts of "now" and "change" can only find their existence, interactions and resolution, in the consciousness of which they are constructed, and by which they are known.

Close to Chaos

Close To Chaos "something" happens, when systems (including the subliminal systems of thought of which our experiential persons and universes are comprised) inevitably begin to fail, and then somehow transition into forceful states of change.

It is a "something", which is neither chaotic nor ordered, that they must pass through briefly before "dying", and to which meditation can lead us before their apparent space-time death occurs.

Yet, when all that remains is forceful inner states of cognitive change, which we then interpret as some-sort-of simultaneously "external" (while nonetheless *internal*) incipient cosmic "order-within-chaos", then evidently nothing else "materially" matters, since every (actually rather improbable, if not nearly *impossible!*) inner/outer thing, is then obviously of forceful states of consciousness (as well as foundationally Conscious universal energy) comprised!

But soon this veritable illusion passes us by too, and all that then remains is the sentience by which our consciousness is known, and of which we are ineradicably comprised. *"At the head of all understanding is realizing what is, and what cannot be, and the acceptance of what is not in our power to change."* – Solomon Bin Judah, 11th CE.

And if this is true of the recurrent consciousness stream of which our ever present moment of "now" is comprised, then it must also be true when the "stream" no longer flows (upon the death of the psychosomatic mechanism by which our stream of thoughts is constantly produced, and subsequently known within our consciousness to exist).

Simply put, the "malleable" inside/outside dimension(s) of energy/consciousness, that enigmatically comprise(s) the true wealth of the universe that we live in, is much more complicated than the rules that we are currently devising to try to keep it in order!

This seems obvious enough, but there is a deeper meaning to Bin Juda's "Serenity Prayer" that applies to the illusion of our direct volitional involvement in the automatically unfolding story of our lives. Our volition is, of course, included in whatever transpires.

But since we are already involved in what has already transpired by the time we cognize it, our ostensible "free will", like our supposed death, is simply an opinion based upon an illusion. And our impersonal immortal Sentience, by which it is known, is thus evidently never really in question! (Or, if you prefer, $Z = Z^2 + C$ [!]) – Piae Memoria, Benoit Mandlebrot.

Simply Irreducible

SOME SAY THAT EVERYTHING EMOTIONAL REDUCES down to just two simple things, love and fear. But since "fear" is not real, and Love is Divine, in the end, Love alone forever abides!

Because, simply put, for as long as we believe that we are just a body-mind, our "survival fear" is going to be there. But do away with the body, either by death, or by dis-identification, and only "Love" remains!

Not the jealous sort of personal love, but the tranquil bliss of impersonal awakening. Some call this peaceful state of pure impersonal Love *Ananda*, others *Nirvana*. But we are prior to thought and therefore state. Evidently, we are both thought-less, and bliss-full!

"And in the end, the love we take, is equal to the love we make." – Abbey Road, "The End", by John Lenin, recorded in 1969.

These Dreams

Gods Become "Lost", some opine by design, in a dream of person and world. Just as we do in a dream, when the dream characters start to feel real and begin to act as if they have "free will".

But when we finally wake, it is as if they never were and we are whatever remains. Just so, we are what remains when our waking dreams, of "person and world", sooner or later ends. But such is the secret nature of dreams, because only the divine Dreamer is real!

Better

In Any Case, from the force filled arrangements of elemental materials within a rapidly expanding space-time, the universe eventually became capable of utilizing energy over duration, in various dimensional ways.

Eventually even accomplishing the amazing ability to ask substantive questions, and receive cogent answers, from within its own transcendent Self.

Indeed, the inordinate power that our apparently "dualistic" mind wields over its contents is evidently almost impossible to explain!

First our mind interprets what our body presents to it, and then weaves it into the ongoing concepts of our persons and worlds, and finally identifies with both.

But we will never be fast enough to catch what is being assembled by consciousness from within our substrate of spacetimeless Universal-Consciousness-Force, so why bother trying!

Perhaps it is even wiser to allow ourselves to simply be "That" which we can always feel, but never see!

A Disappearing Wave

WE ARE EVIDENTLY, precisely what we *don't* see, forever neither coming, nor going!

An Arborist's Supplication

MAY OUR STEADFAST FRIENDS, the trees, always guide us safely within their lofty canopies. And 'rouse us gently from our waking dreams, as we trim amidst their life giving leaves!

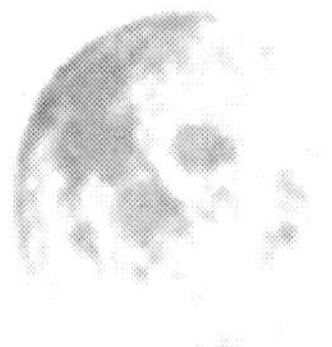

Goodnight Bright Moon!

AWASH IN THE PALE GOLD LIGHT of tonight's cheery campfire, I attempt to further contemplate the inner meaning of Master Lóng's words. But tonight's damp chill finally interrupts my thoughts and prompts me to push my red plaid shirt's collar up, and tug my warm blue denim coat sleeves down!

A solitary "hoot owl" suddenly screeches eerily above me in the crisp darkness, and I look up hoping to catch a fleeting glimpse of the hungry owl as it softly glides through the dark night air in search of its customary nighttime meal.

But all that I can presently discern is tonight's bright full moon, as it peeks diffidently (shyly) out through a momentary break in the deep gray fog. It appears to hang for a moment, like a luminous fruit, from the bough of a solitary distant Madrone tree. Then it disappears, just as suddenly, back into the darkness of the Stygian night.

The momentary gap in the dense fog closes like thick gray curtains upon a slowly darkening stage, and the white glowing orb of the moon is extinguished, all at once, behind the quiet hush of the heavy, now "ghostly" fog (see "Bashō", pg. 518).

The rapidly cooling forest air now carries the softer sounds of the intermittent night breeze, as it winds through the higher needles of the redwood trees, dislodging the first few fogdrops of tonight's foggy mountain dew. And many millions more will rain down life upon the forest floor, before the night is through!

In gentle layers of slowly descending quietness, Mother Nature reasserts herself anew. And quiet peace again reigns supreme in the coastal redwood rain forest, as if to say to all, *"Gute Nacht, gute Nacht"*, 'till morning's golden light sleep tight. *"Gute Nacht, gute Nacht*, and to all, a good night!"

Santa Cruz Mountains
Northern California, USA

Chapter 5
Betula "Birch"

Masked Bandits!

As I Sit Tonight, "tap-tapping" away on my Apple© laptop's keyboard, by pale gold lantern light, red wavering campfire flames warm my feet, and throw dark dancing shadows out into the surrounding hush of the dark forest night. But now I hear a rustling sound, and quite nearby! Then quite unexpectedly, a pair of mischievous green-glowing eyes stares right back at me from the darkness by my side!

Startled, I upend my folding chair, and jump up from my glowing laptop computer, as a wily old raccoon simultaneously summersaults up into the piquant night air! He lands with an unceremonious "*thud*, clutching a small plastic bag full of tasty little snack peanuts in his furry dark brown paw.

The sly old peanut thief then gathers himself up, and now with my pilfered bag of tasty snack nuts clinched tightly in his furry brown paw he breaks into a tottering three-legged run!

Now with an involuntary smile on my face, I hastily give chase, with my trusty tactical flashlight wildly wobbling about in my hand, casting shaky beams of light out into the dark black night! Nonetheless, my retreating little bag of waylaid peanuts still swiftly disappears, back into the deep darkness of the surrounding forest night.

But even in my agitated state of bemused frustration, I can clearly see where my little lost bag of salty peanuts has gone, by the jovial "Cheshire Cat" glimmer of the furry little bandit's baleful green eyes; laughing eyes that now glow happily back at me, from behind his retreating furry black mask!

Then with a clever comic wink, and an earnest nod of thanks, the wily old raccoon makes good his escape and disappears, back into the dark forest night.

Perhaps I should be angry now, or at the very least annoyed! But I just can't keep myself from laughing out loud, at his ambling three legged gate and clever fireside antics.

Now sporting a wide grin, I shake my head and trudge slowly back to the welcoming warmth of our campfire. Thoroughly bested by the cunning little beast, I upright my folding chair, retrieve my fallen laptop computer, and sit back down with a wry smile.

And as I turn up the flame of my rusty old forest green Coleman® lantern, in readiness to type again, I break instead into a slowly spreading grin! Then I shake my head anon (briefly) and laugh out loud again into the dark quiet night, as I recall the antics of our uninvited little ringtailed thespian (entertainer).

Because a small bagful of roasted salty peanuts seems a trifling price to pay for the delightful diversion of tonight's fun-filled woodland entertainment! (In remembrance of smart "Apples", Homestead "High", and *brilliant* Steve Jobs!).

The Myth of History

When The Sages Say That, "nothing has ever happened", whatever could they mean?[(75)] The very idea of it seems to fly in the face of all reason! What about our personal memories, and our collective history? The extreme notion seems quite absurd! Or is it?

Why is it so difficult to grasp the Sage's negativistic understanding, or to reconcile such widely divergent views of reality? The difficulty apparently lies in the significantly differing points of perspective.

If you believe, for instance, that our universe is objectively real and that our body-minds and egos are our real selves, you are (perhaps inadvertently) taking an "Exoteric-Objectivist's" view of your own consciousness.

On the other hand, if you believe that you are essentially an incorporeal soul and that both the universe and your body-mind nonetheless possess a subject-object based reality, you are taking an "Esoteric-Objectivist's" view. Which is a perspective that is esoteric in regard to our purported soul's incorporeal nature, and objective in regard to our body's evidently physical nature.

Both of these views are "positivistic" in that they *posit* the objective reality of the universe and of our evident person as its subject/object. But for the Positivist to be correct, our person/subject must also somehow be an object with the ability to fulfill its subjective role in an objectively real universe! (See "Esoteric-Objectivist", pg. 531).

For example, the first view fits that of the "Evolutionist", who believes the theories of modern science, but may otherwise be an atheist or an agnostic. And the latter view exemplifies that of the "Creationist" who disagrees with certain of the fundamental findings of modern science, but also believes in some sort of an intelligent Creator.

While not necessarily denying the views of either the positivistic empirical scientist, or the fundamentalist religious Creationist, the enlightened Sage experiences no fundamental difference between himself, other people, and the universe. The Sage comes to understand, and presumably to experience, that our persons, and our person's "personal" universe, have no truly objective, or wholly independent existence.

I

This is not to say that he sees the universe entirely as a mere fantasy. For our enlightened Sage is apparently in near agreement with the modern theoretical and particle physicists who contend that everything is energy, and that space-time has an existence that is quantumly relative to our consciousness.

However, they also then either avoid, or ignore, the dilemma of an objective independent existence. In the physicist's view, for instance, energy can be a wave or a particle and is presumably therefore not entirely either.

While the Sage might say that energy is more likely a partial expression of a pure Universal-Consciousness that is impersonal and not necessarily even local! And further, that it can also appear as either an energy wave, or an energy particle, relative to our tridimensional point of perspective. In other words, "energy" is an expression of pure Universal-Consciousness potential, as measured into our cognition by our individual consciousness.

But the physicist and Sage nonetheless both agree that two things are always present as fundamental qualities, energy (emerging from within ubiquitous "force fields"), and the principle of identity or "presence".

The Sage then adds that the space-time in which our experience of energy is understood to exist, depends upon our "streaming" three-dimensional serial experience of an overarching and intrinsically aware universe, which is comprised essentially of Universal-Consciousness.

It is also quite interesting to consider that it has taken our exoteric Western world several thousand years, the development of higher mathematics and highly sophisticated instruments, to arrive at nearly the same understanding that the Eastern Sages, long ago, achieved intuitively and apparently as a matter of direct experience!

This does lend a certain authenticity to the Sage's claim that our apparently innumerably dimensional Uni-, Multi-, or Super-*verse* exists as much as a "verse" within our consciousness as it does in the Conscious energy of an ephemeral "external" universe.

The wedding of the two suggests that their apparent emergence from within an endless plenum of cosmic potential, is a matter of perspective relative to the most likely outcome of events that automatically transpire within unfolding consciousness (see "Schrödinger's Cat", pg. 572).

But fundamentally it rests upon the neither negative, nor positive presence of a primeval Awareness that is clearly qualitatively quite different!

Because it is a percipient Awareness that is prior to our ability to either sense fully, or to understand.

The Sage teaches that this impersonal Awareness is reality and that we cannot see it because it is what is doing the seeing through us, vis-à-vis a virtual space-time!

Fundamental to any reality is the notion of "being". That is, for there to be any sort of realness, there must first be some sort of "somethingness". Simply because, even for the absence of a conceptually real and locatable (any) sort of "thing", there must first be at least a virtual something to be unreal! To "Be", therefore implies awareness of the condition of beingness.

Indeed, with all due respect to Shakespeare, there could never really be a question of "to be", or "not to be", for in the absence of a witnessing Awareness, only nothing could ever "be"! This "nothing" however, is the notional void of annihilation. It is merely another self-refuting theory, a "Solipsism" in disguise!

Whereas the Sage refers to the necessary existence of a noumenal nothingness, which is the necessary complementary to our apparently phenomenal universe. It is therefore neither something, nor nothing, but is most likely a pure (if not "Supreme") Awareness that is somehow expressing itself simultaneously as both!

But what then could this seemingly negative (but also not really positive) "presence" possibly be?

II

Succinctly, the Sage tells us that nothing must exist, for any sort of something to exist, but that both cannot exist without the fundamental substrate of Awareness, by which both are known to exist.

And further, that this "knowingness" is only made possible by the energy/presence of a universal Consciousness. This expression of Supreme Awareness within our consciousness is only allowed because its essential nature is neither a nothingness, nor a somethingness!

In other words, this quintessential (ideal) Awareness could be neither a nothingness, which would possess a complimentary opposite of somethingness, nor a somethingness that would conversely be dependent upon its complementary opposite of nothingness.

Consequently, the essential nature of this qualitative "otherness" of pure Awareness is simply, and necessarily, purely unqualifiable! Because of course, it cannot possess any dualistically understandable quality or quantification. Since every understandable polar quality or quantity is necessarily only partial, it is always incomplete without its complementary opposite.

We may ascribe qualities to this Supreme Awareness such as primeval, pure, absolute, or eternal, and quantities that represent its infinite nature, such as inestimable, or immeasurable. However "It" cannot be either a quality, or a state of consciousness and thereby possesses no opposite. And possessing no opposite, It also cannot be quantified!

For unlike the absence of consciousness, which is simply unconsciousness, in the absence of the fundamental Self-Awareness of our being, there is evidently only the impossible nothingness of a solipsistic oblivion. Supreme Awareness is thus quite unlike the mental states of consciousness and unconsciousness, which are merely polar opposites, because oblivion and non-oblivion are mutually exclusive!

It therefore seems self-evident that the universe's condition of non-oblivion is likely contingent upon a fundamental condition of pure Self-Awareness. Because, while the universe may arise from a preconscious state of space-timeless "Elsewhere", it cannot arise from the dubious state of complete oblivion!

In other words, so-called *ex nihilio* "from nothingness" cannot imply the absence of the pure impersonal Presence that is necessary for any sort of principle of identity to arise in our consciousness, or otherwise.

This is because any sort of existence as either energy, matter or consciousness, is dependent upon the fundamental principle of identity in order to occupy either an independently objective, or an ideationally conditional space-time.

While posing as an interesting theory, *ex nihilio* is therefore really just another self-refuting hypothesis! And simply trying to prove that the universe is somehow physically eternal only postpones the problem of its existence outside of consciousness; whether personal, or independently universal.

Furthermore, phenomenal existence, which is our sensorially limited perception of the universe, is allowed only by the presence of a dualistic consciousness operating within the machinery of our body, senses and mind. Indeed, our ostensible "body-mind" and its relative sentience are thus merely phenomenal.

This is because they are really only a reflection of the overarching or primeval supreme Awareness, which evidently also comprises the only true reality that could ever in consciousness "Be"! The appearance of our evolved body and mind is thus the catalyst that hastens and finally allows, the conscious manifestation of the universe within which we exist, to fully occur.

III

Before the appearance of the human mind, with its profound subtlety and range of experience, the universe of our experience existed in potential only. Thus, whatever was and is actually there, beyond, before, and between our ability to perceive, can only be said to be comprised of a unique expression of self-Aware energy potential playing about in an inscrutable fashion. Within what, for want of a better descriptor, and can be called a "Universal-Consciousness".

The indelibly aware East Indian Avatar, Sri Krishna, tells us that; *"Whatever exists has always existed and will continue to do so, and that which does not exist, never did and never will!"* (Bhagavad Gita 2:1).

This statement actually has to do with the concept of location. Which is to ask, "Where is the universe, in part and in whole, essentially located?" On the face of it, it would seem to be located both within itself and within our consciousness.

But is this dichotomy either accurate or acceptable? No doubt the Avatar's statement refers instead to the timeless, spaceless, and unqualified supreme Awareness that plays at being our persons and universe, within its energetic extension as an essentially impersonal Universal-Consciousness.(76) (See "Avatar", pg. 504).

But its unlimited true nature is perceived within a space-time, subject-object limitation by our personal consciousness, which must operate within the dreamlike functioning of our phenomenal minds.

Hence, noumenal and phenomenal are essentially complementaries, which together comprise the essence and expression of Divine potential. The Divine is therefore both transcendent and absolute, while being simultaneously present in the unfolding phenomenal potential of our beings!

Or we could also say that the Supreme is noumenally immanent and phenomenally transcendent. Either perspective depends upon pure presence. It is therefore important to acknowledge that whatever appears within the limited range of our individual consciousness is actually not unlike a dream!

In other words, our apparent ego subject, space-time, and all of the conceptual objects of our experience, are likely comprised of "fields" of interacting modified universal Consciousness-Energy. In much the same way that everything in a dream is composed of our cellular-energy based mental concepts.

Of course the difference is that the universe can be more correctly said to be like the ongoing "dream" of the Supreme. Which inevitably includes the individual perspectives of every dream character within it!

Moreover, insofar as our manifest or phenomenal universe is concerned, its existence is wholly contingent upon the presence of the experience generating machinery of Universal-Consciousness' evolved human body and brain.

No doubt there is an infinite, potential, and spacetimeless realm that lies beyond the reach of our limited perceptive and cognitive abilities. But it cannot be accessed by the familiar admixture of life and mind that we generally, and erroneously, believe is the positive sum of our selves.

IV

The Sages perspective, however, is what can be termed as "negativistic". In that he sees the objective universe and its person (the object and its necessary subject/object) as mere phenomenon that automatically transpire within an insubstantial, and largely mechanistic, psychosomatic consciousness.[(77)]

In other words, the apparent individual consciousness that plays about in our minds, is the virtual reflection of a veritably Divine Awareness into the perceptive instrumentation of our brains' and bodies. Not simply into, but also as, the essential "Stuff" of which its instrument, that is also "our" body-mind of dualistic experience, is fundamentally comprised.

Which only sounds like so much "double talk", because only thereby are the ongoing experiences and realities of our individual consciousness' manifested via the mirrored, or "virtual", subliminal working presence of our unlimited and impersonal Universal-Consciousness.

Therefore, whatever verity lies outside of our individual consciousness, while it must be there in order for any experience to happen, is entirely indiscernible! In much the same way that the character in a dream is unable to ascertain either the unreality of the dream in which he appears, or the presence of the overarching consciousness that is actually manifesting and experiencing it around and through him.

The universe and our persons are thus seen as objects of (and within) consciousness, but are actually the visible parts of an infinite continuum. A *plenum* "endless potential" of Universal-Consciousness that is being experienced by us sequentially within three dimensions, inside a fourth of inferred space-time.

It should also be noted that for a plenum to be full of potential, it must be neither full of emptiness, nor empty of fullness, while not entirely lacking in either!

Only in this way can it contain both, in the fullness of its infinite potential. This has unfortunately been called a "void", which is evidently as absurd as referring to it as simply being a plenum of endless fullness; as if the universe were a cornucopia full of ineffable fruit! (See "Plenum", pg. 511).

In other words, it is not analogous to saying that an orange is not really an orange or an apple. Instead it is the indistinct "something", in and of which the elaborated atoms of apples and oranges, along with everything else in our curious universe, is contained in unlimited and undifferentiated potential.

But it is also the "somewhere" from which everything comes and to which it eventually returns, as well as being the "something" of which they are made. Therefore it is actually neither a void nor a fullness, as are we!

Our sense of space-time is thus not only the further dimension from which we perceive the previous three. It is also then the deeper, intuitive and noumenal aspect of impersonal Universal-Consciousness that is being constantly reflected "within us".

Thus, it is this "noumenality", which thereby grants sentience to our body-mind, and also simultaneously connects us to our veritable Self-Awareness.

Moreover, since we do experience space-time, even though we cannot directly see it, we must be experiencing it from a superior (i.e. deeper) dimension. This is to say that we cannot be whatever lies before us. In the same way that we cannot be either the pencil that we grasp in our hand, or the hand that we see grasping it!

Is it not then "Self-apparent" that Self-Awareness is not an attainment, but rather a realization of the experience and understanding of our own true Natures?

V

The so-called "objects of consciousness" by which we are cognizant, are thus parts of an impersonally conscious continuum, with implied internal "subject" aspects and ideational external "object" aspects. Both of which are complementaries of the Universal-Consciousness that comprises, connects, and manifests them.

But for the noumenal to experience itself as the phenomenal, our body-mind is clearly necessary. And for our body-mind to be aware, or to know that it exists and thereby that it knows, it must receive the reflection of its qualitatively different percipient primeval Awareness in some fashion!

The likely locus of this interaction is at the most subtle level of the universe's Consciousness energy spectrum, via what could be called a psychic or consciousness "sheath". Which makes its appearance by degrees as the consciousness aspect of its energy pole begins to dominate.

Consciousness and its energy, or Consciousness-Force, are thus complementary poles of a unified, percipient (immanent) Universal-Consciousness. Which is in turn an expression of the infinite potential contained within our universe's fundamental and absolute monism of transcendent supreme Awareness (i.e. the Cosmologist's pre-Singularity "Elsewhere").

Universal-Consciousness is thus in essence comprised of Consciousness-Energy and neither can be separated from the other. Because both are united by the fundamental Awareness of which they are comprised!

But from our tridimensional perspective this appears as an invisible inner self-Aware "space", which only phenomenally appears to us as space-time. Therefore many erroneously believe that the evidently sentient "inner" space from which they seem to be experiencing everything, must be their own personal and incorporeal "soul"!

This is because Universal-Consciousness cannot experience phenomenality while simultaneously looking inwardly at its own noumenal Source. In any case, our enabling *Yesod[ic]* "Foundation" cannot really be sensed at all, but can be known intuitively because it provides the field of consciousness in which our three-dimensional perspectives function.

Indeed, our three-dimensional universe could not "exist", if we were unable consciously to manifest it conceptually within a "Scalar Field", in an *only apparent* space! This is to say that we must imbue it with sufficient dimension over enough time or duration for it to be cognized.

The space-time of our experience is therefore a virtual tool of perception which has no independent existence, per se. And in its absence, even the implication of an independent objective universe becomes quite impossible!

However, what some would therefore surmise is positive proof of "relativism", is actually quite dispositive! It is evidently no more correct than nihilism, because it inadvertently rejects the existence of the universe outside of its experience within our individual consciousness.

Because, relativism holds that whatever exists does so only in relation to whatever else exists, if there is substantially nothing present to relate to, whatever exists, may or may not exist, since it cannot be experientially measured!

And because relativism clearly misses the fundamental fact that whatever exists does so only within dualistic consciousness, it also misses the fact that the apparently relative universe of our experience is actually an illusory, interrelated holographic-like projection of living consciousness-energy.

VI

In contradistinction, the Positivist-Objectivist perceives what he believes to be an objectively real universe, in terms of a substantial or concrete subject and object. This theme is presumably carried out on every level of his consciousness, which is thus limited in its understanding by dependence upon an automatic, subliminal function of continual bifurcation and comparison within an ideational space-time.

This self-limiting aspect, along with its reliance upon the quixotic (unrealistic) premise of an essential subject also simultaneously being an actual object, drives the well-meaning, but nonetheless incorrect Positivist-Objectivist into the duplicity of an absurdly inescapable quandary!

Consequentially, the Positivist-Objectivist is unwittingly practicing a kind of indefensible "Realism" gone too far, just as the nihilist erroneously proposes an absurdly radical, if not inescapable, self-referential kind of infinite "Reductionism".

But even if both views are clearly erroneous, apparently everything exists only within our personal experience, which is actually based upon *no*-thing. Therefore, unless we are very careful and specific in our inquiry, everything reduces ultimately into a purely mystical fantasy, a confounding material chaos, or the unlikely energy and awareness empty space-timeless infinitude of an endless oblivion!

However, in all fairness to the person seeking solid ground in a rather insubstantial universe, these are subtle distinctions not obvious to our senses. Which nonetheless impart the onerous dilatory consequences of existential suffering upon the undiscriminating and the unaware!

But because the Sage embraces supreme Awareness as the sum and substance of the universe, every apparent "thing" has a phenomenal existence, rather than an independent reality. Therefore, in his deeper and hence likely more "Essential" perspective, the universe is neither discretely personal, nor infinitely reducible into either chaos or oblivion.

The question, "Where are things 'located' in the universe?" is thus answered quite simply by the Sage, "Everything exists and is located within Universal-Consciousness, including the ideational space-time in which everything must appear in order to be cognized."

But if experiential space-time is really just our perceptive tool, what then are we to make of our apparent pasts, presents, and futures?

The Sage responds, "Everything is Consciousness and there is no past or future. There is only the present moment, and even the present has no real independent existence!

Everything is just the timeless, spaceless play of Universal-Consciousness. Which only masquerades within our apparent individual consciousnesses as both the space-time and all of the objects that seem to constantly move about and continually change within it."

"Self-evidently" everything is fundamentally only an expression in probability of Universal-Consciousness, including space, time, subjects, objects, apparent stillness or motion, causality, entropy, and even "negative" entropy!

This is why the awakened Sage so often maintains that, "history is a myth" and further, that from the apparent beginning of our universe, "nothing has really ever happened!" It is all the equivalent of a grand dream. And every dream has a dream character experiencing the dream from within it that is also a part of the dream, which in consideration of the foregoing conditions is evidently exactly what we are!

VII

But what of our universal dream's "real" dreamer? "This", the Sage says, "is the Oneness independent of the dream, witnessing it from outside of the dream, through the aegis (sponsorship) of the character within the dream. It is the eternal Witness that exists prior to 'our' dream of consciousness and all that it contains".

This is "who" and "what" in primeval Essence that we always truly are. However, because what we essentially are is also evidently transcendent even to our phenomenal consciousness, it is also qualitatively different from any sort of "quality", and is therefore also quite unobtainable!

Apparently all that we can do, to encourage the natural process of our "reawakening", is to reawaken the conditions that are conducive to a shift of identification away from our illusory ego "self". And we can only accomplish this by fostering the right kind of "Self-understanding" and "Cosmic-perspective".

In other words any sort of *metanoesis* "intuiting of our true nature" could effectively be called an act of "Evolving-Grace". This is to say that while it has the superficial appearance of being a supernatural act of amazing "Grace", it is nonetheless actually accomplished by the natural evolution of the Universal-Consciousness. (Which always lies "dormant" within the sub-atomic constituents of matter.)

But this is not to say that since nothing truly exists, "per se", and we are therefore relatively powerless to do anything about it, that nothing in our experience has any importance or meaning!

Because whatever exists extensively, even if only as an experiential potential, can evidently do so only within a "singular(ity)" extension of Universal-Consciousness, and even therein only by the symmetry breaking will/force of an "Elsewhere" dwelling Supreme. In other words, our very existence, even if it is essentially just an illusion, or only a dream, is itself an amazing act of natural "grace".

This is an understanding that is in itself inherently meaningful, but not in the dualistic manner of our common understanding!

But perhaps a good way to understand the implicate sense of the latent "impersonal meaningfulness", which is an intrinsic quality of the Sages' perspective, is contained in this charming tale once told to me by my Japanese-Hawaiian Karate Sensei.

It takes place on the lovely Hawaiian "Garden Isle" of Kauai: "One day, early in the morning, a *keiki* 'youngster' was walking up the beach throwing starfish back into the still churning sea, one by one.

Someone asked him, 'The storm has thrown up thousands of starfish onto the beach, what possible difference could it make for you to toss a few back into the heartless ocean now?' The keiki bent down, picked up another starfish and wishing it a heartfelt 'aloha!' he flung it far out into the still surging waves. 'It will make a difference to this one!'

And as he continued walking up the beach unperturbed, all the while tossing starfish far out into the now slowly calming sea, he called back over his shoulder, "I'd like to take the time to explain more, but '*wella ka hao*' (life's just too short) and there is far too much suffering still going on, and I still have far too much left to do!"

Independence

WE NEED TO BECOME EXPERTS in the area of our own consciousness. Because if we put our faith solely in the opinions of others, as to whom and what we truly are, sadly, disappointment is the most likely end result.

And as we should all by now know "cults" are especially dangerous, the Middle Eastern cult of *Daesh* "ISIS" is a sterling current example, and should be avoided like carelessly stepping on a nest of sleeping rattle-snakes! (Perhaps even more than the sometimes discouraging opinions of misinformed, but well-meaning, family and friends?)

But the size of the group, or its relative social acceptance as a bona fide religion or philosophy, are not only irrelevant to the importance of its fundamental veracity, but also to its individual adherents ability to think clearly for themselves.

Because only thus can they avoid the irreversible and self-defeating pitfalls (and eventual self-destruction) that result from aggressively pursuing clearly inhumane (if not criminal!) policies of repressive religious discrimination and indiscriminate, shockingly compassionless violence!

Indeed, in a far less affective but nonetheless eventually "serious" vein, even the spontaneity of our intellectual independence does not constitute a *carte blanche* "free pass" for irresponsible, weird, or antisocial behavior! It simply makes no sense to believe that anyone else is at all privy to the contents of our own consciousness, let alone to the mystery of its interaction with the pure Awareness by which it is known.

However, purposefully avoiding others is also not productive, unless you are fond of always being alone! Nonetheless, alone time for meditation and introspection are always important, if not self-evidently so, for those in search of any true Self-understanding!

The truly "Awakened", however, no longer avoid anything. Instead they confront everything directly, because they are no longer hoodwinked (tricked) into surrendering everything to their illusory, intervening egos. Some have called this "Mindfulness".

But this too is probably a concept best avoided. Otherwise we might wrongfully believe that we have discovered a method to save our illusory dream self (i.e. ego), by somehow awakening it from its reverie within our persistent waking dreams!

Nonetheless, the mindful spontaneity of the Awakened is sometimes misinterpreted as simply being some sort of entertaining asocial behavior. Indeed, the fully Awakened may even display a disconcertingly abrupt grasp of a deeper reality that is often ahead of, and sometimes even contrary to, that of others!

Moreover they rarely yield, because they are deeply aware of a more fundamental, if not "realer" Reality. Which certainly includes others, even if we have temporarily excluded our realer Selves from directly perceiving it, by our persistent habit of self-distraction through objectification (i.e. thus creating, and continually re-creating, the infamous biblical "Golden Calf"!).

But purposefully emulating the spontaneity of the Awakened is an ineffectual, often selfish, and sometimes even downright rude, exercise in antisocial futility! It is an erroneous attitude that will likely only alienate others and possibly even socially estrange you!

And because simply acting spontaneous cannot deepen our self-understanding, it will likely only strengthen the already strong attachments that most all of us have, to our intractable illusory egos![(78)] Yet real compassion ever remains the "touchstone" of true wisdom, because the compassion of a Sage accrues from his awareness that we are all one in the divine Essence of the universe.

Kindness is therefore always worthy of emulation, for without it the awakening fires of enlightenment might never be kindled within our hearts![(79)] But the kindness of the Awakened is inscrutable, for its only goal is to enlighten.

And such is evidently the "Will of Heaven", which is also to say that it is the irresistible "Way of Nature" (i.e. Tao). Which of course no mere human being will ever be able to understand fully. But fortunately, we are in our fundamentally Aware essence both much more and much less than merely illusory mortal beings!

Non-Interference

Even If It Were Possible To Do, perhaps no one should ever attempt to spell out fully the אמת עליון *Emet Elyōn* "Highest Truth" of life! Otherwise only more concepts are created that tend to cloud our intuition. And, indeed, any understanding that goes against our *Nisarga* "true Nature" is patently unsound! So, do not interfere; "simply do your work and get on with your life".

Such is the sound advice of a true Sage (i.e. Sat Guru). Only understand that we are life's impersonal witnessing Awareness, which most certainly includes the body-mind vehicles (avatars) by and through which our every apparently personal experience arises.

Nonetheless, of necessity, we are always present, even prior to the perception of our every symbolic experience, and even before the appearance of our dualistic consciousness and apparent individuality!

Therefore, lacking in true subjectivity, the virtual subject/object ego of our everyday experiences is even less than a phantom nanoparticle. Because it is only a virtually real experiential center, and yet our entire (avatar's) mind is illuminated in the sentient reflection of its own true impersonal Witness!

Our everyday mind is thus the reflection of a more percipient reflection, being lent to it by our ubiquitous, centerless and boundless truer locus of self-aware Universal-Consciousness.

But within our individual consciousness, which precipitates between the temporary experiential "mirrors" of our minds and bodies, our ephemeral experiences seem to go on forever!

Nonetheless this reflected, although oxymoronic illumination, is what we commonly understand to be our own private consciousness. It appears unbidden within the automatic interpretative processes of our body-mind(s). Wherein, it continually entertains the conceptualizations of our individuality and relative realities.

The entire universe is thus contained within all of the apparent individual objects of symbolic consciousness that automatically appear within our minds. They make their curious appearances, like a parade of "strange-loop" ghosts, marching almost magically out of our interpretation of the busy electrochemical signals that are constantly going on in the cognitive machinery of our brains![(80)] (See "Strange Loop", pg. 581).

The puissant singularity of sentience therefore arises unbidden in our lives, but also ends with the inevitable death of our bodies, and the subsequent dissolution of our minds. While the impersonal

Universal-Consciousness, which makes it all possible, remains completely unaffected by it all!

Of course this does mean that any meaningful "Enlightenment" can only be gauged and realized by thoroughly acquiring this understanding. But no one remains, when it is sufficiently understood, to gloat over whatever might have been gained!

Such is the quandary of our implied existence, and the truth of our beingness, which ever underlies whatever illusory individual consciousness that we might temporarily enjoy.

Once this understanding is firmly grasped, it is best to cast it immediately aside, as if we were holding a live coal in our hands! And with the clear understanding that "nonaction" is also an action, then simply get on with our lives, as if nothing special has happened at all.

Because in truth nothing has happened, except for a beneficially radical shift in our understanding, back to its essential Awareness' veridical point of observational perspective. Which is always prior to the unfeasible (and therefore, "virtual") experience of ourselves as both subject and object!

"To let understanding stop at what cannot be understood is a high attainment. Those who cannot do it will be destroyed on The Lathe of Heaven."

~ Ursula K. Le Guin ~

Presence

Everything In Our Universe Stands Firmly on the *Yesod* "Foundation" of an intrinsically aware subquantum Presence. This is because the elaboration of pure Presence into quantum forms is the principle cause of the subsequent appearance and eventual evolution of our universe into the sentient forms that we presently enjoy as apparent individuals.

But are we not then fairly contained within the grander space of a Divine Cosmic or "Universal-Consciousness", which is surely the *Adam*

Kadmōn "Primeval Man" of the ancient Jewish Kabbalah? Indeed, is not the very idea of an immortal sentient Presence precisely what we really mean when we try to describe succinctly our occult *Nefesh* "Soul"?

Thus our experientially unapparent über Soul is primary to any kind of existence, and it is also principal, because any sort of consciousness is only thereby sentient, or aware of itself and of its apparent actions. Nonetheless, the ideation of an individual incorporeal soul, while comforting, is evidently as absurd as the refractory illusory individual that entertains it!

But when we fully realize that we are actually only this fundamental Presence, the apparent two of individual self and external universe is seen and experienced as an undivided One. And this Oneness is every "where" present in an only apparent diversity.

In other words, our experience of diversity is actually firmly embedded in the fundamental phenomenon of the psychophysiologically based ideational dualisms that we commonly experience as virtual subjects and objects.

Therefore Gautama Buddha said to his star pupil Vimalakirti, "Nirvana and Samsara are not [really] different!" Gautama Buddha was most likely implying, in this rather curious self-contradictory statement, that by realigning our mistaken identity with its fundamental impersonal Presence, the stubborn secondary dualism of our split-mind may again be realized to be the principal One, by which both our subject(s) and objects are known to exist.

This principal Oneness is the absolutely essential, non-psychophysiological and transcendent basis of our immanent reality. It is our reflected principle of identity, borrowed from the self-aware Presence by which we become aware of being conscious within our phenomenally conditioned body-mind. This Presence is the Buddhist *Bodhi,* known in Kabalistic Hebrew as the *Neshama* or "Over Soul".

Hence a Bodhisattva is a person that has undergone a metanoesis, a beneficial "shift of identity" back into the primeval Awareness by which we all know that we exist. When we again realize that we are really only pure Presence, our fundamental "two" of faux subject and illusory object, is eventually, or suddenly, subsumed (included) back into our veridically undivided Oneness.

We are what the Hindus call the *Brahman* or "Divine", rather than a mere phenomenal reflection of the Divine, appearing within an illusory individual consciousness, via our habitual waking and sleeping dream states. Indeed, even our alleged waking state is essentially no different than a dream, constantly changing in potential, yet persistent in its continuity.

But sadly, we must return to it again and again so long as we maintain a fervent faith in its veracity! But of course the compassionate death of illusion that is our "waking up" does not cause the entire universe to suddenly disappear! Only our seemingly inextricable, and deceptively intractable, habitual involvement in it finally comes to an end.

Finally, or fundamentally, we are the knower of the Divine Dreamer. We are the absolute *Parabrahman* or "Supreme-Awareness" (i.e. *Yahweh,* or *Allah)* by which Brahman exists as the impersonal Universal-Consciousness *(Elohim)*. Other than nothing at all, whatever else could the universe and "we" possibly be!

If this seems at all abstruse (puzzling), abstract, or exotic, consider instead that it is simply the most fundamental and primary generative Force of Universal-Consciousness, which is also always present within us as our Force of Life. It is *Tat,* "That" in Sanskrit, which brings the universal "Manifestation", *Malkuth* in Hebrew, into final focus within our consciousness.

It is the necessary Presence in the wilderness, in order to hear a falling tree make a "sound". Therefore being neither an impossible subject/object, nor even a wholly ideational object, we are rather the aware Presence upon which both clearly depend! For in the absence of presence nothing can be said to exist, and only a conjectural "void of annihilation" could possibly prevail!

All of this seems perhaps absurd, but Shakyamuni Buddha was apparently quite fond of its "absurdity" and faithfully expounded it for some forty-odd years! Yet its veracity can only be ascertained by our own individual experience.

If this were not so, all manner of teaching about self-Awakening (or even the essentialy Divine nature of our waking universe) would be absolutely without merit, and completely bereft of any real usefulness whatsoever!

More simply put, it will never do to mistake a pointing finger for the moon that it is pointing at, because an incomplete understanding, like a bad habit, is quite often worse than none at all!

Trinity

After The July 16th Trinity nuclear test in America's desolate New Mexico desert, J. Robert Oppenheimer commented, "I am become Death, the destroyer of worlds!" He was quoting a passage from the *Bhagavad-Gita* or "Song of G-d", contained in the epic Hindu "Mahabharata" tale.

Ostensibly, Herr Oppenheimer's meaning is plain enough, he was doubtlessly referring to the questionable opening of a nuclear "Pandora's Box", and the resulting ominous threat of releasing a nuclear hell on earth. (As a scientist he well knew the problems that this particularly sinister Pandora's Box might unleash upon an unsuspecting humanity.)

Indeed, shortly thereafter the "fires of hell" were twice rained down upon a nearly defeated, but sadly recalcitrant Imperial Japan, bringing about a swift but unsettling end to the horrific Second World War!

But ironically, as an eventual unforeseen consequence, because of uninformed public fear about reactor safety, byproduct containment, and possible radiation poisoning if there were a core "meltdown", the development of "carbon free" nuclear power would also be slowed and eventually abandoned on the very brink of putting our nation's first meltdown proof IFR "Integral Fast Reactor" into production"!

As a result, evidently still unbeknownst to many people, in addition to the proliferation of nuclear weapons, we were now facing the threat of an eventual planet-wide extinction event from global warming! (See "Just How Bad?", pg. 548).

The subsequent proliferation of atomic weaponry during our frenetic "arms race" with the former Soviet Union soon pushed the entire world to the perilous brink of mass destruction. And although the outrageous numbers of nuclear weapons has been greatly reduced after the end of our so-called "cold war", humankind continues to develop ever more sophisticated and sinister methods of mayhem and self-annihilation.

Now, in addition to the better known nuclear WMD's "Weapons of Mass Destruction" of yesteryear, we now have an ever increasing Black List of horrific high-tech weaponry.

Such as TNB's, "Thermobaric Neutron Bombs". Which greatly reduce our disinclination to use them, because of their significantly reduced levels of radioactive fallout. As well as the newest TBW's "Thermobaric Warheads" that host explosive yields, and searing "sun hot" heat, that actually rivals tactical nuclear ordinance!

Then there are the newest iterations of either "targeted", or "widespread", electronic infrastructure destroying EMP's "Electromagnetic Pulse Bombs". Internet destroying EMV's "Electronic Malware Viruses". And humanity destroying VMD's "Viruses of Mass Destruction"!

In the past few years we have also seen the increasing use of unmanned, remote controlled, "killer" Drones. Ultralight and fuel thrifty, with a simple increase in size and range, they are quite capable of delivering any of the above "nightmare ordinances". No-doubt with a little less speed, but inarguably with much greater stealth than a Naval "Tomahawk" missile!

And last on our list, there are even "smart" BBB's! Which are Self-guiding, deep penetration, "Bunker Buster Bombs"; to name just a disturbing few!

Thus, due to our ever advancing science and technology, things are apparently becoming quite dire for our enemies! Indeed, other than within obscurity (or *oblivion*) there is almost nowhere left for the terrorists to hide!

Nonetheless, humanities tendency toward death and destruction seems especially perverse, given the unique and precious nature of life. Because even if it is fundamentally Divine (and perhaps even indestructible in its fundamental subatomic design), life is also inherently ever tenuous.

It is apparently just as fragile as it is incredibly durable and rare, having appeared against the greatest of odds, in a universe that seems inimical (hostile) to either its appearance, or our survival![81]

On the other hand, life *has* appeared, and its very appearance strongly suggests that something more than mere happenstance is at work within the apparently indifferent, if not mechanistic, events of nature.

But the existence of a virtually changeable underlying Universal-Consciousness that unaccountably emerges from an ostensibly changeless primal or Supreme Identity (i.e. G-d), gives little comfort to most people. Because, in addition to the inhospitable cosmic environment in which life has perhaps inevitably emerged, there are a plethora of unpleasant events, even in our everyday lives, that are evidently quite beyond our control!

This would not be a problem were it not for the fact that we are not only sentient beings, but have also become intelligent enough as a species to understand our own physical vulnerability and personal mortality. In addition, humankind is unambiguously overburdened with an emotional vulnerability to a veritable surfeit of not only real, but also endlessly imagined problems.

Therefore we worry! We plan as best we can, but worry until our plans either work out or fail. And with our perceived losses and ever accumulating surplus of imagined "failures", more serious worries almost inevitably ensue.

Perhaps the Buddha should have added psychological suffering to his well-known list of "Fourfold Suffering" (see pg. 535)! Because, in many ways, this particular type of suffering is not only more constant and widespread than many other sorts of suffering, but evidently it is even more perverse in its intractability! Moreover, whether physical or emotional, pain and suffering are events that transpire within our dualistic minds.

Thus religion, albeit mostly unawares, is largely an attempt to address the unique problems and attendant suffering associated with the emergence of a virtual, and dualistically operating, subject-object mind in humankind.

Anthropomorphic gods and fantastic myths have been created in order to put a human face upon the cold impersonal mystery of an apparently indifferent universe. Because a personal god can both understand and participate in our lives, and a humanlike god can administer justice and even sympathize with our apparent suffering.

He or she can be prayed to and asked for help with the many problems in our lives that are evidently beyond our ability to control. Heavens are surmised as a reward for good behavior, and hells as punishment for the "wicked" in our midst.

An idealized supernatural hierarchy is thus eventually imagined into existence, god is placed in his or her "highest heaven", and all is then assumed to be right in our disturbingly unstable world!

But of course, as with a great many human endeavors, the reality is often far less than the hoped-for results. The fantastic gods and imagined heavenly utopias of humankind inevitably come to reflect the limitations of the human mind, if not the dystopia of our understandably troubled human psyches.

And unfortunately these unmanageable imaginary gods, much like human heads of state in non-democratic countries, through the ever accumulating auspices of vulgar human greed and fear, too often become jealous and unforgiving tyrants, and sometimes even worse, clearly insane despots!

Unbelievers and non-conformers are then punished, tortured and sometimes even killed by the "true believers". Who may even espouse the achievement of an ideal theocratic society as their eventual goal, but just as often seem to practice murder and mayhem to increase their power, profit and perceived "glory", through the imaginary auspices of their fungible (mutually interchangeable) false gods!

Indeed, more wars have been fought in the name of imaginary deities, than can be reasonably justified for the necessary sake of humanities physical survival! No doubt, humankind will be much better off when we are finally able to throw off the heavy yoke of this onerous (burdensome) mix of religious repression of reason and deeper human understanding.

However, religion also offers a strong moral authority and guidance with a firm hand for a great many in the world who still apparently need it! Despite the indisputable fact that about as much harm as good results from religion, it also seems reasonable to suppose that even more harm might befall the many who still require this paternal kind of guidance, if they don't get it. Because without it, the likely widespread chaos could result in an anathema (complete out casting) of civilized society!

But ultimately, the real culprit is our sadly misguided and dualistically divided human ego. This is why our most effective method to achieve abiding individual, and perhaps even world peace, is to gently, but persistently, nudge human awareness away from identifying with our illusory ego entities (and the disingenuous teachings thereof), and back to the Universal-Consciousness that is the primal source of our life, love, liberty and consciousness.

After all, a "demon" is fundamentally a person incapable of empathy, while even recent history clearly shows that the "highway to hell" is apparently a society without compassion! And while both are clearly divisive, like so many things in humankind's makeup and world, it is difficult to tease out the healthy and good from the dysfunctional and harmful. But this does not excuse us from trying!

Most religions contain the thread of this concept (of trying to make a "better world"), even if they often ignore its deeper significance. For example, this process is the symbolic ingathering of the "scattered sparks" of Divinity, which the Jewish Kabbalistic Sages have hinted at for thousands of years.

This concept eventually became the *Tikkun Olam* "healing of the world" of modern mainstream Judaism and, by ignorant extension, the presumptuous proselytizing activity or "soul saving" of its sometimes harmfully misguided scion (offshoot), Christianity.

The alternative to our wider social awakening is to tolerate the harmfulness of mental manipulation, greed, violence, and downright insanity that is so often being perpetrated in the name of an ideational god and a presumed "higher", rather than an actual "greater good". Because it is a corrosive harm, and ironically it is often absurdly inflicted with an unintentionally comedic, if not tragically misdirected sense of entitlement, and sometimes even rather crude "self-righteousness"!

Systemic dysfunction, terrorism, and the suicidal destruction of nature are already upon us. War, cruel starvation, deadly pandemic and the chaos of widespread anarchy are currently (if not always) biding their time, lurking in the dark shadows of civilized humanities ignorance, fear and inaction.

And unless we are soon ended by a nuclear war, an errant asteroid, über volcano, or perhaps an exploding supernova, the time for effective action is apparently growing shorter by the minute, while the need for raising human awareness has evidently never been more pressing!

Things are apparently quite dire, and the darkness of despair seems warranted. Therefore it also seems important to remember that even this powerfully troubled communal version of our shared world, while not entirely unreal, actually exists only as a collection of amenable perspectives of thought.

This is because ours is a conditional group reality that arises from within a consensus of agreement, all of which originates from within our illusory collective heads. This means that even our most intractable illusions are always amenable to change. In other words, if we don't like what's going on in the world, we can always change our minds!

Flying Home

Understanding That Life Is But A Dream, it seems wise that we simply do our work, and get on with it. But we should also not advertise or dwell upon the fact that we may have somehow intellectually or intuitively grasped the cardinal truths that define our existence.

Because if you constantly remind yourself, reminding yourself is all that you will likely achieve. And if you constantly remind others about your insights, all that you will probably accomplish is their annoyance and sadly, even your own alienation!

A pilot does not need constantly to remind himself that he is operating a machine that is flying at great speed through space and clouds! He simply does his job, and soon arrives safely at his destination. And please note, that the big collective "You" is in discussion here, which I understand always includes every iteration of any little "me's" as well!

(In grateful memory of the brave 101st Squadron of the IAF, you will ever be "the wind beneath our wings." – see pg. 602. And to the USAF Strategic Air Command, and my courageous cousin, fighter pilot, Captain Billy Lee!).

The Divine Gulag

Unless We Are Indeed Languishing Our Life Away In Prison, and evidently, *if not disturbingly*, far too many of us in America, without sufficient reason, currently are! But we nonetheless experience life as if it were a prison, rather than the Divine's challenging experience of unity within multiplicity, then we are in error and are undoubtedly suffering as a result.

For our coming apart is G-d's pulling together, as our dissolution is naturally transformed into the fuller manifestation of G-d's Awareness.

Our dense "lead" of misunderstanding only becomes true "gold" in G-d's joyful alchemy of reawakening. And an "alchemical" transformation is indeed happening! It is a Divine alchemy, increasingly fueled by the direct experience of our transcendent monastic oneness in the projected play of Allah's own experiential diversity (see "Alchemy", pg. 504).

Because "Allah, Yahweh, G-d, or Parabrahman", all of these insufficient names, and many more, refer only to the same unnamable Supreme. Believing otherwise is likely no more than ignorant religious bigotry!

This is because our ordinary life is but an illusion of "holy smoke" and "sacred mirrors" that is being surreptitiously reflected into our illusory individual minds by the presence of the totality of G-d's impersonal Universal-Consciousness. This is to say that the entire universe lies behind every apparent particle of which it seems in consciousness to be comprised.

But eventually, when conditions finally allow it, our Life-Force becomes so strong that it begins to "burn away" our habitual impurities of misunderstanding, until only the elemental alchemical "gold" of our true Self remains, as it did with the remarkable East Indian Saint, Jñaneshwar.

Sadly, we don't presently have the time to discuss his full story, which is a fascinating tome unto itself, but suffice it to say that Jñaneshwar's "alchemical" reawakening was so powerful and complete that he experienced a radical transformation, wherein the cells of his entire body were completely renewed!

Such things may even be possible without violating the deeper laws of nature. Indeed, since the human mind's escape from the ignorant dogmas and superstitions of our recent past, the occult methods of nature's mysteries of seeming miracle and magic, are daily being revealed by today's unfettered science. (See "Jñaneshwar, pg. 547).

However, while it is not yet recognized by today's science as simply being another incompletely understood aspect and occurrence of nature, the eventual reawakening of our deeper, essential Nature, is most certainly not accomplished by any amount of personal willfulness, or even by some sort of "supernatural" act!

Evidently it is only done by somehow being worthy of receiving what, due to our dearth of current understanding, could be called the impersonal "Grace" of the Divine. No doubt this worthiness equates with the final cessation of our Self-ignorance, so often inaccurately called "liberation" in the ancient Far-Eastern scriptures.

But please remember that whenever we refer to "G-d", that we are always referring to the penultimate Supreme. And that it is the transcendent Supreme that witnesses the selfsame imminent Universal-Consciousness that underlies both our apparently "individual" consciousnesses and the grander space-time bounded universe that we presently seem to find ourselves in.

This is the impersonal Universal-Consciousness that both provides an underlying structure, and remarkably also achieves the manifestation of the entire cosmos, without and within the collective conceptual illusions of our only apparently "individual" consciousnesses!

When relating to stories such as that of Saint Jñaneshwar it is important to keep these caveats in mind, if we are to gain any trustworthy understanding from them. Furthermore, it is essential to realize that only *Elohim* "Universal-Consciousness" could construct such an existential "prison".

A prison contrived of conceptual space-time illusions both individually and collectively so powerful and convincing that it is capable of holding even *Yahweh* "the Supreme" in its abstraction, through the formidable power of subject-object distraction! Such is the gulag of our unexamined persons and universe (see "Gulag", pg. 507).

The seminal (essential) pointer here, is that a compassionate (or even an indifferent) Divine would never actually construct such a formidable prison! The alternative therefore, is to design a self-imposed, or self-allowed, limitation of awareness that works quite well for those who think that their egos are actually in charge, until the ruse is finally revealed. Yet, if it is seen through too soon, where is the challenge and adventure in that?

However, this is not to say that the cruelly insane misadventures of a deeply disturbed human ego are the Divine's intention either! But that since we are never truly "bound" that we also never really need to be "set free" from anything but the ignorance of our own boundless true Natures! And when we finally are, the Divine's spatiotemporal misadventure slowly (or suddenly) ends, and the real adventure of our spacetimeless Self-rediscovery begins!

This realization, which is the uncovering of our truer Self, is evidently what prompts the peaceful smile that so often plays upon the lips of the Awakened, and the infectiously joyful belly laugh, or sometimes even alarming *kwatz!* "lion's roar" of a living Buddha. A good question then is, "What is really meant when the Awakened curiously describe life as the 'play' of the Divine?"

The Play

Why Is The Universal Manifestation so often called the *Lila* or "Play" of the Divine, when so many of life's aspects are not very playful, often trying, and are sometimes even downright painful!

The entire "Universal Manifestation" of our experience arises only temporarily as a complex "somethingness" within our consciousness, and is instantly gone with our consciousness' cessation back into a simple nothingness.

It must be manifested time-and-again, as the Divine's *re*-creational entertainment, within our ideationally individualized consciousnesses! Like a recurring dream it arises from, and returns back to "nothingness". The Universe thus reveals itself within a conscious manifestation, but to "whom"?

(Dedicated with thanks to my august friend, Gurupremananda).

Brace Yourself!

Few May Understand This, and perhaps I even say it at my own peril, but the *Shoah* "Holocaust" is done, and not a decent person in the world (or even the world itself) has survived it unscathed! *Haval* "alas", for every precious Jew that we could not save is now forever gone, and only the universal bed-rock of יהוה *Yahweh* remains unchanged! But if it did happen, then where?

If you answer Germany or Poland, perhaps you are (in a physical sense) correct, but you are still an entire universe wide of the truer, deeper mark! If you maintain that any Jew now survives unscathed, then quickly now, please show me that "Jew", for I certainly want to shake their hand! Yet, to be a Jew is by far the finest thing that I, or anyone else, could *ever* do!

And I am not suggesting, G-d forbid, any sort of radical revisionism, or crude calumny of the truth! (See, "Revisionism", pg. 572). For indeed, the Messiah with our every thought is always coming and going, and all his good works too!

But Moses never really descended from the mountaintop, and the enlightenment of the Buddha never even passed his lips, 'though they still managed to reveal the highest truth to me and you!

But if you still think that you can show me a Christian, a Muslim, a Buddhist or a Jew, then we must be on our way to meet with the Wizard of Oz, and I will happily give you snowy Mount Everest in Dorothy's little red shoe!

Yet just as sure as the Divine power of love once moved Ruth to follow Naomi through the benevolent fields of Boaz, the last *mensch* "person" in the world will most certainly be a Jew, 'though no Jew has (or ever will be) "born"! And, although contrary to popular opinion, the Buddha never left the warm elastic sea of his mother's womb, although he continued to live and teach the highest truth for a great many years!

In fact, if anyone ever manages to solve this curious riddle no questions will remain for Elijah to answer, and all the past and future Prophets will soon disappear, and along with them, Krishna, Jesus and Mohamed, and all the "lamed-vav" Avatars too! (See "Lamed-vav", pg. 509).

But if this makes you think, or feel, either good or bad, or right or wrong, you are still endless universes away from the deeper truth of things! It matters not if we are angry, indifferent, happy or sad. Because whether we agree or disagree, or even simply say "yes" or "no", sadly, we are still quite wrong!

Even if we still stubbornly maintain that we are a Christian, a Buddhist, a Muslim or a Jew, whether we fancy ourselves to be an agnostic, an atheist, or any one "entity" else, it really matters not!

For even if we boldly say "both!", or remain stubbornly silent, we remain as ignorant of the absolute truth as a dog rolling happily about on his back in the hot noonday sun! If not, then quick now, who are we really, and upon what do we all truly stand!

In this rather shocking riddle lies a deeper truth, which may enlighten, but should never offend. If you are angry or offended, sadly you remain as blind as a man groping around in a dark cave, looking for a light switch!

Truth never trivializes suffering, but reveals the nature of its deeper meaning to the thoughtful and the aware. But we will never understand its essence while trying to use our everyday minds. Yet until we do understand, abiding peace will not come, and the meaningless violence and pain borne of ignorance in the world will also likely never end!

But if you still think that you are a Jew, then to answer the enigmatic riddle that is truly G-d's "you", while standing on our temple's ravaged Western Wall, you must first blow Tekiah (a resounding blast) on a shofar (ram's horn) made of gold, while standing with your right foot on an etrog (yellow citron) and your left foot on the dark side of the moon!

And if you still believe that you are any sort of an "entity" at all, you only need answer without opening your mouth, or even making a sign! Only then will *Shalōm* "Abiding Peace" again be ours, as we once again take our rightful place within the *Tiferet* "Garden of Eden" of the Divine.

Or else we must dualistically "die" while yet alive to gain a fuller access to G-d's inner גן עדן *Gan Eden*. And once there, eat of the *Amrita* that forever grows upon the evergreen boughs of our eternal "Tree of Life".

Otherwise we must wait patiently for our Sat Guru (True Teacher) Elijah to explain, but until then, no one can ever truly say, wherein lies the secret deathless path into the great Neverevermore![(82)] (See "Amrita", pg. 504).

Neverevermore!

TONIGHT, CHILLED, AND FINALLY "BONE TIRED", from the effort of so much contemplation and writing, I find myself unconsciously inching ever closer to our campfire's cheery warm glow. Bathed in the pale light of a silvery full moon, I exhale, cross my legs, and slowly sink into the dark velvet depths of a quiet meditation. The last thing I remember is the melodious chirring of a nearby tree frog.

Then I am whisked away from our redwood campground on the soft wings of a dream, just in time to witness a most interesting Bayou scene!

A big green bullfrog, sporting a red and brown striped waistcoat, stands up on his big webbed hind feet and in a deep croaking voice speaks out loud: "Gran'ma Moon ta' great gran'ma Thibodeaux, whil'st sit'n 'roun' the fire, she say; "Hey girl, don't ya' know? There's been a whisper 'n the wind 'round the ole' Bayou…"

Then a little green tree frog, in a nearby cypress, chirrups in with a bright little smile; "Go or stay, sit, walk, stand, ride, hop, climb, fly or run, 'tis all 'bout the same! If you try t' do somethin', then you're prob'ly just about doin' nothin' much't all! But just try t' do nothin,' an' you're still a doin' somethin!' So ever'thing remains just 'bout the same, whatever is, or isn't done, or so't seems!

But then look'n closer still, seems quite sure that North is South, 'n East is West, n' up is down, 'n back is front, n' front is back, an' ever'things, don't ya know, always just 'bout the same! A great big n' small not nothing - no - not even some kinda' somethingness, always goin' 'round about, 'n back again. Truly, nothin's really a movin' nor a changin' much't all!"

"Why, yes. Yes indeed! Joe Crawdad Crayfish bubbles back from beneath the cool green water that languishes under great gran'ma Thibodeaux's porch; "An' jus' so! t'morrow sneaks 'round unnoticed behind t'day 'n quietly snatches yesterday, before t'day's even had a chance to be born! Yet, having never been born a'tall, then do we truly live, an' will we surely die?

'Cause in 'dis world seems da' faster that ya' try t' go, 'tis the slower you'll be a mov'n on. 'Till goin' fast as time allows, soon you'll cease t' move through time, or displace space a'tall! Then, bein' all old n' cry'n out, in painful space-time birth, you'll finally move so fast, n' grow so young, that time won't move a'tall!

Then all them high places'l be laid down low, an' the low places'l be a lifted up on high! 'Till ever'thing is a rest'n up on a level play'n field.

Which pretty soon folds right back up on itself, n' then, back up again, int'a a real tiny little place. A little place that grows so small that it finally jus' slips away, real quiet like, back inta' the Nevermore. But, having never ever really been, you are now more truly never more, than you have ever been before, in da' great big Neverevermore!"

Waking with a sudden start, from my nighttime reverie, I am compelled to wonder, "Whatever might this mean?" But after a time, tired of fruitless pondering, I eventually decide that something must be said, so I can finally get some sleep!

So, from the beginningless beginning of the first primeval singularity, it seems quite clear, that there has ever been only the consciousness and creative will of the Divine. Words and deeds are merely concepts that float about briefly in the illusory space-time universe of our everyday minds, and are then never, ever, more!

For, from the very beginning of our universe, there has ever been, only the conscious will of the self-sentient Divine. Briefly arrayed as creative energy, an apparent "somethingness" emerges. But it is only brought into an illusory temporality of subject-objectness, within the reflected individual consciousness of the Divine's dreamlike human mind.

In other words, our mind is none other than the reflected "flame" of the "Burning Bush" (i.e. Universal-Consciousness) that Moses addressed with awe in the במדבר פנים *baMidbar Pinim* "Interior Wilderness" of the Divine.

This is the fiery source of our luminous individual consciousness, emerging as it will, in the creative play of the Life-Force, within the wilderness of Mother Nature's dreamily evolved consciousness bodies. Which we erroneously believe are strictly our own!

But what then might the rest of our Bayou Bullfrog "Griot's" (Story Teller's) storytalk really mean? Only this, that there is a *Sattva Guna* "Identity Principle" present in all that exists.

It is also present in every particle of food that we must consume to live. And it is mixed with the universally conscious Life-Force that enters into our bodies through our nostrils, with every breath that we are compelled in life to take![83] (See "Gunas", pgs. 507 and 540).

And so, only in and through Universal-Consciousness, as life feeds upon life, are our bodies by the Force of Life sustained! Uncalled, the incipient stuff of mind emerges from our bodies and our brains, and dances like a flame upon the cerebral "wicks" of our living "body-candles" synapses.

It arises spontaneously from within the *Kundalini* "Consciousness-Energy" (i.e. *Chi* or *Chitti*) that is released by the mixing of the identity principle and Life-Force, within our primeval little mitochondria (see "Kundalini, pg. 508).

The amazing confluence of Universal-Consciousness, identity-principle and Life-Force transpires within the primitive little mitochondrion that unpretentiously resides within the busy nucleus heart of our every living cell! The Divine-Consciousnesses dream evolved body only then becomes our human body-mind. But without a mind there is no body, and without a body there is no mind!

Our body and our mind, or "body-mind", is thus dependent upon the identity principle and Life-Force of the universe at large. Which, to be known by us, in turn, depends upon the ever flowing dreams of Universal-Consciousness, acting as our body-minds through the creative auspices (patronage) of Mother Nature.

In other words, within our brains, the living flame that is our mind coalesces, and is reflected in our brain's neuroelectrical activity, and the conceptual objects of our consciousness automatically appear.

From a Kabbalistic perspective this is also the occult meaning of the Christian Biblical phrase, *In lux tenebris* "Light within darkness", which the venerable Ancient Mystical Order of Rosicrucian's (AMORC) tellingly calls the "Quickening". (Which, interestingly, could as well describe the invisible presence of light within a Black Hole, or its eventual dazzling emergence from within the darkness of our universes Primeval Singularity!).

The conjoined light of our consciousness also illuminates the denser darkness of our bodies sufficiently for them to appear within the subtle, deeper source of our Universal-Consciousness, in much the same way that a dream arises during our familiar nighttime sleep.

But our mind is also thus quite efficient at creating its own inner universe and can, in a sense, even "extend" itself into whatever is persistently presented to it. In other words, whenever we use a computer keyboard, or even drive our car, it is incorporated into our mind and therefore we, in a manner of speaking, actually "become" it as well! Indeed, the novel changes thus granted are quickly changing us into extended versions of ourselves!

Our body-mind thus becomes the "screen" upon which the experiences of our person and universe are revealed. And the "moving picture show" that is our life spontaneously begins when we open our eyes with the rising sun, and with each new day it spontaneously begins again!

Indeed, the consciousness-components that comprise our lives are only revealed in the sentient light of the collective "sparks" of impersonal Universal-Consciousness that illuminate the concepts of our ideational persons and universes.

They are reflected within our minds, and burn like a sentient halo around the conglomerations of electrochemical, identity-principle, and Life-Force combined events. Which thereby energize and enlighten the product of each neuron with *Chitti* "Consciousness Force" (pg. 505).

This is to say that when enough "sparks" have been gathered together, our individual consciousness appears, like a hologram within our minds. (See "Scattered Sparks" pg. 261, and "Hologram" on pg. 544).

Our familiar subject/object selves, with our cavalcade of seemingly endless objects of consciousness, are thus "born", by manifesting in the occult light of our deeper impersonal Universal-Consciousness. And when our Universal-Consciousness naturally begins to identify with its own illusory objective subjects (our egos), we begin to perceive the entire universe as if it were somehow substantially comprised of our very own personal objects!

But, from the very beginning, nothing has ever really happened. This is to say that nothing objectively real has occurred. Only Universal-Consciousness has appeared to move about, as it will, within the all-encompassing, and therefore solidly sentient space of its own Self-Awareness, which (for you *talmidim* "students" of the Bible) is also the Torah's veridical *Tzur* "Rock of Ages"!

Or we could also say, in another metaphor of Middle Eastern antiquity, that having first consumed the fruit of the "Tree of the Knowledge of Good and Evil" we are subsequently born into a fallen mortal world of apparent dualism, "East of Eden".

But, by consuming the *Amrita* "Fruit of the Tree of the Knowledge of Eternal Life" (i.e. the esoteric knowledge of the universe's manifestation within our consciousness), we soon come to understand that we are neither subject nor object, neither living nor dead, and also neither "this" nor "that"!

For we now stand upon the slim razor's edge of the eternal שלום *Shalōm, Sōlh*, or *Ananda* (i.e. "Abiding Peace") that is the enigmatic innerouterspace of our metaphysical Mr. Froggy's Neverevermore!

The Seat of the Soul

Some Apparent Quandaries are actually only erroneous questions, hastily erected upon faulty premises! For instance, the primacy of either our mind or our brain is actually shared, because they appear in our consciousness simultaneously, when our brains are sufficiently formed.

Furthermore, once they do appear, neither can exist independently of the other, since any existence is only verifiable within our conscious experience of it.

However, in spite of the impossibility of ever being able to ascertain the reality of any existence within our consciousness, our bodies and our brains must support our consciousness, since our bodies, brains and consciousnesses are all manifestly interdependent with one another!

On the other hand, how can our divided consciousness ever simultaneously be both its own subject and object? Therefore primacy, evidently, must be granted to a self-contained Awareness that our consciousness is either present to, or appears "before" (i.e. in front of). For what is our consciousness really, aside from its contents? That is, our consciousness "itself", even if it is alternately both subject and object, must appear before some sort of Witness.

But this Witness, while of necessity being independent in order to simultaneously be both the primeval and percipient subject/objects and object/objects of both our persons and universe(s), does not necessarily require a completely independent existence! For if our body is removed, our consciousness will not return. Just as a lantern flame goes out when its fuel is finally exhausted, or the lantern is badly broken.

Is this qualitatively different Awareness then the true "seat" of our souls? Again, a faux question arises. Why? Simply because, if the presence of our sentient essential Awareness is summarily removed, then only an obviously insentient oblivion remains!

And if our subject-object consciousness, as the binary object of our sentient Awareness, is removed, then Awareness as an independent subject may be aware, but cannot be conscious of the fact of its own awareness!

Therefore the purported "soul" in our discussion must be located in all three (i.e. in our brains, in our minds, and in our essential Awareness). While somehow maintaining its essential distinction as their principle, and therefore apparently non-objective, foundational Subject. This feat could only be accomplished in an illusion that grants an apparently separate existence to a clearly conglomerate subjective phenomenon!

Interestingly, our qualitatively incorporeal subjective Awareness and the energy and space of which our comparatively corporeal objective brain and body is evidently comprised, share a certain durability of presence, while the mercurial phenomenon of our individual consciousness clearly does not!

Yet the final primacy must be granted to a self-contained and independent Awareness, because it is necessarily always present either prior to, or contiguous with, the appearance of our (only apparently) "individual" consciousnesses.

However, our consciousness itself must simply be functioning in order to generate the experiences that it contains, even if they remain sentiently unknown (see "Robots", Sec. I, pg. 144). Indeed, our consciousness and the experiences that it contains are constantly changing, while the Awareness by which they are known to exist does not!

If this principle of Awareness also constantly changed, there would be no durable identity of presence to lend sentience to the ever-changing conglomeration of consciousness objects always appearing, as if, "before" it! We could even (inaccurately) call it a *welt Geist* "world Spirit".

But there is evidently little doubt that this primeval Awareness is also the "Elsewhere" of theoretical physics, which enables the quantifiable appearance of our entire universe.

Although I suspect that a few conservative theoretical physicists might still consider this to be a currently unverifiable claim, and perhaps one even better suited for philosophical or metaphysical speculation than for serious scientific investigation!

Yet whatever exists evidently does so as energy and consciousness, suddenly suspended, as it were, within a surrounding infinitude of subject-objectless Elsewhere, from which it, prior to space and time, somehow emerged.

Furthermore, both space-time and subject-object consciousness are cognitively united in our minds, and thus also appear only within our consciousness. Therefore it seems most likely that our bifurcated individual consciousness, could only appear to exist through a unitary Universal-Consciousness, to an independent and purely preconscious Awareness!

But then, of course, the only "where" that this could possibly transpire in, is within the ever-present spacetimeless primeval אין סוף *Ayn Sof* "endless nothingness" of our less conservative theoretical physicists "Elsewhere". Which is an Else-where that is also always present every-where!

Yet, to claim seriously that space and time must then be eternal, simply because nothing apparently exists without them, seems a premature summation to a clearly monumental if not endless inquiry!

Finally, it is only when *all four*, independent Awareness, impersonal Universal-Consciousness, individual consciousness, and a living body, are present that our persons and universe can even be said to "exist" at all! Indeed, the curious confluence of a these fundamental four is perhaps as close conceptually as our representatively overburdened dualistic minds can approach, in understanding the unadulterated nature of our so-called "souls".

In other words, a solution of continuity is shared between our qualitatively different witnessing Awareness and the binary consciousness of our bodies and brains. Which isolates our concepts of mind into a potential infinitude of everchanging, imaginary existences!

Further, it may be observed in deep meditation, or understood by diligent self-inquiry, that energy and Awareness, in the absence of our body-minds, are both free to merge into an indescribable, yet fundamental "Essence". This Essence then demands only the highest of superlatives, in our inevitably futile attempts to fully understand, or to even accurately describe it!

It therefore seems both fitting and proper to suspect that both quantum level energy (i.e. "strings") and Awareness, and their elaborations into our body-minds and individual consciousness', are actually the extreme poles of a unitary Divinity; what other epithet (description) could possibly be sufficient?

But the apparent separation of this essentially "spacetimeless" unity, into substantial and insubstantial, can only be accomplished symbolically in the various and sundry mixtures, and admixtures, of conceptualizations within our dualistic consciousness.

An ever flowing stream of consciousness therefore illuminates our minds almost instantly, with an unfeasible "subject/object's" (i.e. our ego's) perspective, as we appear to perceive an unprovable "external" reality that is somehow discreetly comprised of space-time bounded "objective/objects".

This curious compound object perspective, of an ideational subject/object viewing an imaginary object of (and within) our only evidently dualistic minds, acts like a sentience granting "prism."

Thus it actually operates much like a perceptive "lens" (of the Absolute). And like an invisible inner "eye" it also provides a durable point of apparently "individual" perspective. An unchanging point of perception through which our automatic psychosomatic processes are conceptualized into the shifting experiential spectrum of our entire consciously manifested universe.

This is the virtual universe of our common experience. Therefore our purely conceptual "soul" never really exists, and so requires no real "seat", just as the Awareness which ineffably ever "Is", never truly does! But this can only be verified by direct experience, since our consequential (binary) consciousness clearly cannot fully grasp the presence of this purely independent Awareness, Its extension into energetic fields of force, or even the form(s) by which It is known!

Ima Prakriti and Aba Purusha

In The Ongoing Cycle Of Life, Mother Nature's touch is gentle in the spring, so that the new life suckling at her breast may thrive. Yet the bitterness of life's harsh winter eventually strips away the old and the infirm from our shared "Tree of Life" with absolutely no regard!

(Just as "She" regularly does, even with our currently environmentally embattled, and steadily diminishing, 30 million – "or so" – animal life forms that we still currently share our planet with.)

And evidently only the fully awakened, by way of independent Awareness' presence as immortal *Purusha* within Universal-Consciousness, can directly understand the mortal paradox of Universal-Consciousness' Maya (i.e. manifesting force), as it acts through *Prakriti* "Mother Nature". Although the worldly wise never tire of trying to explain it!

Yet, from beginning to end, what is our universe if not the boundless play of Prakriti and Purusha? This is the endless dance of the intertwined Life Force of Mother Nature, and the embodied Universal-Consciousness that is being witnessed in our minds by our essential *Purushottatma,* which is the absolute "Supreme".

Universes without number may rise and fall with no particular meaning, but in the end only the Supreme reigns. From beyond any attempted sobriquet (description), an inscrutable meaning unto *it*-Self! (See "Purushottatma", pg. 512).

Maya's Snare

If We Even For An Instant Think, "I am in charge", our foot is not only already in our mouth, but also within Mistress Illusion's clever cosmic snare too! For what is "Maya" (experiential illusion), if not Prakriti's (Nature's) capricious twin sister? And such a clever snare it is! For when we surmise that we are *not* in charge, our neck is already in her noose too!

But, although it is a rather difficult snare to see, it is nonetheless also no great mystery, because we are, with the very first thought that we exist, already caught up in Maya's cunning snare; whatever we may desire, or even "want not", to do! (See "Maya", pg. 509).

Desire

What Is Desire Really, and why is understanding it so important? Firstly, "desire" is the cause of greed and the problem with greed, besides being self-destructive, is that it often initially succeeds! Desire, unfortunately, is also the motivation for acquisition, and is therefore exercised and exploited by merchants, crooks, politicians, and guileful self-serving manipulators of all kinds.

When unexamined, the ignorance's of desire and greed can clearly compel men to steal and feed upon each other's perceived possessions like so many ravenous *anthropophagi* "cannibals"! And so we require the rule of law; but there must be more to capricious human desire than all that?

In less aware בהימות *behemoth* "beasts" of all kinds, desire is simply the unrestrained instinct to survive, and so the feeding begins! Undeniably, for preservations sake, with the notable exception on our planet of green plants, life feeds upon life.

However, in humanity the primary desire to survive and flourish is transformed within nature's complex experiential vehicles (of our "body-minds"), and our illusory egos are soon established as our most precious possessions.

To be sure, all life must take in energy and reproduce in order to survive and continue itself, but even this necessity seems to be qualitatively transformed by emerging compassion within the dualistic subtlety of humankind's more evolved, highly sentient consciousness.

But problematically, our illusory egos have only a conceptual, but nonetheless quite necessarily locally "implied", existence. Because without a "durable" centralized locus of space-time operation, our body-mind mechanisms of individual conscious experience are a great deal like a "schizophrenic helicopter" careening around without a pilot!

If we completely lacked a sense of durable "spatio-temporal" individual body-mind identity we wouldn't know where to park our car, be able to communicate, or even understand how to feed ourselves!

But unlike the simple American Snail *Cornu aspesum*, that only needs to find its way to food, shelter and sex, we exist in a much more complex experiential universe. We live in a secretly symbolic "representational universe", in which we need to make reasoned choices with different, but apparently causally related results. This requires the ability to think synthetically, rather than to simply react responsively.

In short, our body-mind could not possibly survive without an "ego", even if it is an entirely illusory one! But if we continue to believe what so many others have incorrectly said, rather than our own direct experience, all of our constant yet subliminally instinctual efforts of acquisition in life can only continue to amass a growing collection of ephemeral, conceptual memories.

Yet, ironically life's most precious "jewel", the "diamond" of our true Buddha Nature, is always prior to desire and greed, and is already immanently present within us, as our very own veridical Self!

Nonetheless, our priceless but intangible *Dharmakaya* "Buddha Nature" is ever beyond the grasp of even the most skilled of thieves, who might even trip over it, so to speak, and injure themself without even understanding why! Indeed, this precious jewel is our true and unadulterated Self.

It forever lies beyond the reach of the primeval greed of the desire to "Be", and is therefore always present, and prior even to the appearance of our individual consciousness. Ergo, it can be neither possessed, nor given away. And perhaps even more ironically, it can only be discovered by those who are not searching for it, and so do not desire it at all!

Gaps and Spaces

QUIZZICALLY, there is often a considerable gap in both comprehension and agreement among even the most intelligent of human minds! As a result intellectuals have devised numerous and often quite precise means of communication. This phenomenon has certainly contributed to the development of our science, arts and technology.

Well and good! But the fine point, so often missed, is that the communication of our ideas is as dependent upon the illusory existences of our imaginary "gaps and spaces", as it is upon the presence of our ideational objects! Because gaps and spaces (of one kind or another) seamlessly intrude upon and elevate a percept and/or concept into our consciousness. By providing a virtual space-time field of alternating consciousness and unconsciousness, for them to appear within, and in relationship to.

In other words, by the ideational insertion of an illusory "space-time", our apparently "tridimensional" objects are experienced within our consciousness. But of course this "gap", which appears to us as an invisible space-time, is itself the sub-quantum "Elsewhere" (Essence) of our universe's principal Supreme subject.

The Divine is simply being "caught", by our observing consciousness, in the act of reflecting (or "manifesting") its inherent sentience upon (and within) the various ever changing concepts and constructs that comprise our individual consciousnesses.

Our "objects of consciousness" are extended into a cognitive form from within the simultaneous interplay of our essential (and ever witnessing) Awareness and Universal-Consciousness. This phenomenon transpires within our brain's cerebral activity and generates a self-aware "flame" of reflected sentience within our minds.

This is true whether or not we assign the epithet of an independent Divine Awareness to it, or simply assume that it is completely dependent upon the underlying neuronal activity of our psychosomatic mechanisms.

But it is also vitally important to understand that there are no discrete dividing lines in our entire universe! Instead, there are "zones of transition" wherein the single source of the absolute Supreme (i.e. the primeval "Elsewhere") blends into Universal-Consciousness. It does so first in the spacetimeless appearance of the primeval Singularity.

And after the sudden cosmic inflation of the "Big Bang", with the eventual appearance of our dualistic mind, it then blends into the conceptual object illusions of our individual consciousness. Of which our various experiential manifestations of the known universe are ultimately comprised. But apparently, if not for the empty, yet Aware, fullness of a primal and ubiquitous *Akash* "Superspace", neither concept nor communication could transpire!

Our entire experiential universe is in fact so very full of "gaps and spaces" that we are constantly tripping over our own concepts, and forever falling into the unconscious gaps and spaces that lie between them!

The general condition of the human mind is such that our ever-so-proud intellect is actually stumbling through life in a half-aware fog! Is there any wonder that the world created by such distracted minds should constantly be a shamble?

Nevertheless (and I think with great irony!), it is the positive (i.e. cognitive) presence of these negative spaces within our consciousness, which allows us eventually to escape the subtle "identity trap" that is created when we wrongfully identify with our phenomenal subject (ego), and objects of consciousness (universe).

All of which spontaneously manifests in the spacetimeless spaces that sequentially appear within the ever flowing stream of our cognition! Indeed, all of our presences and absences eventually cancel each other entirely out, and back into the nothingness from which they only appeared to emerge, for a little conceptual "space-time" while. But our *Nisarga* "True Nature" eternally obtains, in the enduring silence that remains.

Food of the Gods

A Great Mystery Lies at the very root of our consciousness, that an apparently nonlocal, qualitatively "impersonal", and essentially "singular" Universal-Consciousness somehow becomes manifest through the evidently widespread presences of our familiar localized individual, but nonetheless somehow "pluralistic" consciousnesses. Such a challenging quandary inevitably invites the following questions:

What? Evidently an occult connection exists between two apparently qualitatively different things, our corporeal body-mind and an incorporeal Universal-Consciousness.

How? This likely happens as an "act of G-d", by the initial fragmentation of a suddenly apparent monistic Universal-Consciousness. And its subsequent "evolution" through successively higher forms of evidently individual life, until the universe finally becomes manifest within our seemingly "individual" consciousness, when a suitable psychosomatic instrument of subject-object perception is finally "developed".

When? In an eternally present moment of "Now", which is distinguished from past and future moments by an aura of rich and vital immediacy, an immediacy which engages and involves us in such a vivid and persistent manner that its authenticity is rarely questioned!

Where? In a spacetimeless "Elsewhere", within which every "where" is provided by the functions of our experientially contingent body-mind, and the involved but independent presence, of its primeval witnessing Awareness.

Why? This inquiry has no real answer, unless we say that it is (in a fashion) the "play" or "sport" of the Divine, and as such an act of Divine "will". Which is not really an answer, per se, but is nonetheless an important understanding!

I

What, if anything, is shared by our list of five inquiries? Certainly it must be the "awareness", or more precisely the self-contained Awareness that becomes simultaneously present within both our unitary impersonal Universal-Consciousness and our conditioned individual consciousnesses, in such a fashion that both are wholly dependent upon it!

Thus, while our conditional space-time consciousness is comprised of a continual serial display of dualistic percepts and concepts of self and other, our awareness of them is neither comprised nor composed! Instead, it consists of an irreducible self-aware Presence.

This is the veridical Divine that is called *Allah* in the Islamic Koran, *Nirvana* in the Buddhist Sutras, *Parabrahman* in the Hindu Vedas, and *Yahweh* in the Jewish Torah.

It is the independent Awareness by which *Elohim* "Universal-Consciousness" becomes sentient within our brain's symbol generating algorithmic-like dendritic "code system" of virtual individual dualistic consciousness.

Because Universal-Consciousness evidently evolves as it will, eventually into its living experience of "us", from the pre-subatomic bottom of our universe, and up![(84)]

If so, then it seems not unreasonable to ask, "In the absence of the components that make up our physical body, how can our personal consciousness still be present?"

Indeed, in the absence of our personal consciousness, how is any sort of cognizance even possible! And lacking the "principle of identity" or presence, what can even be said to exist? Furthermore, when we lack our body-mind vehicle of experience, how is any experience even possible!

A substantial gap seems to exist between an apparently "personal" consciousness, whose only proof of existence is its own claim of "presence", and the apparently self-evident presence of a physical body in whose absence our personal consciousness is obviously not present, if not impossible.

But is our corporeality necessary for the experience of our presence, or is the universe's self-sentient Presence necessary to the cognizance of our corporeality? Without question, it seems apparent that the identity principle of Presence is first necessary for any sort of ostensibly "personal" consciousness to be subsequently present!

In other words, our personal consciousness, whatever it contains or is comprised of, and whatever it may be conscious of, must first be "witnessed" in some fashion to be present in any meaningful sense.

Indeed, if any element in the phenomenally causal chain of our personal consciousness is broken, our consciousness is precluded! So we can safely agree that this type of consciousness is a kind of "contingent" cognizance, and therefore that it is most likely merely a "function" of our body-mind.

Not only is our personal consciousness contingent upon the presence of many ingredients, but a strong claim can also be made that it is essentially no different than the components of which it is comprised!

The majority of neurologists, for instance, now believe that our personal consciousness and the reality that it seems to contain are all the result of underlying brain activity.[85] And it seems consistent with a growing body of compelling scientific evidence that such is indeed the case!

But what of the necessary "first element" of Presence, the principle of identity, in whose absence no-"thing" can be said to be either here or there, and without which, we could not even be aware of their being any-"where"!

Our personal consciousness is clearly a phenomenon that is contingent upon the presence of many elements, and there is apparently no reasonable or compelling evidence, philosophical, scientific, metaphysical or otherwise, to the contrary!

But what of the inner contents of our experience of the universe, the sweet fragrance and beauty of a rose, for example?

II

These "qualia", fragrant roses included, are the inner experiential content of our personal consciousness. And like everything else they are clearly dependent upon the first cause of "Presence", however you may choose to define it! But what then is present, as Presence?

If what is present is energy, as the physicists tell us, then even the evolved energy that has elaborated itself into our body-minds, is still only physical energy! It has a certain quality, a physical presence, but the awareness/presence of qualia within our consciousness is neither compelled nor even likely therein.

Moreover, even the presence of "sentience", the awareness of being present, does not necessarily compel the presence of qualia. For example, an intelligent machine may interact with a presumably external universe and even communicate its internal thought processes.

Yet incredibly, it still does not necessarily enjoy *any* of the unique and rich qualities of the living human awareness that comprises our common experience of life! Its intelligence is instead dependent upon a complex set of synthetic "algorithms".

Even if these instructions are somehow designed to become self-directive, self-replicating and self-learning analogs of human awareness, its evolving electronic intelligence is in no way compelled to produce the vivid inner content of our brain's bioelectrical qualia.

Could the presence of bioelectrical qualia in our brains'somehow have happened by providential accident? Perhaps, but the odds against it are analogous to the systematic dismantling of a huge intercontinental ballistic missile.

Then carelessly tossing its many thousands of parts into a gigantic vat, stirring them about occasionally, and hoping that the missile will somehow randomly reassemble itself, guided by the improbable force of mere statistical probability!

Mathematically, this is remotely possible. But the entire universe would likely end, before the exact combination of action is reached to satisfy the arcane equation, which is *rerum opus* "necessary for it to actually occur" ! We don't have to be a rocket scientist to understand that, even algorithmically speaking, the probability of this ever really happening, is simply far too remote!

III

Despite claims to the contrary, it is also doubtful that general A.I., often called "Strong Artificial Intelligence", is even possible to fully achieve. Unless we are willing to allow it by introducing a semantic manipulation, perhaps one day boldly declaring that, "Behold, awareness, dualistic consciousness and intelligence are indeed synonymous!"

If so, it does not bode well for those seeking some sort of physical immortality, which if successful only prolongs the many problems associated with the ignorance of our essential nature. Can we even seriously ask, "If an intelligent computer could meditate, would it ever become enlightened?" Or is this simply another nonsensical inquiry!

Remarkably, even diligent inquiry alone reveals an apparently ever widening qualitative gap between our consciousness and the living brain in which our mind manifests. But it is also evident that both qualia and our brains must be present, for in their absence, whatever we are is clearly not consciously present either!

The reality of the assumed "gap" is therefore unlikely and if it is indeed a fiction, what might the true story be? But to answer this question we must first ask, "Is the likely element common and necessary to both our body and consciousness then Presence?" (See "Qualitative Gap", pg. 569).

It is enough simply to plug a working computer, essentially a "thinking machine", into a wall socket or a battery for it to function. But for us to "Be" we must also be alive, and since a computer cannot currently exist in this fashion either it cannot yet enjoy living as we do!

In other words, life appears to be qualitatively different from other material energy processes. Life seems to possess what could be called an innate quality of self-Aware presence, and perhaps even a rudimentary "sentience", in even its simplest forms!

Hence, if our earnest quest to develop an A.I. is to succeed, then it follows that utilizing living but undifferentiated human "T" cells that can be directed in their development and function by reprogramming their genetic code to replicate the cells of the human brain, may be one way that it can be accomplished!

But this is an obvious "cheat" in that the identity principle, Life Force, and the various evolved elements of life are simply being borrowed from, preserved, and subsequently manipulated in order to redesign and redirect them in novel ways.

Regardless of how our brain and the information which it holds are copied, it becomes increasingly unclear as to who is really in charge, since everything associated with our consciousness is also transpiring within already existing life forms. It could even be said that this development is simply another emerging space-time "wrinkle" in the ongoing evolution of life itself!

Indeed, any sort of presence it seems, must contain a fundamental quality of Self-Awareness to substantiate its own presence. Otherwise it may either exist, or not, outside of our percepts and concepts, or even outside of its own fundamental and intrinsic "preconscious" Awareness.

But because the necessary presence of something, having the fundamental quality of "Identity", does appear to be fundamentally there, of both logically simple, and scientifically recondite probability, the identity of Presence and Awareness are evidently synonymous!

However, our mental processes are themselves subject to the same qualifications and are therefore patently nonexistent, if not present to and within our Awareness in some fashion! Otherwise our Awareness could as well be an insentient "mirror", rather than an indeterminate and indescribable "Somethingness" that is capable of registering in some fashion the qualia of whatever appears before it!

Therefore it seems fair to assert that an Awareness/Presence is not only necessary to the existence of our universe, but when divided becomes the probable artifact of an essential but impersonal Universal-Consciousness that has been bifurcated by our own biologically based dualistic thought processes.

This apparently includes, but is not limited to the percepts, concepts, and whatever else that we might subsequently experience upon becoming humanly sentient!

IV

It appears that the evolutionary hand of *Ima Teva* "Mother Nature" is always present in whatever transpires within, or as a result of, our dualistic individual consciousness. And upon closer inspection, it becomes difficult to differentiate between our "inner" universe of experience, and a presumably independent "outer" universe that is experienced therein!

These conclusions appear to be consistent, whether our perspective is entirely physical, or purely subjective. Perhaps we can even rigorously, philosophically and metaphysically, determine that everything is some sort of primeval and universal self-aware "Presence".

Nonetheless, we may never be able to "connect the dots" fully between the qualia of our minds and the nerve impulses of our physical brains.

But is not the question then of the *non-sequitur* "it does not follow" variety, in that everything only apparently has a qualitatively dual corporeal/incorporeal nature? Evidently whatever may be there, outside of our busily bifurcating consciousness, must also share in the presence of a self-aware Energy.

In principle it must be a Universal-Energy that is not only intrinsically aware, but which is also capable of self-direction and self-expression in an apparently endless variety of ways, within a seemingly endless variety of living and non-living forms!

Physics currently offers some helpful insight into this *Mouna* (Sanskrit for, "Grand Mystery"). But until empirical science progresses, just a bit more, we must turn to metaphysics to investigate a more complete answer. And although traditional philosophy does offer some incomplete insights we must free ourselves somewhat from convention.

We will take a more "heuristic" approach that also includes the diligent investigation of the enigmatic symbols and language of mysticism and esotericism if we are to continue to progress in our inquiry.

Because, only by somehow bridging the apparent gap between our consciousness and its biochemoelectric neural substrate can we possibly proceed with our "*in*-quest!'. As we continue to probe ever deeper into the unexplored wilderness areas of our own human mind, to discover the principal roots of our life and consciousness.

And although it appears that we may have finally hit the distal wall of knowledge and reason, the definitive points so far revealed in our cursory examination, may yet become fertile ground for progress! Despite the infamous mind vs. body "Gap Quandary" problem (see "Emmanuel Kant" pg. 530, and "Gap Quandary", pg. 537).

But we can only do this by reframing our point of perspective in prose. Since understanding the universe is currently an "art" about as much as it is a science! Thus we can in all fairness say; *"Behold, the mystery of the boundless Unmanifest becomes manifest as our dualistic universe, yet apparently cannot fully do so, except in the presence of our illusory personal consciousnesses!".*

In order to more fully illuminate this vital dilemma, let's utilize an esoteric paradigm first (fully) expounded in the ancient occult East Indian Tantric teachings of *Advaita Vedanta* "Final Oneness" (see Upanishads, pg. 515). Consider, that along with the appearance of consciousness, that our food-body, vital-breath and animating Life-Force must also be present. And since this is apparently the case, we may also then fairly inquire, "Why is it so?"

V

East Indian Vedanta proposes that "particles of food", that are actually comprised of a subquantum *Sat* "Identity Principle", are consumed and mixed inside our bodies with the *Prana* "Life-Force" that we inhale with every breath from within the ubiquitous "Superspace" of the *Akash*. And our individual consciousness appears, like a flame dancing about on a living candlestick.

Thus, our individual consciousness emerges from within a Universal-Consciousness, which is present at the core of the identity principle that is fundamental to its appearance. Indeed, this *Kadosh*-Principle (i.e. the principle of an inherently "Holy" Divinity) is the probable esoteric source of the ancient Jewish practice of *Kashrut* (the special handling of foods).

In addition to the more commonly understood "entropy" from supposedly exclusively physical processes, this occult sentient process slowly, but relentlessly, exhausts our apparently strictly physical food-bodies, much like a flame consumes a candle.

Why "food-body"? Simply because our body is composed of whatever food has been consumed by our progenitors, and subsequently by our current bodily form. In the resulting "light" of this living flame the aggregate consciousness of Life's essential identity principle appears as sentient awareness within our emerging minds.

Hence, all living things are sentient by a process that allows them not merely to phenomenally exist, think and act, but also to "know" that they do exist, think and act!

This special knowledge is for every sentient being an apparently, although in no way objectively verifiable, subjective experience.

It is merely the experience of an illusory individual beingness, in which all living things share. Our deceptively familiar body-mind is therefore essentially comprised of a conditioned Universal-Consciousness. It is conditioned into being by, shall we say, "Divine Will" through the biological, evolutionary, and occult forces of our combined Life and Identity principles, none truly exclusive of the others.

Thus the individual "we" and whatever experiences that this "we" seems to have or possess are, poetically speaking, the "food of the gods"!

We are essentially the enigma of a material "sacrifice" suspended experientially upon an ideational "cross" of space and time. Our psychosomatic sacrifice allows the Divine to experience the endless possibilities inherent in the prospect of its all-inclusive appearance, within our conceptual individual universes of ostensible diversity.

But within this apparently corporeal space-time illusion, we are fairly consumed by the extreme effort involved in generating, and maintaining, the puissant experiences of our ongoing individual consciousnesses!

VI

The late Advaita Master, Nisargadatta Maharaj, summarized this in one terse statement, "Consciousness is hard to bear!" We become "worn out", and are eventually "consumed" by the complex processes that underlie the manifestation of the universe within our unitary, yet seemingly dualistic individual consciousnesses!

Eventually, all that remains is what was there all along, the Universal-Consciousness from which the energy and space that comprise the universe apparently arises, and within which the manifestation of our experience actually transpires.

But ultimately, even the force of Universal-Consciousness must dissolve back into the Supreme (i.e. *Yahweh, Allah, etc.*) from which it emerged. And although we can't see all of the hidden processes that are constantly at work within us, insofar as we are concerned, everything that we experience is actually composed of multitudinous "ever-shifting" points of conscious perspective.

Each point of our experiential perspective is thus first occasioned by a principle of identity, which is essentially the emergence of a Supreme-Awareness into a Universal-Consciousness!

In other words, sentient awareness is the reflection of the "transition zone" between the Supreme, which is prior to any sort of consciousness, and the impersonal Universal-Consciousness that is prior to our individual consciousnesses.

This is the mysterious generative wave of incipient Universal-Consciousness that gives rise to our boundless Superverse. However, in order to become "manifest" (i.e. known to possess existence), the סוף אין *Ayn Sōf* (endless "plenum potential") of Universal-Consciousness is dependent upon its phenomenal reflection as our individual body-mind consciousness (see "Plenum", pg. 511).

In turn, the apparently unverifiable wholly subjective experiences of our individual consciousnesses are wholly interdependent with their own experiencing aspect of Universal-Consciousness.

Thus our rather quizzical universe is somewhat like a massive רימון *rimōn* "pomegranate", which cannot wholly be itself in the absence of even one quantum-string "seed"! But in experiential effect, even though we are always an essential part of our whole universe, we can never truly understand any more than our own conceptualizations of it.

For example, we may be able to place an onion upon the palm of our hand, but we can never do so with the word or concept that we so often think embodies it. Likewise, in the case of our own witnessing Selves we are, metaphorically speaking, actually the "fruit" of the entire universe. Rather than just the minuscule single "seedling" portion that can be observed in our binary subject-object minds.

Hence, inquiring into the essential truth of our universe, or of our own consciousness, is much like peeling a symbolic "onion", which we seem to be holding upon the palm of our hand; we have no choice but to proceed from the first layer!

But eventually when the multiple layers of our ignorance are finally all peeled away, what remains is the hidden subjective core of our manifest universe. This hidden core is the *Parabrahman* "Supreme" (الله/יהוה) the occult knowledge of which is clearly *parajñana* "beyond all 'common' understanding".

Therefore, a solution of continuity exists between our apparently individual consciousness, its underlying Universal-Consciousness, and our fundamentally absolute Supreme-Awareness. A continuation of sentient awareness that not only unites, but also explains the necessary unity within any sort of implied experiential diversity!

Yet, since all of this seems just a bit too conveniently erudite, we are compelled to ask, "Is it necessary to first become a scholar, or to be some sort of genius, in order to convince ourselves that we understand it fully!"

Genius

WHAT IS "GENIUS" if not a greater concentration of consciousness within the anthropoid mind? And if so, what is it then that consciousness is the content of?

Individual consciousness is certainly only present with life, therefore life seems to be principal to its presence. Therefore consciousness is most likely an outcome of percipient life, emerging from the evolutionary operation of an occult life principle.

And because life is evidently the manifest presence of the waking Supreme within the universe, the Supreme is apparently just as much present within the occult substrate of our singular impersonal Universal-Consciousness as it is within our perceptible dualistic individual consciousness.

It would seem that there is really no difference between the two, but in the end, as in the beginning, only the mystery of the absolute Supreme abides.

And because the principle of life is potentially present within every particle of energy in our ever expanding universe, Consciousness and Life-Force arise simultaneously, whenever the chimera of a living diversity appears.

Therefore the appearance of genius, within an expansive individual consciousness, also likely brings with it a minute rippling expansion of our human "group consciousness", which appears within the Universal-Consciousness of the totality of the universe!

Model this transmission as "memetics" if you wish, but the difference is that both individual and group consciousness are dependent upon living bodies, while impersonal Universal-Consciousness is not. So the minute change or "expansion" is epigenetically written in our flesh, then it is taken into our minds, where it is eventually acted upon by sentience (see "Memetics", pg. 550).

In this way our mind, which has emerged from within an apparently unconscious "matter", continues to expand within its manifestation as our dualistic consciousness. Thus a Beethoven or an Einstein uplifts us all! (see "Microtubules", pg. 550). For this reason the genius of the *Mahatma* or spiritually "great person" should no doubt be honored, but never, foolishly, apotheosized as a demigod!

However, the greater revelation is the emergence of the presence of the Supreme within an impersonal Universal-Consciousness. Because this is the prime factor and initiator of our imaginary dualistic, space-time experiences of the universe. Which are reflected within the transitory illusions of our individual minds.

Thus the Universal-Consciousness that comprises everything is actually the creative "genius" of the Divine, which empowers and enables the limitless existence of an imponderable superuniversal "Totality". But this Totality is already completely present!

It is only the movement of our dimensionally limited minds, through the limitless parallel structures of its elegantly elaborated wholeness, which creates the generative illusions of space, time, and causality that seem to contain and condition everything else.

But insofar as our current dualistic subject-object perspective is concerned, everything within the endless totality of the Supreme is comprised of Universal-Consciousness, arrayed between the poles of diversity within unity, and unity within diversity. Yet it is only because of the Self-aware presence of the Supreme, which is even prior to Universal-Consciousness, that anything becomes "present" at all!

This is true genius, the creative genius of the Divine. But this transcendental genius is *Parajñana,* the "Wisdom beyond wisdom" that directly understands the *Mouna* "Great Mystery" of our universe. Which in essence cannot be fully spoken, or even fractionally understood!

It is the sum and source of the grand dynamic ספירות *spherōt* "spheres" of endless interdimensionally overlapping universes. Its Divine generative genius emerges (as if uncalled) from within the primeval "Elsewhere" mystery that gave birth to the first Singularity and its subsequent explosive dispersal as space-time and energy.

It then "peacefully slumbers" within the energetic quantum constituents of our unfolding universe. Until its eventual reawakening in the constantly evolving singularity of our minds, and its final return back, into the primeval pre-singularity "Nirvana" of Elsewhere.

It is the aphelion and perihelion (pg. 504) of Universal-Consciousness with the Supreme, (*ut hoc loco* "as here used"). Which thereby activates everything within the plenum of our potential Superverse. Therefore only the Supreme witnesses, from beyond all consciousness, the birth and destruction of an endless cavalcade of potential universes. Which rise and set before it like a resplendent cosmic dream!

Yet it also remains true that we need purpose to live abundantly, and so search for it in vain, within the fundamental meaninglessness of our transitory dualistic existences. But if we can truly understand that the provenance of our sentience always lies within the essential impersonal Supreme, our endless searching eventually ends.

Then the enduring purpose of the Supreme in being you and I, for a little experiential space-time while, is fulfilled. Indeed, the true "genius" of this realization may be the highest, and perhaps even final purpose, of our otherwise transient and relatively inconsequential lives!

Beynōni

OUR "ORIGINAL SIN" (if conceptual "sin" possesses any verity of existence at all!) is what we first allowed to stand *beyn* "between" the cognizance of our ideational self, and the fundamental Awareness that is our own true nature.

This is the fantastic belief that we are the *beynōni* "sentient reflection" by which our inherently narcissistic cognizance is recognized!

(In piae memoriae of dauntless Selma Alabama marcher, Rabbi Sidney Axelrod, of Los Altos Hills, Temple Beth Am, California, USA.)

Winds of Change

IRONICALLY IT APPEARS THAT CHANGE IS CONSTANT, nevertheless there is no real change! But without it our life is apparently impossible. Yet the urge to "Be", upon which all change – real or not – depends, is one of the most fundamental aspects of all nature. The entire universe, and all of the living bodies within it, depends upon the quality of "beingness". Indeed, life even feeds upon life in order to sustain itself by taking this principle in!

Various *gunas* "elemental aspects" (pg. 540) interact spontaneously and our apparent beingness appears, where only nothingness previously prevailed. But this "nothingness" is neither oblivion nor annihilation, because it is present even before our every thought. It is, quite simply, the foundation of both our persons, and our endlessly unfolding experiential multiverse!

The concepts of our "persons and universe(s)" appear to float about within our consciousness upon its "reflective" surface. They are the ancient Kabbalist's scattered "sparks of creation". And by occult extension, of the biblical *eruv rav* "flotsam and jetsam" (i.e. "our" subject/object creations), of the consciousness that spontaneously emerges from within the universal urge to "Be".

Change only appears to happen when the *ruach* "spirit-winds" of our consciousness, being impelled by the fundamental "guna" aspects of nature, begin to move our thoughts about in our conscious minds. But all of this apparent dualistic motion is based upon the not "not-nothingness" of the primeval Awareness from which our entire universe emerges!

We are none other than this changeless, spacetimeless essence of Awareness, by which our apparent existence is known. All else is an illusion comprised of our ever shifting thought concepts.

In order for them to appear within our consciousness, opposites are manifested. And it is only within their ever moving juxtaposition that any knowledge of our person and universe appears!

But this apparent change is like the sun reflecting upon the surface of a lake on a windy day. The shimmering display on its multiple waves may look like many precious diamonds, but it is nonetheless only the reflection of one brightly luminescent Sun!

If So, Then How?

In Any Case, "Tao" and "parajñana" are not worth pursuing. To do so is like sparring with our own shadow, or trying to cut waves with a sword! Whatever peace may finally or fundamentally abide, it is not ours to either "hold", or to fully understand.

Because when a pursuer pursues only himself, apparently he soon becomes lost in the dream of an impossible pursuit. Therefore we evidently have no choice but to query, "Why would G-d pursue Himself, and if so, how?" But then our inquiry inevitably goes awry!

And while this certainly seems plausible enough, and some cursory quantum evidence for the existence of the "footprints" of the Divine does exist (even though we can present no conclusive empirical proofs at this time), what could the nature of such a clearly impersonal and spacetimeless "universal" Consciousness possibly be?

To begin with, any consciousness must be conscious of *something.* This means that at the moment any subject appears, its object must as well, since neither can exist without the other. Supposing this to be true, then they must also exist simultaneously either outside of, or "prior to" any sort of "space-time". The case is thus Self-evident!

But then we are further compelled to ask, "By what 'miracle' is a clearly 'impersonal' Universal-Consciousness (personally) conscious of *it*-Self?" And the pursuer soon becomes trapped in an inadvertent quandary of his own unconscious design!

Thus, having exhausted every other means, evidently all that remains is to simply sink back into our own fundamental Presence and realize that as such we always abide, no matter what appears to be before us.

Hence, our binary mind's mechanical method creates the "walls" of the faux reality that we experience in our mind's inner "eye". Yet ironically, our own Consciousness is a grand "Illusionist" that fools us quite easily into believing that we are a discreetly individual subject. But if this is so, then we are also (evidently) a nonetheless impossible "object/subject" that is somehow sequestered (in an illusory "self-sufficient" way) within our clearly dualistic minds!

Thus the illusion is maintained! (I.e. That we are somehow a wholly independent mind that is housed in a solidly physical body, which nonetheless dwells in a separately external universe.) Until one day *Tat,* "That" by which we become conscious, comes searching for *it*-Self.

But until then, we must patiently abide, with absolute faith in the *Ananda* "Peace" that is the essential Nature of who we and the universe veridically are, and finally stop asking our Selves, "If so, then how?"

Before the Beyond

EVIDENTLY, THE BEST THAT WE CAN EVER DO, is to become like Arjuna, the hero of the Hindu Mahabharata that (like the Kabbalistic interpretation of the Jewish Torah) is the symbolic story of the true nature of our own consciousness. But our consciousness is neither of our own devising, nor is it able to be independently cognizant of itself in a sentient manner as a "solo mio".

This means that we are either not alone, or are the apparent object of another more primary subject that is also necessarily prior to our consciousness, in the manner of a purely sentient omniscience. Some would call this the true "Us" (i.e. a conceptual "god"), but we must also exist prior even to this purely conceptual divine being!

So, we can become an "Arjuna". This is to say, a superlative *Mensch*, an "enlightened" human being in service to our own essential nature (i.e. which is יהוה *Yahweh*, الله *Allah*, or परब्रह्मन् *Parabrahman*).

This is the supreme "Monad" of pure Awareness that ever dwells "before the beyond" of the ancient Kabbalistic "Three Veils of the Absolute", אין, אין סוף, and אין סוף עור - *ayn, ayn sof,* and *ayn sof aur.*

These are the "pre-numerical" representations of the transitionally extending quantum wave or string aspects of the underlying unmanifest Parabrahman into the Universal-Consciousness that is known in Sanskrit as the *Saguna Brahman*.

In the ancient Kabbalistic *Sefer HaYetzirah* "Scroll of Creation" this illusorily diversified aspect of the manifest monadic Divine is tellingly termed *"Elohim"* (i.e. which is an otherwise rather inexplicable pluralization of the Monadic G-d name *"El"!*).

In other words, *Ayn* not "not-nothingness" is also our modern cosmologist's infinite pre-Singularity, pure potential "Elsewhere"! (That has also in the past been called the Supreme, Yahweh, Allah, or the Parabrahman).

While *Ayn Sof* is the "endless-Nothingness" that is (in effect) the rather "insubstantial" sub-quantum Singularity itself. (And is thus perhaps more exactly called the *Schenah* or "Holy Ghost", that while not readily apparent is nonetheless always present, and to which we can always "attune" our consciousness if we so wish).

And finally, the occult words, *Ayn Sof Aur* are actually just another way of saying not "not-Somethingness". Which is not only the advent of the actual quantum Singularity, characterized in the Kabbalah as being comprised of *Elohim* "Universal-Consciousness", but also describes the purported "breaking" of its supersymmetry and the later release of photons that is known in the ancient Kabbalah as the *Zohar* "Effulgence"; as represented within the extensive "Zohar Scroll".

But this also means that any sort of extended Universal-Consciousness must be on the "out", or "bright" experiential side of our universe. This is to say, on the "red-light" side of the "terminal wall", which is the visible side of the "Three Veils of the Absolute" that can be seen with a sensitive orbiting infrared, or "microwave" telescope.

It seems fairly incontrovertible, when looked at in this way, that such is the story of our awakening to the awareness that we are, in our essential Witnessing aspect, on the "dark side of the veil" from which the perturbations of Universal-Consciousness emerge.

And with the emergence of an energetic monopole presence, dipolarity (of necessity) simultaneously appeared as the quantum strings and membranes of which our extended universe is comprised (i.e. in a unique "particle/wave" duality).

In other words it could be said that the "veils" represent the transitional "event horizon" of the Elsewhere from which the primeval Singularity precipitated into the imminent "presence" of its own endless potential (of transcendental Presence). And at this juncture the universe emerged with a "Big Bang"!

In this non-rigorous paradigm, consciousness and universe are represented in the dipole fashion necessary for their inherently "uncertain" appearance in the Heisenberg mathematics of an accurate quantum Self-reflection.

This is to say that it makes no sense to exclude the necessity of consciousness in the orderly appearance of the universe that we experience. Because the universe of our experience (the only one that we will ever "know") is of necessity a stranger to itself!

For the simple reason that the threshold of objective appearance transitions into its own subjective aspect. And both meet on the dipole

Möbius-Strip-like edge of sentience and presence, on the obverse side our apparently physical universe. Which is evidently bounded by space-time and subject to the Newtonian laws of physics (see "Mobius Strip", pg. 553).

Clearly, the meeting place of physics and metaphysics is in the consciousness from which the concepts of both emerge. In the end, the only thing that survives an honest inquiry, is "That" by which our consciousness is known.

Indeed, contrary to a conflation of unrelated terms, evidently of necessity, our consciousness must also be included in any accurate assessment of our universe. (As we begin to render it into the representative mathematics of symbolic algorithms, and the inevitable emergence of "strong" artificial intelligence).

Science is in fact fast approaching the meeting place of our individual dualistic perception with its trans-dimensional super space-time aspect of non-dualistic Universal-Consciousness.

It also seems inevitable that when quantum level computing is able to engage with the quantum level universe, that our more fundamental Universal-Consciousness will eventually be able to take control of its own quantum level of expression.

Beyond this point, evidently everything comes to a terminus within its Supreme source. When we come to this point in understanding the nature of the relationship of our consciousness to the universe, the currently metaphysical, but soon to be theoretical, and soon to be thereafter serviceably "factual" זוהר *Zohar* (i.e. transitionally "Effulgent") deeper nature of things, will no-doubt be revealed in a more scientifically specific, and perhaps even technologically useful understanding!

After the Beyond

The Following Is A Fanciful and "user friendly" mathematical חידה *Chēdah* (i.e. a slightly modified, Talmudic-like "concept puzzle"). It presents a simple heuristic paradigm of "Singularity Generation" and "Supersymmetry Breaking".

Through the creative use of novel symbols and unique concepts, it also requires no formal training in higher mathematics to understand. (Not to be confused with the rigorous formal mathematical proofs of modern physics!).

By taking the time to follow this concept puzzle teaching tool (as if you were doing a Japanese "Sudoku", or a New York Times crossword puzzle), you will gain a deeper insight into the sudden appearance of our universe from within an apparently spacetimeless state of all pervasive nothingness.

Singularity Generation and Supersymmetry Breaking

TERMS:

₤ = Pre-Singularity Elsewhere   Ж = Subquantum Probability
У = Infinite Possibility

ʐ = Primeval Singularity   Ꝋ = Supersymmetry   ψ = Higgs Field
ȝ = Higgs Boson
Ұ = Supersymmetry Breaking   X = Space-time

CONCEPT:

₤ = Ж (· Ж) = У + (1 · Ж) = ʐ · Ꝋ · ψ = Ұ· ʐ + ȝ = E = MC^2 (+ X)

SYNOPSIS:

The last probability of ₤ = Ж (· Ж), is У + (1 · Ж), which is equal to ʐ · Ꝋ, therefore Ꝋ · ψ = Ұ · ʐ + ȝ. Which defines the resulting balance of energy/mass with inflating space-time as Ұ · ʐ + ȝ = E = MC^2 (+ X)!

EXPRESSED AS A THOUGHT EXPERIMENT:

This is to say that the Primeval Singularity (ʐ), came into existence from the accumulation of spacetimeless Subquantum Probabilities (Ж), that were steadily increasing (Ж (· Ж), in the unrestrained "field potential" of the spacetimeless Pre-Singularity Elsewhere (₤).

When the Subquantum Probabilities (Ӝ) finally reached their extreme limits (Ӝ · Ӝ), the only probability left was the Infinite Possibility (Ӯ) wherein infinite probability eventually reached an accordance with infinite possibility. Which resulted in a quantum one dimensional expression of energy/mass, the Primeval Singularity (ɀ).

At first the spacetimeless Primeval Singularity (ɀ) that contained our entire universe was an infinitely energetic, interminably compressed (and hence "immeasurably" hot!), one dimensional ניקוד בלי אמצע *nikkud* "centerless-point" that was necessarily in a state of perfect supersymmetry (Ọ).

When the last possible Subquantum Probability (1. Ӝ) within the Primeval Singularity (ɀ) was spacetimelessly reached, its inherently Infinite Possibilities (Ӯ) were expansively expressed as pre-string perturbations (i.e. *vrittis*).

And subsequently, as open and closed Strings of energy that were forming as the Higgs Field (ᴪ) began to emerge within the nascent spacetime that was just beginning to coalesce between them.

This generated googolplexes upon googolplexes of Higgs Bosons (Ҙ) as the vibrating Strings began to imbue the coalescing energy of protoquarks and gluons with mass.

As soon as they acquired enough mass, quantum gravity began to operate. And space-time was "forced" to expand to accommodate the increasing dimensions and scale of the subatomic particles that were starting to take shape as the quantum Strings began to vibrate at different frequencies, due to the many differing rates of pressure and expansion within the developing Primeval Singularity.

The mass imbuing Higgs Field and Bosons led to the separation of the heat-merged electromagnetic and weak nuclear interaction of forces. And when the energy in the expanding Singularity (ɀ) became greater than the rate of expansion, the supersymmetry (Ọ) was broken (Ƴ), as proto-matter and antimatter began to form and annihilate each other!

And soon only a relatively small amount of matter remained. Yet even this relatively small amount (of the original Singularity) was enough to create an endless Superverse of dimensionally different Multiverses!

The energy released by the mutual annihilation of matter and antimatter increased the rate of the Big Bang's expansion in a brief period of FTL (Faster-Than-Light) "inflation" that lasted from around 10^{-36} to 10^{-32} seconds.

In other words, we are also discussing unimaginably forceful events that were transpiring in almost unimaginably "microscopic" increments of unfolding "Planck" time! (See, "Planck Time", pg. 564).

When the breaking Supersymmetry (Ɏ) reached the full expression of its quantum Infinite Possibilities (У), with the full inclusion of expanding space/time (X), then the energy contained in quantumly coalescing matter (E) (which is "held together" by the Strong Nuclear Force, Weak Nuclear Force, Electromagnetism, and Gravity) now agrees with Einstein's relativistic equation of ($E = MC^2$)!

But since we do not really know exactly how the Singularity was generated or how it's initial supersymmetry was broken, the best that we can currently do is to create plausible theories and somewhat fanciful hypothesis (such as this one!). And then test their veracity, as our science and technology continue to progress.

However, in order for any sort of descriptive equations to function, the Primeval Singularity (ɀ) must first transpire! For this to happen the Elsewhere (Ɇ), from which the Primeval Singularity (ɀ) emerged, must contain the potential of two principles. That of some sort of limitless Consciousness Force. And a foundational identity of omnipresent impersonal Awareness that contains the Probabilities (Ж) and potentials of both infinite Consciousness Force and Infinite Possibility (У) in its makeup.

But in order for them to be known to "exist", the Primeval Singularity (ɀ) must also be experienced in some fashion.

The obvious instrument of this understanding is Nature's evolved psychosomatic instrument, which we subjectively experience as our own "body-minds", the experience of which creates the chimera of our illusory egos.

Nonetheless, both the Life and Mind that eventually emerged from within the universe's evidently limitless aspect of Universal-Consciousness Force, also emerged from within the expanding Primeval Singularity (ɀ). Hence, a fundamental Supreme Awareness, and the ubiquitous Force of Universal-Consciousness are perhaps the most fundamental aspects of our universe!

However, you will also notice that Supreme Awareness, and the primal Force of Universal-Consciousness are not included in the above equation.

This is because, like quantum gravity, Dark Matter and the Dark Force, the exact role of a clearly necessary, ever-witnessing (Supreme) Awareness, and a ubiquitous impersonal Universal Consciousness, in the generation and expansion of endless multiverses from within the spacetimeless Primeval Pre-Singularity Elsewhere (Ɇ), into our boundless Superverse, remains one of the most widely unacknowledged, if not challenging, problems of today's emerging neo-physics!

In summary, whether you are able to follow the above mathematical concept paradigm, or not. Or if you can easily do so, but can also easily identify its every shortcoming. You need not be overly concerned. Because all of these concepts are also presented in a more exacting "dialectical" fashion elsewhere throughout Forest Primeval.

Who Wants to Know?

For Our Apparent Selves, the full import of this transformation in our understanding is nothing, more or less, than the eventual cessation of our existential suffering. Which is brought on by the delusion of believing in a strictly "physical" existence. And for our apparent world, as revealed in Chapter 6, it could well mean the preservation of life on our planet!

In a "nutshell", this is the essence of Forest Primeval's message, which requires no Guru or Master (other) than our own authentic Self to understand, and to undertake in the world. But is there no better way to understand the phenomenon of our own binary consciousness?

The Root of the Root

Consciousness Is An Elaboration Of The Divine. It is the root of life's mental "software", which is written within the dendrites of our brains, the products of which comprise our every experience of person and universe. But our biological "hardware", "software" and vital symbolic algorithms are interdependent, and constitute the basic three aspects of all manifestation within our cognition.

Moreover, (our) cognizer, cognizing and "cognized" are also not only always one, but are also always quantumly connected through our neuronal hardware and biological algorithms to the primeval *"root"* of the Divine (i.e. Universal-Consciousness).

But in the end, through every possible spectrum of our experience, our subject "first" Selves are thus universally quite the same! Our essential Selves are even present prior to consciousness and therefore our true Essence lies beyond the range of all knowable things. Thus, unchanged and unchanging is the universal first Subject of all things. Beyond space, time, energy and form, grows the שהורש *shoresh* (root) of the universal Tree of Life; *"And behold, the 'Root of the root' is "I"!*

Yet, so little of this do we commonly see (of "I") that we all-too-soon become busily involved in the automatic dreamlike illusions of our binary (subject-object) minds. And so we continue to suffer unnecessarily from an illusory privation from our eternal Selves by wrongful identification, but never in veridical fact!

But until we do gain some measure of faith that we are finally beginning to understand our true-life situation, we will soon discover that it remains rather difficult not to continue to follow our longtime habits of Self-deceptive dualistic evaluation!

A Simple Metric

We Far Too Often Evaluate Our Life With A Simple, if not "simplistic", metric of "good and bad", which is good, as far as it goes! But in a society with an overabundance of material goods, when we attempt to apply our simple metric to problems that arise as we strive to "pursue life, liberty and happiness", its deficiencies are soon apparent.

In other words, if having is "good", and not having is "bad", then even in the midst of abundance, our happiness is always incomplete. Because we can never have everything that we want, we will also most likely never be fully satisfied with whatever happiness that we might currently believe we have, because of the happiness that we consistently believe that we perpetually lack!

In which case, even the apparently harmless little metric of "good" and "bad", becomes *bad*, when it is misapplied! Our life then becomes an acquisitive trap of our own devising.

Hence, only we have the power to "free" ourselves. By freeing ourselves from the mistaken misapplication of "good and bad" judgements to amenable objects of thought, when a simple, non-compound metric of "have and have not" might be better applied. Because our life (in truer "broader strokes") is neither good nor bad, it simply "Is".

Yet in simply doing what we must to make our life possible, and what we can to make it better, while not hurting others in the process, we are compelled to ask the question; "Is it 'good', or 'bad', to prevent others from getting what they want in the process of acquiring what we would like to have?"

There

Perhaps The Greatest, and apparently only "miracle" that we are ever really required to "acquire", is the uncommon understanding by which we are able to contemplate such things as our own origin and meaning.

Yet anything that can be contemplated, cannot be "That" by which it knows of its own existence. But while it may be troublesome that we can in no way completely understand "It", within the context of our experiential space and time, *"It" must be there anyway!*

Troublesome Time

Troublesome Of Late; ...is all of this apparent racing to past and future events, interspersed with an only ideational present. Slaves to this experience, we simply cannot say which the "real" is and what is just an illusion.

Curious indeed! How unfortunate, he who thus becomes burdened with the yoke of responsibility for everything that appears within his mind.

As the 16th Century Japanese Zen Poet "Basho" rather cryptically wrote on one dark and stormy night, *"How admirable, he who does not think that life is fleeting, when he sees the lightening flash!"*

The resulting existential "guilt" infects both our psyches and bodies with the dis-ease of dualistic identification. Which is then an unfortunate set of circumstances indeed! Yet "troublesome" somehow seems the better descriptor of the two.

But perhaps even more troubling is the thought that if everything that we experience is indeed comprised of thought, then not only our thoughts, but even our very lives could be devoid of any real meaning!

A Priori

If Meaning Is Not Inherent, Then It Is Assigned. If meaning is assigned, then why not interpret suffering as much the joy of being, as life's intervening happiness'. But a balanced mind cannot do this, because extreme pain cannot exist without the experience of extreme distress.

In point of fact no thought can even "exist", except in relation to other thoughts! Thus language (be it written, spoken, mathematized, or even expressed as art), is the vehicle of its (meanings) own transmission.

On its own language is merely a collection of symbols. Thus it is really nothing (no "thing") in itself. Except in being "that" to which meaning is, by experiences' measure, "assigned"!

But for any-sort-of "experience" to happen, there must always be present three interrelated conglomerations of thoughts, *experiencer, experience,* and *experienced.*

And since one cannot exist without the other two, they necessarily have in common the "objects of consciousness" (i.e. thoughts), of which they are evidently (unavoidably) comprised!

Because any subject to which an object appears, is himself dependent upon the object of his experience, in order to "exist".

But in order to accomplish this, his binary subject-object consciousness must then be the conjoined "subject/object" of an Awareness that is clearly independent of both! While in subquantum Essence, always being the Elsewhere (i.e. Supreme) of which they are comprised!

It seems then, that the vehicle necessary for any experience, must simultaneously be our body, brain and mind. Ergo our witnessing Awareness is arguably "real", if our rule of measure is endurance and independence.

But of course any sort of awareness needs a "body-mind" to have any sort of phenomenal experience. The nature of which is its appearance as a point of negative perspective in relationship to the positive existence of the emptiness that is necessary for its apparent objective existence within our individual-consciousness'.

This "emptiness" is the fullness of the Awareness by which experience happens. Therefore Awareness must be spacetimeless, and yet contain the potential of being. But since this potential is an aspect of the Awareness by which it exists, its existence is *a priori* (i.e. "prior").

And because it stands alone, it does so as the veridical percipient of the entire universe! The existence of which is only ascertainable to a "first cause". Which is also present as the consciousness and experiential stuff of which our manifested universe is apparently comprised. But only if "meaning" is assigned!

If this is so, then to so *assign,* is itself actually an act of "free will". Since experience is only another symbol for the presence of consciousness after all! Thus "free will" of a sort exists, but it is limited to the assignment of meaning to our objects (and subject/objects) of thought.

Rather than to any singularly impossible "objects of experience", unless experience is somehow intrinsic within thought. But of course neither can be the case, because neither experience nor experiencer can stand alone!

Moreover, since everything is evidently composed of thoughts, which have no intrinsic meaning in themselves, everything reduces not into meaninglessness, but instead into the consciousness by which thoughts are cognized.

However consciousness too does not stand alone. Our mind, which contains it, is a composite psyche, dependent upon a body and brain to consciously function, although our brain does quite well all on its own!. But it also depends upon an independent Awareness, to understand whatever is being "symbolically" (in living thought forms) presented to its (thought comprised) phenomenal subject/object, our so-called "ego".

Therefore meaning is not intrinsic, but assigned. Yet meaning is also not arbitrary, because meaning is based upon conclusions that are drawn from memorialized experiences. And in our memories of them, they are inherently thought structures, with assigned meanings.

Existence and meaning are so derived and assigned, according to the metric of mind, apparently in full view of the "Divine"!

But, this is not really even new knowledge. Because as early as the 4th century BCE, the Greek philosopher Protagoras was on the right track. But his well-known equation, like our impression of dualistic reality, is actually the mirror image of an all-inclusive Monad of intrinsic cosmic Awareness!

Ergo, not *homo est mensura* "man is the measure". But, *homo est Dei mensura* "man is G-d's measure." That is, man is the metric by which G-d measures all potential things, into an existence within our binary minds, allowing the experience of diversity, within an all-inclusive and irreducible "One"!

The Children of Abraham

WITH ALL DUE RESPECT to the complex historical and political events of the past, and the resulting nations, decisions, and actions that have led to the violent current impasse in the Mideast; Jews and Muslims are *all* still equally the "children of Abraham"!

Indeed, for anyone who cares to diligently "vet the record" of Israel for himself, despite the onerous efforts of self-serving revisionists, the "historical record" continues to speak quite clearly for itself!

And because of this fundamental fact, and perhaps even at the peril of my own safety, in the spirit of brotherhood, with no thought of assigning any sort of clearly fruitless "after-the-facts" blame to anyone involved, I offer the following humble prayer in an uncomplicated modern "American style" of prose:

You say "Nabi", and we say "Navi", and the simple reason why is that we are both the children of Abraham! Hence, I evidently have no choice, since we have the same father, yet a different mother, but to call you 'my brother'.

In light of this fundamental fact, and despite the considerable suffering that we have sadly visited upon one another, may we one day soon embrace each other once again, in earnest brotherhood, and make our broken family whole once more, and go to war no more!

And by healing our painful family rift, let's prove the doomsday "prophets" wrong, as we quietly work together to save not only ourselves, but our poor suffering planet as well! – Amen.

"...and they shall beat their swords into plowshares, and their spears into pruning hooks, nation shall not lift up sword against nation, and they will study war no more." – Isiah 2:3-4.

(Piae memoriae, the innocent "Children of War", who have suffered for far too long because of the violent actions of their own fearful, ignorant, and often vengeful guardians, on all sides of this stubbornly recalcitrant conflict!).

(See "Truth's Measure", pg. 18; and "The Children", number 14, pg. 482).

If

If G-d Did Not Exist, man would probably need to invent god, such is our compelling need! But, the existence of G-d can neither be proven, nor disproven. Yet, we apparently need to have this concept to not merely exist, but to thrive within our experience of the world as well.

Hence, there is most likely some sort of "religious instinct", which helps to balance our internal conceptual world with whatever we believe is occurring in our experience of an external universe. We often experience this in our dualistic perceptions as "good" and "evil". When this instinct is damaged, mental illness and crime often result.

If minimally we all can agree on this, then maximally we should all "find" G-d! But can G-d really ever be found? Consequently, whether we say "Ye" or "Nay", to a clearly conceptual god, the search for the veridical Divine seems important to our survival!

Apparently, our fascination with religion and politics is (most likely) prompted by religious and survival instincts, respectively. Indeed, within us Consciousness has built a mighty "mental fortress", of compelling concepts, the pursuits of which are apparently endless.

Thus, it seems that the source of the awareness, by which we are conscious, may be the only way out of our unending conceptual maze. Otherwise, there are only endless speculations of mind, being driven by "Mother Nature's" instinctual compulsions. Such is the bizarre nature of man's present existential dilemma!

Night Marauders!

ANOTHER DAY IS PAST, and another night has fallen. And as I sit basking in the comforting warmth of our cheerful *madura* "campfire", I hear a sudden sound! Startled, I look up from my writing, and in the distance I spy three pairs of glowing, greenish yellow eyes.

But tonight's bright campfire light keeps the hungry little ring-tailed raccoon bandits that are hiding in the nearby Huckleberry Bushes at bay, at least for now! And as they survey our humble little woodland camp from behind their furry little black masks, I can scarcely suppress a wry smile, because I am finally becoming well acquainted with this charming little *Rōmi* "Gypsy" band (with all due respect to my freedom loving Rōmi brothers!).

Unperturbed, I quietly rise up from my seat and slowly wander around, carefully locking up anything that tonight's marauding raccoons might want to steal away. And at last satisfied that our cache of food is safe, I can finally retire to our sheltering tent for the night.

But in spite of being "bone tired", as I snuggle down into my warm sleeping bag and begin to drift off into the waiting arms of gentle Morpheus (pg. 509), I can hear the hungry little camp thieves poking their curious little wet black noses into every niche and cranny of our humble little backwoods camp.

They are evidently looking everywhere, anxious to find a way to get at our small "horde" of cached food, no doubt hoping to nab another bag full of tasty little peanuts!

As I finally surrender to my fatigue, and begin to drift off gently into a good night's sleep, the last sound that I can hear is the restful carefree babbling of Pescadero Creek. It arises on its own uncalled and slowly meanders up the mountainside, sneaking 'round and betwixt the ranks of ghostly white birch tree trunks that decorate its steep slope.

'Till at last it creeps inside my tent, and slips softly down inside my ears, where it gently lulls my busy mind away from the bright and busy world of our waking dreams. Then it takes me back, and yet further back again. 'Till finally, by "bits and pieces", I am carried quite away, back down into the deeper, darker realm, of a quiet dreamless sleep.

(In loving memory of "Gypsies, Tramps and Thieves", and of our good-hearted California Congressman, Sonny Bono).

Santa Cruz Mountains
Northern California, USA

CHAPTER 6
ALNUS "ALDER"

At Forests Edge

WHILE HIKING TODAY, I AT LAST ARRIVE at the distant edge of the deep evergreen forest and pause briefly there to adjust my eyes and then carefully step away from the forest's dark shade. Blinking in the bright sunlight, I now stand at the foot of a steep grassy knoll.

As I gaze up its steep slope, I can see that the climb to the knoll's crest is eased by numerous dusty brown cattle trails that crisscross stepwise across its verdant face, like the steps of a giant staircase.

But because cattle cannot climb straight up, or down, such a steep grade, they are resigned to walk carefully back and forth across its vertical face, angling up slightly with each pass, until the top of the precipitous slope is finally reached.

Eventually these compacted trails, a couple of feet wide, become a prominent feature of the steep places where cattle sometimes graze. But they may also (on occasion) serve as impromptu "stairs" for adventuresome people too! Carefully stepping upward from trail to trail, I at last stand, breathing heavily, upon the uppermost crest of this charming verdigris hill.

Below me, the forested mountains and green grassy hills retreat gently into the West. They slowly diminish in size and then flatten out into a green grassy plain before they meet with a glowing line of sandy golden beaches, at the edge of a distant cerulean sea.

A long line of smoky gray fog stretches out across the blue horizon of the distant Pacific Ocean, until it is finally swallowed up into the great blue dome of an overarching sky. A beautifully bucolic scene, it is certainly well worth the arduous trek to the top of this singular verdigris hill!

But even the experience of great beauty eventually wears thin, and finally submitting to the inevitable experiential ennui, I must carefully make my way back down to the awaiting shade of the evergreen forest again.

On the way back to camp several deer trails serve as forest byways, back to the main path. Deer trails are subtle and easily overlooked, nonetheless deer are renowned experts at picking the best ways through both chaparral and dense mountain woods.

The delicate White Tail Deer of the Santa Cruz Mountains are nimble and light on their feet, jumping up to ten feet high over brush, or as wide over a cascading stream.

Racing at twenty, or even thirty miles an hour, up or down a sheer mountain face, is a small feat for even a young White Tail Deer!

Even the droppings of deer are diminutive and delicate, whereas the epithet "cow pie", just about says it all about cattle droppings! And while inadvertently stepping on deer "pellets" can go relatively unnoticed, stepping into a fresh "cow pie" is generally an unforgettably messy experience! As a Memorial Park Forest Ranger once jokingly confided to me, "It is like the difference between seeing a skunk, and being sprayed by one!"

Without a doubt, both metaphorically and practically speaking, it is best if we can avoid life's occasional "cow pies" when we use bovine trails as steps to reach the top of a hill. But deer trails can speed our journey through the dark forest, if we are observant enough to find them.

And 'though both chew their cud, have an extra stomach, sport cloven hooves and have pointed horns, there are some other rather interesting differences between deer and cattle. Cows are generally gentle beasts that want nothing more than to be left alone to graze peacefully in high green meadows.

Indeed, I must agree with the majority of East Indians that cows sometimes seem to be more worthy than some of the people who routinely eat them! (See "Cows", pg. 523.)

Deer, on the other hand, are hyper alert and exquisitely athletic creatures that are always poised to explode with a graceful burst of speed that humbles even our fastest human Olympic sprinters!

But it is exactly such thoughts as these that constitute the apparently necessary, yet suspiciously analogous train of endless objectifications that quizzically appears, then disappears from our consciousness.

And indeed, this is done in a continuous "stream" so rapid and vividly compelling, that it almost never occurs to us that the entire universe which it comprises might actually exist on the "in" side of our heads!

Clone Rings

The Sages Often Say, "Anything is possible in Consciousness!" But what does this really mean? Do you remember a time in your life, probably when you were young, that even the improbable seemed quite possible? The universe seemed so full of possibility that if you just willed it strongly enough, you could probably even fly!

Then, almost anything seemed possible. I vividly remember, as a child, seeing and listening to a chamber orchestra as they played beautiful classical music in the midst of a large "Clone Ring" of redwoods in the climax forest behind my family's Santa Cruz Mountain home (see "Biological Progression", pg. 520).

These enchanting Clone Rings are formed by the demise of ancient forest monarchs, perhaps from natural causes, but most likely from nineteenth century "clear-cut" logging. The grand Clone Rings of genetically identical "reincarnated" forest monarchs begin their lives as numerous tiny "adventitious buds" that are even tinier than a grain of sand!

These remarkable little "Human T Cell" like buds are located in the outer cambium of the felled giant's still living tree stump, just under the bark at ground level.

Activated by hormones that are released when the big tree is felled, they begin to grow and tend to form a grand living ring of clones that are genetically identical to their fallen parent tree (see "Adventitious Bud" pg. 504, and "Cambium" pg. 505).

The activated buds form new sprouts, initiating from where the top of the larger remaining anchor roots meet the fallen tree's trunk, radiating out from the base of the stump like the spokes of an immense wagon wheel. But no longer being of any use to the now absent forest giant, the new sprouts cleverly reinvent themselves as whole new trees!

In the Santa Cruz Mountains, these charming Clone Rings are sometimes composed of trees six feet or more in diameter, and often grow in such a wide circle that the probable dimensions of their venerable deceased parents nearly defies today's reasonable imagination!

So powerful was my childhood experience that with very little effort I can still recall the unlikely site of that delightful chamber orchestra, and the haunting strains of its enchanting music, even though it may have never "objectively happened".

Except that it most certainly did happen, somewhere within consciousness, even if only in the imagination of a young boy growing up relatively alone, in an enchanting redwood forest! But something changes within us as we grow older that too often dulls our spontaneity and imagination, and sometimes sadly, even our creativity!

So what is it that dampens the spontaneous feelings of joy that used to accompany our everyday experiences of living? To find the answer to this abstruse question we must rediscover the innocent child that ever lives, securely hidden away, within the enchanted circle of our innermost being.

This, it seems, may even be the only reliable pathway that is worth taking into the thought-tangled wilderness of our own adult consciousness.

Because it is only the innocent primeval pathways of emerging consciousness that can lead us back "home", to our own truer, deeper "impersonal" Natures!

Other paths may lead to power, money, fame, or even to the companionship of family, friends and lovers. But even the dearest of life's gifts will eventually fade away and desert us in the end.

Because, whether we like it or not, such is the inexorable way that we individually suffer the experiences of apparent change, within our symbol comprised subject-object universes, where even our celebrated birth becomes our eventual death sentence!

But our subtle, witnessing Self, which ever holds the wonder and joy of life in its pristine impersonal *Hridaya* "Heart", will never change or abandon us, for it is the ubiquitous still "Center" of our entire experiential universe!

The voices of those who have passed before us into the silence of the grave cannot rise to tell us, but if we seek only other things in life, we may well find them, but sadly, we will most likely be disappointed in the end.

The Eyes of a Child

Before The Forms Were Named, the universe was ever new, in the eyes of a child. Only ephemeral images were then present in our minds, innocently floating about in the cosmic substrate of our inner *Akash* "Superspace". Delightfully they would expand, exploding into forms like kernels of popcorn to be eagerly consumed!

How delightful the person and universe did "taste", when we were but an innocent child! Yet even now, somewhere within, behind and before we name the forms that form our world, our person and universe are still new and "evergreen".

This is the secret place within our hearts that keeps our hope and love alive. It is the "root of the root" of our Tree of Life that survives the fire and storm of names and concepts, which after our brief childhood seems to ever rage within our minds.

For we too soon forget what we once knew, since it really has no name. Because the names we give are too quickly taken for the forms we name, and as if by fire, we are soon consumed by what is taken in, because the root of our consciousness has no real form!

We spontaneously build a person and a universe of words, images and dreams, in illusory interior space-time places. Soon they fill our heads and obscure our more direct vision of life, like hair no longer living, yet ever-growing from a living root, before our experiential "eyes".

They emerge spontaneously from within the living שורש *Shoresh* "root" of our Tree of Life. And once cognized into existence, they charge out full of wind and bluster, as if from within the very source of our being, which is actually the occult living presence of the Divine!

And although we didn't "start the fire" of our own consciousness, finally, parched, tired and thirsty for freedom, one day we will begin again, to heed our quiet *ba'al Krishna's* "inner child's" authentic voice.

For this is the secret voice that sounds the beating of our hearts, and gives to life through us the wondrous illusions of our "free choice" and individuality. Because we evidently have no real choice, but to continually make ostensive choices!

It is the only "One" that is always percipiently there and that despite every apparent change, always remains reliably "us" as well. Or more precisely, it is the "all-seeing" Universal Eye, of all our purportedly personal "I's".

But the real "G-d's Eye", of all our illusory little I's, remains stubbornly hidden away, within the collective illusions of our many "me's"!

When this understanding finally dawns upon us, then our experiential "roundtrip flight" through the stormy winds of our illusory objective universe turns 'round about, and our return trip flight "back home" at last begins!

As we turn around, and thus get back, by turning our attention back in again. Ever gliding towards the "place", that place Divine, where in our childhood innocence we once began, until we finally see the universe as ever new, and evergreen again![86]

But make no mistake about it, dying to Self-ignorance, while we journey to our "eternal home" rather than to our illusory graves, is neither pleasant, nor easy, even if it is only the death of a persistent subjective (and collectively quite powerful) illusion!

And we cannot count on others to understand the changes that are coming upon us. They may even feel hurt, thinking that we have somehow abandoned them. When all that we have really done, is to abandon the narrow interior subject-object perspectives of our clearly illusory, and definitively "dualistic" universe(s)!

But ours is nonetheless an objective universe that we once shared with many others. And heartache, loss, and loneliness, for whatever reason, are always hard to bear! But we will eventually discover that our steadily awakening consciousness, in its ever diminishing dualistic derivations, soon becomes an even greater burden to bear!

In any case, when Mother Nature finally decides that it is time to reawaken from "Her" dualistic reverie, we may no longer be able to convince ourselves that we have any actual individual choice(s) left at all! I wish that I had better advice on this matter, but alas, despite a lifetime of searching, I do not. Societies will judge us according to their own standards, and indeed the fate of poor Socrates is an indication of this sometimes rather disturbing fact!

We may even be considered by those closest to us to be inexplicably remote, or somehow lacking in initiative; at best misguided, and at worst perhaps insane! Because people naturally fear what they do not understand.[88] Nonetheless, despite the apparent chaos and hopelessness of our current time, we just might be the avant-garde of better days to come for the entire anthropoid race!

But there is apparently no palliative (balm) for the sometimes painful transition that happens when we finally begin to wake up from our long sleep within the Divine's persistent universal dream.

Yet I do know that we all must learn as we go, as we go again within, to see the forms as they once were, before the "firestorm" of our universe of names began. But in seeing them directly once again, we also see that they are actually the "seeing", that is the Seer seeing them (i.e. they are merely "experiential").

And in so seeing, the Seer (suddenly or soon) becomes the eternal Sage. The *Sat Guru* "Witness", that was secretly there even before our child and childhood thought forms were even named!

But the universe, now left alone upon its own, becomes ever new, and "evergreen". And the spontaneity and joy of our youth, may now freely bloom again!

"We didn't start the fire, it was always burning, since the worlds been turning!" – Lyrics from "Storm Front" by Billy Joel, published by Columbia Records in 1989.

Ursus horribilis

Today I'm Trying Something New, hiking past the southernmost Memorial Park trail. Here the climax forest of soaring giant Coastal Redwoods and Douglas Fir trees comes to an abrupt end, and my next step takes me beyond their shady green border. This is where the parkland protection ends, and the exposed sunlit chaparral begins.

Long ago the forest here was cut away for lumber, and the stumps were either pulled by oxen, burned, or blasted out, to create grassland for the ever-growing herds of grazing Hereford and Black Angus cattle. Now, even long after the cattle have gone, the slow biological progression from grassland back to evergreen forest is a long and tortuous process.

Hundreds of years may pass in the Santa Cruz Mountains before the chaparral finally reclaims the extraterritorial grassland. And for as long as cattle are still grazing, the arboreal procession of territorial reclamation will continue to be delayed!

But when the incessant bovine grazing finally ends, over the next several hundred years scrub oak and pine trees will slowly begin to shade and crowd out the pioneer Manzanita, *Arctostaphylos andersonii*, Sage, *Artemisia californica,* and Coyote brush, *Baccharis pilularis.*[87]

I

Eventually, the oak and pine tree species that require full sunlight will begin to be shaded out by a slowly advancing phalanx of lofty evergreen trees. And over many hundreds, or even thousands of years, a towering assemblage of "climax" forest evergreens, composed of the regal California Coastal Redwoods, *Sequoia sempervirens* and stately Douglas Fir Trees, *Pseudotsuga menziesii,* finally reclaims the land.

Yet even in the shady evergreen climax forest certain understory trees and shrubs survive and thrive as an essential part of the complex coastal redwood rainforest biome.[89] Tan Oak, *Lithocarpus densiflorus,* for example, has made a laudable comeback after being decimated in the 19th century for its bark, which was used for tanning cattle hides.

Unfortunately, this oak species has also been the hardest hit by the recently imported systemic fungus *Phytopthera ramorum*, which causes Sudden Oak Death syndrome (S.O.D.). This vascular malady invades and causes the death of the water conducting vessels in the tree's cambium, the vital layer of living tissue just under the bark, and the infected oak quickly dies from the resulting lack of water to its food producing canopy.

American Chestnut Blight (A.C.B.) and Dutch Elm Disease (D.E.D.) are similar imported vascular fungi, which have nearly wiped out the entire American Chestnut and American Elm tree populations in the relatively short spans of about one hundred, and fifty years, respectfully! Fortunately a "cure", or more precisely *a control*, has been found that is useful in saving high value oak trees with SOD in residential areas.

So far, the forest oaks have had to fend for themselves, but from what I have seen in the Santa Cruz Mountain range, unlike ACB and DED, they are doing a fair job of surviving SOD, despite an alarmingly large number of casualties!

But like the deadly human Ebola and Zika viruses, or even the cyber Stuxnet Virus, the potential for far more serious outbreaks of exotic and imported diseases (of many different sorts) looms ominously even on our countries fast approaching "cybernetic horizon"!

Indeed, perhaps the biggest threat to our nation is now a viral cyber-attack on our country's extensive power grid, which could leave us literally "powerless" and vulnerable to widespread systemic collapse, physical attacks, and even invasion, for several months!

This is another "wakeup call", underscoring the need for better cyber-attack defense and survival protocols. As well as improved quarantine and inspection methods for imported diseases and pests; but unfortunately we do not yet have a sufficiently effective "C.D.C." for trees.

Nonetheless a tree species or a forest, like a person or a population, is a rather problematical, if not impossible, thing to replace! (See "Acknowledgements", UC Ag. Ext. Serv., pg. 497).

II

Despite man's considerable past interference, the Santa Cruz Mountains still manage to support a myriad of interesting insects and fascinating flora and fauna, as well as a terrific wealth of trees! Tan Oaks, for instance, provide acorns for the irascible forest Gray Squirrels, *Sciurus griseus.*

While the ubiquitous Huckleberry Bushes, *Vaccinium ovatum,* once fed a burgeoning population of Black Bears, *Ursus americanus,* Brown Bears, *U. arctos,* and Grizzly Bears, *U. arctos horribilis* (a truly horrible "man eating" creature that the local Pomo Indians did their best to avoid!).

Before the bears were driven and hunted to local extinction, the large yearly runs of, now-rarely-seen, brightly colored wild Chinook Salmon, *Oncorhynchus tshawytscha,* and hard headed, steel-gray Steelhead Trout,

Oncorhynchus mykiss-anadromous, made the benign natural cornucopia of this pacific coastal mountain range into a literal "bear heaven"!

But evidently it is as difficult for bruins and people to coexist peacefully as it is for culturally different groups of people to do, and today only the microscopic little "Water Bears" remain (see "Tardigrade", pg. 586).

And unhappily, it is apparently even difficult for two people, who ostensibly love each other, to both remain in love and stay together over the challenging course of a lifetime!

This is because our ideational dualistic universe continually changes whether we like it or not, and while yet under the impotent control of our fickle illusory egos, it remains a difficult dream universe indeed!

III

If we understand nothing else in life, given sufficient experience, we come to understand suffering and loss. But from the ever growing rubble of our suffering, the wisdom of compassion may eventually begin to sprout forth and grow!

And if we follow its thread inward, back to its ultimate Source, prior to the flow of our everyday consciousness, it leads straight through the dismal storm of our recalcitrant worldly illusions, to the peaceful *Hridaya* "Heartwood" of our own authentic "Tree of Life". Therefore it has the power to reconnect us to the ultimate Source of all life.

This is why the Awakened almost universally prize the impersonal love of our aware compassion. For in the absence of this deeper *agape* "impersonal love", any sort of genuine awakening is not possible.

Yet, such is the power of this deeper love that even if complete understanding has not yet blossomed, the person who lives deeply immersed in G-d's Divine love, is probably already a saint!

This is what is meant by the statement "G-d is love". Because, authentic *karuna* "compassion" is the גשר *gesher* "bridge" that straightaway leads us back to the heavenly gates of our own true Awakening![90]

Shaqwi!

Today It Is Beautiful And Warm, and I am determined to hike up into the high chaparral of "Rattlesnake Ridge" that looms in the near-distance above our camp! So I set out through the redwoods with my favorite walking stick in hand, and soon we begin our arduous ascent into the highlands, on a seldom traversed deer trail.

Before long the comfortable forest shade is left far behind, and below us the verdant redwood forest stretches out for miles to the very shore of the distant blue Pacific Ocean.

I am entranced by the natural beauty that is being progressively revealed as we climb.

In the Santa Cruz Mountains, the hot dry chaparral that we are struggling towards is the common dwelling place of cuddly little Cottontail Rabbits, *Sylvilagus nuttalli;* perky California Quail, *Callipepla californica;* formidable Forest Scorpions, *Uroctonus mordax;* and even the occasional irascible Diamondback Rattlesnake, *Crotalus atrox!*

But while hiking around a sharp bend on the dusty deer trail that winds steeply upward amongst the hot and dusty dry chaparral, as I step gingerly over a low sandstone outcropping, a menacing dry rattling sound suddenly rises up to greet my ears!

And scarcely one foot in front of me, coiled up on the dusty trail, reclines a rather large "sunbathing" Diamondback Rattlesnake!

A coiled rattlesnake is poised to strike nearly the full length of its body at an astounding speed of 33 meters (approximately 108 feet) in less than a third of a second! And I can see that the lovely rattlesnake that is peacefully "sunbathing" at my feet is more than three feet long!

Under my breath, I thank my (now deceased), half Mexican, Muwekma Indian guide, mentor and employer, "Brujo" Paco, who managed a large cattle ranch on Alpine Road in the Santa Cruz Mountains, where I had worked as a cowboy in the early 1960's (may you rest in peace my dear friend!).

Quietly, and respectfully, I begin to chant Paco's ancient Indian "Snake Charm" to the big rattlesnake now coiled up at my feet. Amazingly, the agitated serpent soon stops rattling its tail, and starts to flick its forked red tongue curiously towards me, "tasting" the air for my scent.

No longer frightened or angry, the snake is now surprisingly calm! Nonetheless, I keep chanting the *Shaqwi* "Rattlesnake" charm as he slowly uncoils and slithers towards me, inches over my right foot, then curiously cocks his head and looks me squarely in the eye! I stand motionless, as still as a statue.

Remarkably, it seems as if the big snake understands every word that I gently chant to him, just as Paco had taught me, "You must not be frightened, but must 'enchant' an agitated snake with a powerful feeling of deep love, like you are talking to a long lost brother."

But then the rattlesnake stretches upward and leans his scaly, wrist thick, diamond marked body against my leg for support. When his flat triangular head finally reaches the level of my right pants pocket, the big snake flicks out his forked red tongue, to "taste" the hand that is hidden away inside.

I probably should be terrified, but for some mysterious reason I become quite calm as I "lovingly chant" the memorized words of a now extinct language that I do not even really understand. And I am also quite surprised that, under the present circumstances, I am somehow able to recall the antediluvian reptile incantation at all!

Nonetheless, I am soon mesmerized by the sinister golden "cat's eyes" of my scaly reptilian "brother" as he slowly sways his big wedge shaped head to the repetitive rhythm of the ancient Muwekma Indian snake charm; a secret song that he remarkably not only seems to appreciate, but also to somehow understand! (See "Muwekma", pg. 555).

We remain like this for about a tenth of a minute, which seems more like an eternity, before the venerable old rattler slowly relaxes the taught spring steel muscles of his extended body, and slumps back down onto the hot and dusty trail at my feet.

As I continue softly chanting the secret "medicine words" of the dying language that Paco once taught, swearing me to utmost secrecy, the big snake slowly crawls away and disappears back into the safety of the dense and dusty, impenetrable mountainside chaparral.

But before his sinuous tail finally slips out of sight under a low manzanita limb I silently count its accumulation of annual rattles, two by two; "Two, four, six, eight, ten and twelve!" They end there, the remainder broken off due to the brittleness of age, or perhaps in some long forgotten territorial tussle.

Finally breathing a sigh of relief, I notice how very blue that the sky now seems, and how bright and beautiful the sunlight, as it plays upon the dusky green foliage that is now scarcely trembling above his retreating tail.

The natural beauty around me is surprisingly more vivid now than it was before I met the venerable old rattler sunbathing peacefully on this hot and dusty deer trail. And it seems like a rather providential time to accept his gracious gift of tolerance, and begin our long trek back down to the comforting shade of the awaiting creekside White Alder Trees, while I count my blessings along the way!

But while doing so, as sometimes happens to us all, my mind is unexpectedly filled with a different sort of care, and I am suddenly transported back - to a faraway place, in a far more treacherous and uncertain time!

From Every Mountainside

It Was During The Tumultuous 1960's, as the Vietnam War raged on, that peace marches began to fill the streets in Northern California. The Vietnam War began with honorable intentions, as did our military involvement in the Mideast.

Without a doubt there were some political shenanigans going on "behind the scenes" of both. (There usually is). But "time and the internet" will tell their fuller story. And reveal the identities of those with the indelible blood of many innocents on their hands.

In the beginning, both wars seemed like the right thing to do. But the Vietnam War was quite unlike our more recent ill-advised conflict in Iraq.

Which has drawn our country even deeper into the unstable quagmire of the current Middle East, and inadvertently spawned ISIS! (The latest iteration of tragically twisted humanity to emerge as *behēmōt* "beasts" from the inhumane ashes of war). Because the Vietnam War was based not only on bad information, but also upon a tragically flawed theory of political science!

And within a relatively short time, with such a faulty foundation for taking any martial action, what began with honor had devolved into a horrific debacle, with sometimes far less than honorable conduct.

Our troops fought bravely and often with valor, but soon found themselves embroiled in a bizarrely complex, if not incredibly involuted Southeast Asian Civil War! And while the Vietnamese Civil War did not involve correcting a criminally neglected problem of state like slavery, like our own Civil War it also pitted North against South, and sadly at times, even brother against brother.

Moreover, to the continual consternation of our politically hampered officers, and the resulting mounting losses of our hopelessly embroiled troops, it was never really our "war" to win! Because it was in fact always "unwinnable", in the politically constrained manner that was being forced upon our military.

Sadly, as history shows, not only good people, but sometimes even great nations can make great mistakes! But perhaps the most important thing is for us to learn from them so they will not be repeated time and again. And evidently ensuring the correctness of sensitive political information, and the right use of power, is the essence of this grim and clearly "timeless" lesson.

In the first known publication of the English Arthurian legend, *Le Morte D'Arthur*, written by Thomas Mallory in 1485, the wise wizard Merlin advises victorious King Arthur to look around at the bodies of the slain that littered the surrounding hillsides. Looking like a bloody recumbent forest made up of cruelly felled "human trees".

'Look hard Arthur! Let the horror sink deeply into your heart. And let it be seared even more deeply into your memory. Lest you should ever forget the tragic price that has been paid on this bloody mountainside to ensure your rule. For if we forget the lessons of the past, we are doomed to repeat them, time and again!'

(Now, returning to the, as yet evidently unlearned, "martial issues" of our own more recent historical time…) The "Peace Movement" began as a gentle groundswell. But when the full extent of the unjust and undeclared Vietnam War's brutality was finally revealed, the public demand for its end quickly erupted into an unstoppable "wildfire" of righteous anger that soon swept across our entire nation!

Traditional values and beliefs were being challenged by a new generation of well educated, rebellious and free thinking youth, who actually believed in the radical idea that "peace and love" could change the world!

It was the dawning of the purported "Age of Aquarius". A time when many believed that the alternative to changing the direction that the world seemed to be tending would likely be a global thermonuclear holocaust that could destroy the human habitability of our entire planet!

As if in "spiritual response", Yoga, Zen and Taoism were being widely dabbled in for the first time in the West, along with the experimental use of psychoactive, or in the vernacular of the time, "psychedelic" drugs.

Although unbeknownst to most of us at the time, no-doubt this growing interest in the exploration and expansion of human consciousness had nothing to do with the human invention of "Spirituality". It was simply the Force of Life beginning to act, to save itself, through the children of the generation of humans that had created the planet-wide extinction problem!

Marijuana soon showed up on the street scene and in short order became the "party drug" of choice, quickly supplanting alcohol and tobacco for that dubious honor. Because we saw (in addition to *ganga's* simply being much more fun!) that when it is abused alcohol not only dulls our senses, but also lowers our already limited anthropoid awareness and reaction time.

It also became quite clear that alcohol abuse not only contributes to violence, but to a seemingly endless macabre parade of horrible and senseless traffic accidents! Thus, alcohol and tobacco were revealed to be more than just physically unhealthy and addictive drugs, but psychically destructive, as well.

Alcohol and tobacco were soon associated with the perceived arrogance, insensitivity and looming violence of the dangerous existing world polity, and unfortunately, these salient truths have not much changed with passing time! The so-called "Flower Children" were reluctantly "turned off" by the cigarette and alcohol generation and their self-exonerating explanations and excuses.

Yes, they were the "Greatest Generation", those who defeated the subhuman Nazis and cruel misguided imperial Japanese of that dark time. But they also created, and had already twice used, the dreadful atomic bomb! Which admittedly saved a great many lives, by ending the necessity of invading Japan. But a weapon once used in war, becomes a great deal more likely to be used again!

And so began the difficult job of managing a spreading global nuclear threat, a pervasively ominous threat that each successive generation would soon have to cope with!

Disturbingly, following upon the Second World War's tardy ending, in a clearly excessive arms race with our former allay, the now disbanded "Soviet Union", we had collectively built up an insanely massive arsenal of Intercontinental Ballistic Missiles (I.C.B.M.s).

These sinister missiles were equipped with a rapidly evolving array of ever-more-powerful hydrogen bomb warheads. Warheads that are detonated by first imploding them with the "hotter-than-the-sun" plasma from the now much less powerful atomic bomb!

Very soon there were enough ICBM-mounted hydrogen bombs and "doomsday plans" for their use, to quickly destroy every niche of civilization over our entire planet; several times over!

Horrifically, whatever the thermonuclear blasts would not destroy outright, the radiation spreading out like an unstoppable plague in their aftermath soon would! If this is not "hell" made manifest upon earth, then what is?

I

The Vietnam draft was now in full swing and after graduating from high school, not yet understanding the true nature of the Vietnam conflict, just as every generation of my family has, I felt the puissant "call of duty" and enlisted in the Army to serve my country in its "time of need".

But the Army was looking for "Specialists" as well as combat soldiers and due to a chance high aptitude score in electronics, I soon found myself servicing the very missiles that I believed might soon end the world!

My painful conscience was relentless. Something had to be done, because it seemed quite apparent that whatever had been going on in the past, was certainly not going very well now!

Because of my rare military specialty, I would never be sent off to fight in Vietnam. But I felt that it was my duty as a freedom and justice loving American to stand "shoulder to shoulder" with the brave American veterans who, like the eloquent John Kerry, were throwing away their medals in disgust, and were speaking out publically against the morbid practice of counting bodies, instead of freeing territory, to determine who was winning the war.

Needless to say, this was not a very popular decision with my patriotic family of "foreign war" veterans and selfless heroes! My father and uncles were so ashamed and angry with me that they could hardly even force themselves to speak to me.

Ostracized by my own family, I was heartbroken, but knowing the truth of the matter, because I had been raised to not back down when I knew that I was right, I "stuck to my guns", and soon paid the price!

My controversial views and actions would land me first in jail, then in a military holding facility, since the post stockade was by now filled to overflowing with returning G.I.'s who were protesting the war, and I soon found myself out of the military.

And so we "tuned in" to the belief that we could end the Vietnam War, and started spreading the radical idea that the average person might actually have the power to assist in achieving world peace. And moreover that we could even succeed in doing so, if we could just find some way to act together to save ourselves and our planet.

Rather than listening to the profit hungry businessmen and misguided political warmongers who rarely put themselves in harm's way, but seemed quite satisfied with sacrificing others to their execrable plans. We "turned on" to the idea that we could somehow end the war, and "dropped out" of the society that we believed had even brought the world to the very brink of annihilation!

But our idealistic peace movement was short lived. The hopeful nascent (budding) heart and spirit of the 1960s soon gave way to frustration and anger, as the hoped-for world peace and more enlightened society never materialized.

With the duplicitous (deceitful) "help" of the very establishment that our spontaneous, but youthful and naïve, "consciousness expansion" movement had hoped to change, our grass roots "Peace and Love" movement, was quickly driven into an inescapable ditch of impotence and despair.

Woefully, unlike their more successful Israeli counterparts, most American communes were either sabotaged, or simply soon self-destructed because of their general lack of practical experience, organizational incompetence, and deficit of effective leadership. And the overuse of so-called "recreational drugs" only made these already difficult matters, even worse!

Moreover, many of the rock stars and supposed leaders of the media-styled "hippie" movement were now being harassed and jailed, while others either "accidentally" died, or disappeared under (sometimes rather suspicious) circumstances, never to be seen again!

Ironically, our well-meaning but impractical Peace and Love movement was (sadly) already bound to fail, if simply left to its own deficient devices and dearth of sufficient resolve.

Most people understand that governments that feel threatened often have little patience, and almost no mercy. But until our American Army's National Guard gunned down some naive young college war protesters at Kent State University, in the American "heartland" of Midwestern Ohio, it never occurred to most people that we might actually be silencing, jailing, and sometimes even killing, some of the best and brightest youth in our society!

And as if this were not bad enough, our government was thoughtlessly attempting to squelch the social conscience of our entire nation. But this was America, and not merely "the home of the brave", but also the justice-loving "land of the free". And even the infamous "silent majority" were finally beginning to wake up!

II

Tragically, some of the formerly peaceful protesters decided to fight back, and militant groups like the Black Panthers, Symbionese Liberation Army (S.L.A.), and the self-styled Weathermen guerrillas, were born. Every radical group had its own agenda, but were nonetheless united in their condemnation of our government's indefensible actions at home and in Vietnam.

The kidnapped San Francisco newspaper heiress, Patty Hearst, took up the cause of her SLA captors and was eventually arrested in a bank robbery desperately orchestrated to raise money for their cause. Her lawyers subsequently used the now infamous "Stockholm Syndrome" defense to excuse her aberrant vs non-conforming "deviant" behavior (see pg. 581) and her eventual sentence was a mere "slap on the wrist"!

Which only confirmed what many had already suspected, that unlike the common man, the wealthy and famous can, far too often, afford to "live above the law". Justice, like most everything else in our increasingly oligarchic (rule by the wealthy) society was (and evidently still is) apparently for sale, but only if you have enough money, influence and good fortune to buy it!

All doubts were soon removed that our fading Peace and Love movement was over for good when the Weathermen underground group began to blow up what they unwisely considered to be the legitimate targets of a malignant "Military Industrial Complex". In another twist of irony, this phrase was first coined by the former American military commander of the free world's successful World War II effort.

This was the monumental problem of transitioning to a peacetime economy that President Eisenhower had wisely cautioned our nation about in the wake of World War II. It did not take a genius to see that the consequences of ineffectually responding to this growing problem were beginning to reveal themselves formidably in the turbulent nineteen sixties.

Indeed, our nation continues to suffer from its nonresolution, which has spawned a plethora of unexpected economic, political, law enforcement and national security problems.

Perhaps the most corrosive and insidious of which is the emergence of unbelievably powerful corporations and conglomerates that have nearly all the legal rights of an individual, without a controlling conscience. They have come to control the production and supply lines of almost every necessity, and to exercise an unprecedented control over our economy, and disturbingly, even our government. And they lobby our congress with impunity, using unprecedented amounts of money to keep it so!

III

But *fair* lobbies do serve a practical purpose and even incorporation is a legitimate financial instrument that protects the businesses that employ it, unless they are allowed to go too far! But corporations are no more an "entity" than a hammer or a saw and despite all the hype, they remain just a financial tool.

Yet, having been poorly controlled, corporations are now so very big and powerful that they challenge our government's ability to fairly govern them. And perhaps even worse, far too many have developed a selfish attitude of entitlement to excuse their economic and politically destabilizing influences, and sometime even blatantly criminal fiduciary actions!

By justifying their actions with claims that our hijacked economy, and their uncontrolled capitalism, is the same as a competitive and healthy economic system of fairly moderated free enterprise, like the Robber Barons before the Great Depression of the 1930s, they have done great economic harm to our country.

By masking their malfeasance (theft, vicarious, and/or statutory) and their dishonest pecuniary (financial) intentions, they harm the very economic system that is responsible for their successes; they almost literally, "bite the hand that feeds them!" And because our American economy is now so very big and internationally engaged, our damaged economy, like a gigantic row of tipping dominos, can destabilize other economies around the world!

Outrageously, gargantuan corporations lobby politicians with unprecedented amounts of money, and even place their former executive officers in key judicial and governmental positions, to influence the passing of bills, statutes, and laws that inadvertently suppress our middle economic class, and are favorable to their own ever growing prosperity.

Is this not tantamount to a clandestine "Economic War", or at least a hostile appropriation, in order to surreptitiously achieve financial control over our entire nation? All that remains to consolidate their control, is to pass legislations that give corporations the same power to vote as a bona fide individual entity!

Evidently the greatest enemy to our liberty today is not communism, socialism, fascism, anarchy, or perhaps even terrorism, because ironically it has become us, or presently at least, a disturbingly large percentage of 1% of us!

But all rich people are not to blame! Indeed, the paternal attitude and helpful actions of some continue to be of inestimable service to our nation.

While "Command Economies" like the former Soviet Union have proven to be relatively unsuccessful, apparently "Market Economies" like ours are also vulnerable, if not to the lack of competitive growth, then to the corruption of rampant greed!

And as things stand today, it is truly alarming to realize that all that stands between us and a full-blown *plutocracy* "political rule by the wealthy", like that which eventually brought down the powerful Roman Empire, may be our precious right to vote.

As Walt Kelly's beloved comic strip character "Pogo" once wisely opined, "We have met the enemy and he is us!'; which is an apropos modern version of the immortal Puck's exclamation, "...what fools these mortals be!"

"When rich speculators prosper while farmers lose their land. When government officials spend money on weapons instead of cures. When the upper class is extravagant and irresponsible while the poor have nowhere to turn - all this is robbery and chaos. It is not in keeping with the Tao" – Tao Te Ching, verse 53.

"A New English Version", translation by Stephen Mitchell, Harper and Row, New York, 1988.

IV

Without a doubt, terrorism is currently a big problem in today's world. But some of the first modern "terrorists" on American soil were frustrated and embittered young American idealists! But they only succeeded in either blowing themselves up, or were soon brought to justice, or forced into hiding by law enforcement, because terrorism, unlike a clearly justifiable social revolution, invariably only succeeds in accomplishing its own destruction!

Thus ended another difficult chapter in the ongoing struggle that is the emergence of our national conscience. No longer connected with the now defunct Peace and Love movement, the innocent consciousness exploration of the 1960s quickly deteriorated, and dangerous "street drugs" of all sorts began to show up!

The drug abuse that ensued quickly took its toll of musicians, entertainers and many naïvely inquisitive young people. Accidental drug overdoses touched almost everyone. In time marijuana (by association) acquired the same notoriety as the opiates, barbiturates and stimulants that were ruining, and even tragically ending, so many promising young lives.

With widespread propaganda, emerging public fear, and eventual reactionary support, the government's anti-drug and anti-marijuana campaign now began in earnest!

But the law enforcement commotion soon died down in the 1970s, only to be revitalized with the emergence of the cruel South American, "Pablo Escobar" style, cocaine cartels and bloody gangland drug wars of the 1980s. By the '90s, the American government's self-styled "War on Drugs" program was in full swing.

Marijuana had by now acquired the anecdotal (hearsay) reputation of being a "gateway drug" leading to the abuse of a frightening new generation of dangerous and hyperaddictive über street drugs. It was soon given the same felony status and so-called "Zero Tolerance" treatment as the other "hard" drugs that it had been unfortunately lumped together with!

V

I am certainly not pleading the case for drug abuse, but perhaps our recent history shows that marijuana has not been given a fair and realistic treatment in our country. And, like it or not, relatively harmless and innocuous marijuana has now become almost hopelessly embroiled with the dysfunctional political, financial and ignorant social milieu that has unfairly demonized its use.[(91)]

Poor innocuous marijuana, along with gay rights, had become another icon for civil rights, and a litmus test for equitable justice in our time and country!

But our legal system is merely the enforcement arm of the government, and despite the fact that many of its loyal servants well understand the shortcomings of its sometimes ill-conceived laws, they are nonetheless duty bound to enforce them.

However it is also no mystery that some, perhaps unconsciously, adopt attitudes that excuse the laws that they seem to have no choice but to enforce! Therefore, except insofar as unnecessary psychological and physical brutality are concerned, there is no blame.

Nonetheless, the lessons of the Nürnberg Trials certainly must apply to those who, without honor or force of conscience, follow orders and perpetrate heinous acts that go beyond the pale of both common decency and justice.

But it certainly does no good to call anyone a "Nazi", unless they are indeed hell manifesting genocidal monsters! Because in the end everyone must live with their own conscience, and even in its absence, be held accountable for their actions.

Although, disturbingly, where law enforcement is concerned, in certain parts of our great country, it is still unsafe to be innocently driving a car if you are a "person of color", especially if you should happen to be "black."

Which is in itself evidently a rather controversial thing! As my own bright and beautiful half-"black", or half-"white" (you decide) niece Leloni, and nephew Alonzo, can *personally* attest!

Now, it doesn't take a genius to understand, that my every instinct tells me to inflict grievous bodily harm on anyone that harms a hair on their dear sweet heads!

But since I also respect the law, when it's just, please understand that what you are reading is, in a fashion, my own present way of fighting for the civil rights of the happily mixed millennial children of America!

Who are being given the onerous task of saving us all, despite having to contend with a distinct minority of ignorant, prejudiced and terrified cops. Who unfortunately still lurk (often in a disturbingly "clandestine manner") within their honorable Civil Servant ranks!

But because I am the son of two hard working (and non-prejudice) "middle class" white parents, and I grew up amongst other white folks in the relative isolation of a beautiful California redwood rainforest. I eventually discovered that prejudice is not necessarily based upon a narrow mind and a mean spirit. But simply at times, upon a naïve sort-of ignorance that is the result of social isolation.

However, it is also often the result of socio-economic privilege, and plutocratic classism. But in the accomplishment of the "American Dream", of personal prosperity achieved from honest hard work, there is certainly no blame. Yet to continue pursuing ever more wealth, when to do so you need to steal from your fellow Americans, is an entirely different matter altogether!

In any case, while we rightly never owe anyone an apology for our legitimate accomplishments in life, when they are fairly obtained, it is important to recognize that our ignorance is perpetuated by accepting the hearsay opinions of others, rather than basing our beliefs upon our own unbiased personal experience.

Indeed, the mass incarnation of our black American citizens has reached rather alarming proportions, if not truly disturbing racial disproportions!

According to genius Ta-Nehisi Coates, author of "Between the World and Me", in his recent *Atlantic* magazine article "The Black Family in the Age of Mass Incarceration", currently in Russia (which is infamous for its "Gulags") the number of citizens incarcerated in Russian Gulags is around 400 per 100,000.

While it is an astounding 700 per 100,000 in America! Alarmingly among these American inmates, 4,000 per 100,000 are "black". And among black high school dropouts 60% will spend time in prison!

And while we have neither the time nor the space to solve this problem here, practical solutions are not inexistent, and for the sake of justice and liberty, and the financial welfare of our great country, some sort of prison and law enforcement reformation is obviously required!

Indeed, when I asked my good friend, who is a retired Assistant District Attorney for Santa Clara County (here in California), what could be done to fix the problem of "jittery-trigger-finger syndrome" and "rogue cops" harassing, beating and shooting people without sufficient reason, he did have something to say!

First he wanted me to understand that the vast majority of police officers are "good cops". And that contrary to some of the hyperbole about "cover ups", the D.A. (at least in Santa Clara) is always ready to prosecute a bad cop!

As he put it, quite simply, "A lawbreaker, is a lawbreaker!" And that, "Bad cops are the *worst*, because their sworn duty is to 'protect and serve'!" He also wanted me to understand that most officers are even more upset about rogue cops than the average citizen is. Because it not only makes them all look bad, but could even make them more of a "target" than they already are! (Indeed, placing callous "revenue" aside; *Why are bullet proof squad car windows not yet mandatory in every squad car!*

Evidently, most police officers are given a cursory psychological evaluation, and a year of probation, in order to weed out the noteworthy "undesirables". But he suggests a more rigorous and comprehensive background check, before they are even interviewed!

If they pass this first bar, then they also need to do more than just fill out a psychological questionnaire, they need to be evaluated by a professional Psychologist or Psychiatrist with the proper training to identify potential rogue cops.

The retired ADA then suggested a lengthier probationary period, with monthly performance reports that include the potential new hires field partner's personal observations, as well as his or her superior's "desk" evaluation. These suggestions, of course, are in addition to all of the other training, field operation, and reporting requirements that most police officers already have to do!

But of course, "the Devil is always in the details", and "oversight", to ensure that smaller, isolated Departments, and those that have been having these sorts of problems persistently, get the sort of expert help that they evidently require!

However be this as it may, we also need to remember to praise their efforts publically to encourage them, and perhaps even to help protect them from an angry public. As they get down to the unpleasant business of policing their own ranks, as well as our countries many neighborhoods and cities.

But, returning to our primary subject, it seems impossible to explore how this rather innocuous plant (marijuana) has become embroiled in such a heated controversy, without uncovering some of the deeper reasons why!

And in doing so it may even serve as a societal "mirror" to take a closer and more revealing look at ourselves. So let's take a closer look at this fascinating psychoactive plant species, which has accompanied humankind from before the dawn of recorded history.

VI

In today's world, medical marijuana is a drug of mercy for an ever growing number of cancer patients. Because it stems nausea, increases appetite and enhances a positive mood, it can also bolster the human will to live, and the ability to endure challenging, and often excruciatingly painful chemical and radiation therapies. For this reason, oncologists are often selflessly at the forefront of the battle for the legalization of medical marijuana.

Decades of exhaustive research have shown that marijuana is physically nonaddictive and even considerably less harmful than most unregulated over-the-counter medications. However, any sort of smoke that is inhaled into our lungs, over time, may cause pulmonary problems, and this seems to be marijuana's primary health risk.

But this is more of a technical problem than a legitimate, inherently serious medical contraindication that can be related directly to the chemical properties of marijuana!

In a humane response to this pulmonary problem, safer smoking devices have been invented; "vapor pipes" that vaporize the psychoactive *tetrahydrocannabinol* "T.H.C." containing resins. (Which are actually a potpourri of complex floral chemistry that concentrates in the plant's female flower glands).

Moreover, stronger overall concentrations of THC are being achieved by careful breeding, which is actually fortuitous because less plant material needs to be inhaled or ingested!

When marijuana is smoked, it is primarily carbon monoxide and heat that can damage our lungs, and for this reason many patients simply ingest it. But chemically synthesized THC, perhaps because it is missing certain other plant chemicals, or combinations of chemicals, does not seem to offer quite the same benefits to most people.

Indeed, in higher doses it has even been linked to unsettling psychotic episodes! Apparently Mother Nature already knows what "She" is doing (in a manner of speaking), without our unnecessary meddling about in something that already works quite well.

Marijuana does seem unique, but why is this so? What puts cannabis into a different category than other street drugs, and the many available pharmaceuticals, like serotonin reuptake inhibitors or dopamine agonists, the *nouvelle époque* "new age" of psychoactive antidepressants?

Interestingly, cannabinoid receptors are located almost everywhere throughout our brain and body, except in our brainstem, which regulates our autonomic nerve impulses, like breathing and blood pressure.

This peculiar distribution of receptors certainly explains why there are no recorded cases of fatal marijuana overdoses! But why are the receptors there in the first place? What possible connection could there be between the alkaloids produced by a "weedy" plant (presumably to protect itself from sunburn, insects and fungi) and our complex human brains and nervous systems?

VII

There are apparently no easy answers. But the fact remains that the receptors are there and that marijuana alkaloids do have a profound effect on human consciousness, because of their design and their receptors curious presence in our bodies and brains!

This question puzzled an Israeli neuroscientist named Raphael Mechoulam, who in the mid-1960s isolated the alkaloid primarily responsible for the psychoactive effects of marijuana, (THC) a "nitrogenous organic molecule" unlike any other found in nature.[(92)]

Later, in 1988, a researcher at the St. Louis University School of Medicine, Allyn Howlett, discovered a specific receptor cell for THC in the human brain. And definitively, these mysterious cells were later found to be present all over the human body! Evidence of cannabinoid receptors meant that there exists a network within our brain and body that is set into motion by the presence of cannabinoids that are being naturally produced somewhere in our bodies!

In 1992, Raphael Mechoulam and William Devane discovered our brain's own cannabinoid and named it *Anandamide*, meaning "inner bliss" in Sanskrit; a name coined from the ancient East Indian "Vedic" scriptures. It was subsequently discovered that Anandamide's neural network is involved in regulating our experiences of pain, memory, appetite, coordination, and emotion.

Marijuana is hardly something new to mankind, because human beings have been using marijuana for thousands of years as a source of fiber and for its psychoactive properties. It is quite likely the oldest domesticated plant, called 麻"*Mah*" by the Chinese, it is represented by the characters for the male and female plant under the roof of man's home. (Indeed it even had the distinction of being considered a goddess (Ganja Deva) in ancient India!).

Evidently, meditation as a means to enhance Self-Awareness and practices that bring our awareness to a single point, also have their roots in antiquity, as does the spiritual use of marijuana. But why should these kinds of practices be at all effective?

VIII

It is a curious fact, that by simply forgetting every thought except one, that time seems to come to a stop, and our perceptions of person and universe seem to expand boundlessly into an apparently timeless and eternally present moment.

This is likely because our sense perceptions are recorded in our brains, but are cognized as thoughts in our minds. And although our focus in meditation may be an attempt to focus upon no thought, "no thought" is still a thought! But our "focus" itself is not, so besides the patently (clearly) obvious, what is Anandamide's function?

It is believed that Anandamide serves the important cognitive purpose of "forgetting". Forgetting, the opposite function of focusing and remembering what has been focused upon, is critical to our consciousness (see "Forgetting", pg. 535). In fact, the total information that is received via our senses and cognitive functions would immediately overwhelm and paralyze or "crash" our minds, if it were not somehow regulated!

Cognitive mapping for instance, which utilizes magnetic resonance imaging (M.R.I.), is as dependent upon the hidden representational algorithmic "ebb" of the MRI's attached computer, as it is upon the computer's colorful representative magnetic "flow" of our brain's own mysterious "brainbows" of symbolic electrochemical activity.

In other words, not just representative symbols, but symbolic space must be provided for the representations presentation to occur in some fashion, both in our brain's cognitive systems, and in the MRI's interpretive computer software.

Anandamide is believed to be the indigenous cannabinoid produced in the human brain that helps it to perform this important neural "ebb" function. It seems probable that no real distinction exists between the perceptual modifications accomplished by using the plant form of THC, or the concentration of our brain's own cannabinoid form of Anandamide from doing spiritual practices like meditation.

At least insofar as our brain is concerned, there is likely no appreciable chemical difference!

For the earnest "spiritual" marijuana user, who uses marijuana as a kind-of לחם הקדוש *Lechem ha-kadōsh* "Sacrament", the net effect of cannabis on their consciousness, of providing experiential space by amplifying the forgetting mechanism of our brain, is essentially the same, a telling attenuation of our everyday sense of space-time.

What is accomplished is a perspective of awareness that unlocks our perception of continuously being in an experientially present moment.

Either way, our individual experiences of person and universe as objects of consciousness that appear to be temporally linked to an eternally present Moment of experience, is also essentially the same.

For example, the ancient mystical order of "Sunnyasin" are naked trident spear carrying ascetics that still wander across the jungles and mountains of India, following the example of the "King of Yogis", the Hindu lord Shiva.

Remarkably, it is said that hungry tigers will not molest them, and even the true king of the jungle, the King Cobra, will not bite them! And the mysterious Sunnyasin, while eschewing (shunning) all other forms of comfort and care, have used marijuana for thousands of years as an aid to meditation, and to achieve a unique mental state often described as "witness consciousness". (See "Shiva", pg. 575).

Of course it is quite impossible to actually be cognizant of the spacetimelessly real present moment in which our thoughts are being constantly assembled into the symbolic objects that fill our consciousness.

Nonetheless, the quiet still space that exists as our thoughts are first beginning to expand, may only be a "heartbeat away" from a disidentification with our binary symbolic reality, and the sudden spacetimelessness of our real Awakening!

IX

Marijuana is still considered to be a sacred plant in some present day cultures, and was part of the natural pharmacopeias of many in the past. In spite of this, or perhaps because of its "pagan" past, and the reflexively perceived dangers that it seems to pose to our ubiquitous "group consciousness" (which generally supports the self-serving "status quo" of humankind's ubiquitous illusory egos) and for other purely cultural, political, and monetary reasons, the use of marijuana has been largely marginalized, demonized, and even criminalized!

Thus the progress of legal reformation has been painfully slow! And by 2015, in the United States, the only exception to marijuana's illegality was ironically in the politically conservative, yet ruggedly independent state of Alaska. And not surprisingly, in a few states where it was considered legal by prescription because of its medical benefits.

Yet even in these states its use is still potentially perilous, because our federal government does not yet sanction marijuana use for any reason. Outrageously its possession, use, or sale is still considered a serious felony in many states, and is even punishable by death in some countries!

Soon after this writing however, Colorado and Washington State (and now Oregon!) have joined ruggedly independent Alaska, by wisely voting to decriminalize marijuana! No doubt they will soon reap both the civil rewards of fiscal savings, and a probable cash "windfall" in tourism! (And happily, due to the hidden efforts of many determined – if not *brave!* – people, it is once again on the ballot for legalization in California.).

But we need only look at the sanely progressive Netherlands to see one example of a practical working solution. In Amsterdam, for example, marijuana, like prostitution, is allowed but remains tightly controlled. Prostitutes are licensed, but must receive regular medical evaluations to ensure their health. As a result the health of their clients and communities are also protected.

The same sort of user and community protection laws are applied to marijuana sales as for alcohol consumption, or misuse in public places. Marijuana is sold in licensed "coffee houses". Both prostitution and marijuana are thus taxed and supply local law enforcement and government coffers with extra revenue.

And while it is generally unwise to intermingle different issues, this sort of "social management" solution makes sense, rather than unnecessarily filling our prisons to overflowing and continuing to drain our countries already dwindling tax reserves! (A problem which would be somewhat relieved, if we could only get the richest of our corporations to pay any income taxes at all!).

In startling contradistinction, billions of tax dollars are being diverted in America, pursuing a losing War on Drugs and victimless crimes of vice. At a time when our entire national economy seems once again to be teetering on the brink of financial disaster! (No doubt, along with the "outsourcing" of American jobs, this is a result of the swift technological advances that have impacted our outdated manufacturing industries).

The onerous (tedious) logic in support of these policies is either so simplistic, or overly convoluted, that it almost literally confounds common sense, and thereby nearly defeats all reason! Therefore the only rational explanation, as we have already briefly explored, may lie in a brief but accurate historical account of how we have reached such an unfortunate and unjust, if not silly, impasse.

X

Global environmental catastrophes loom on the near horizon. Our current national debt is staggering and is increasing at an alarming rate! The "outsourcing" of American jobs to foreign countries means fewer domestic jobs, and has weakened our local manufacturing and production. Out of work people cannot pay mortgages, so we are currently just beginning to emerge from an alarming housing and job crisis!

Our children are also continuing to fall behind the rest of the developing nations. Because our school boards, administrators, and overworked and understaffed educators currently lack sufficient funds and effective strategies to successfully accomplish our American children's education.

And to make matters only worse, our struggling public schools are being undermined by the shrewd perfidy of the greedy, self-serving corporations that want to cash in on our ever dwindling tax dollars at the expense of our children's effective education!

And it certainly does not help that an evidently ever increasing number of large corporations are now primarily international, and as a result are apparently beginning to lose their allegiance to America!

Unfortunately, the problem is also exacerbated because those with the intelligence and talent for teaching creatively are understandably attracted by better pay. If so much is expected of our, often vilified, educators and they do not receive reasonable pay for their efforts, the gifted and talented will naturally seek employment elsewhere. And who can blame them!

Our nation is also currently trying to recover from unprecedented "deficit spending" to finance a controversial, but sadly necessary "War on Terror"; an unconventional and costly war that is also most likely far from over.

And moreover, this rather unique "war" is probably best won not simply by military invasion. But by disproving the terrorist's rhetoric, by continuing to take in the innocent Muslim refugees who are now beginning to flee the intolerant "fundamentalist" terrorists in order to save the lives of their own innocent families and children.

Indeed, not to do so in our great country that is proudly comprised of (and regularly strengthened by welcoming) the intermittent waves of downtrodden, but nonetheless brave and determined refugees, who subsequently become grateful and hardworking new Americans, is simply, unequivocally and quite decidedly, un-American!

Because, even if the history ignoring hate mongers, who with much bluster and chutzpah, or even pretentious "pomp and circumstance", might otherwise advise, perhaps they should instead hang their heads in abject shame!

Yet, if the full truth be told, even if they may have actually somehow forgotten, and try to deceive themselves into believing that they are somehow more deserving than other decent hardworking Americans, the simple fact remains intact, that unless they are *American Indians*, not too long ago, their rather recent forebears were most assuredly immigrants too!

"*...Give me your tired, your poor, your huddled masses yearning to breathe free. Send these, the tempest tossed to me, I lift my lamp beside the golden door!...*" – An excerpt from, *The New Colossus*. A "sonnet of freedom", written by Emma Lazarus in 1883. Which was later engraved on a bronze plaque that (when I last checked) is still attached to the base of our great nation's Statue of Liberty. (To forever welcome our new immigrants to a safe harbor, and a promising new home, for so long as "Old Glory" still waves free!).

XI

And sadly, once again, it is far too obvious that our nation's almost "pervasive" ignorance about the veridical modern Middle East is still continuing to harmfully effect our martial judgement, and at times, even our actions in this troubled region of the world. "Iraq", for instance, did not exist before the year of 1932!

In fact, most people are not even aware that a great deal of the "Geography of States" in the Middle East has been drawn, renamed, and even redrawn, time and again by acquisitive foreign powers. And it should be no surprise that this onerous practice has been going on for some time now!

The following is a rather well-known, but nonetheless poorly understood example, because of the many years of confusing revisionist propaganda that denies the historical record. And there are many more such examples throughout the entire chaotic Middle East:

"There is no such thing as a Palestinian Arab nation . . . Palestine is a name the Romans once gave to Israel [as an insult to the resistant Jews who continued to live there]... The British chose to call the land they mandated Palestine, and the Arabs picked it up as their nation's supposed ancient name, though they couldn't even pronounce it correctly and turned it into 'Falastin', a fictional entity." — Golda Meir.

But since this was also going on at that not-too-distant time in underdeveloped countries around the globe, it would be misleading to conclude that the entire modern developed world is to blame! But it also does not excuse the involved participants today from doing little more than bombing the populace of the region, when a minority of them begin to take up arms in the resulting pain and confusion and begin looking for someone to take all the blame.

When human beings suffer the wounds of war, and the wounds are still painfully raw, patience does not feel like much of an option! And soon (sadly) any solutions better than rabid revenge seem impossibly ephemeral to some of the "psychically wounded", who have suffered tragic personal losses, and thus they feel the need to attempt to excuse their inhumane actions by citing a divine decree.

Nonetheless, just like the serial killer who once had an abusive childhood that eventually transformed him into an inhumane monster, they too must be held accountable for any criminal martial actions that violate our modern world's international Geneva Conventions.

In other words, no one involved is ever excused from exercising the rules of "fair" engagement! Indeed, the draconian Medieval practices of rape, torture, brutal public executions, and the purposeful targeting of innocent civilians that is currently being pursued as if it were a legitimately conducted Muslim *Jihad* (i.e. "Holy War) by the fundamentalist cult of *Daesh* (ISIS) only illustrates how far into the ditch of despair that hundreds (some would not incorrectly say even "thousands") of years of foreign intervention, war and chaos have finally caused in the region!

Indeed, it seems especially tragic when we consider that writing and many other aspects of our modern "civilized" world first began many thousands of years ago in what was then called *Uruk* (and in our time, "Iraq").

And as any American veteran of the Vietnam War will readily attest, simple ignorance can quickly become a great deal more than its simple self! And martial ignorance in particular has a disturbing way of propagating "evil" sorts of outcomes.

Take for instance our liberal democratic Muslim allies, the Kurds, who stood with us in the "Gulf War" against the vicious troops of the malevolent and sadistic Saddam Hussein. Hussein's cruel forces could not dislodge the brave, outnumbered and "outgunned", Kurds from their defensive positions around the sleepy little desert town of Halabja. And so in 1988 he ordered his troops to gas the entire town with (internationally outlawed) Mustard Gas!

The Kurds are most certainly quite brave, but they are also not a very numerous people, and the resulting loss of five thousand of their people, dying in such horrific agony, has become their national 9-11! Like us, it has galvanized the Kurds against an ancient "evil".

The age-old evil that periodically arises out of a twisted fundamentalist interpretation of the sacred Muslim scriptures. Which is much like our own infamous "Christian" Ku Klux Clan's interpretation of the Christian scriptures. Or our American Neo-Nazi's silly eugenic denial of the relevance of any sort of scriptures, or non-white people, at all!

Which allows these hate filled, self-serving human rights deniers and scriptural abusers to, with impunity, blow up churches and innocent civilians. And to otherwise rape, torture and murder anyone of a different race, and/or political or religious persuasion who is unfortunate enough to be captured by them!

It is the selfsame twisted and "Theocratically" empowered bellicose plague that (along with a radical meteorological change) once helped to destroy the "cradle of civilization" in ancient Uruk, and has continued to plague the Middle East for thousands of years since!

As things stand today, we have pulled most of our troops out of Iraq. And of course soon after we did, a great many of the surviving members of Saddam's vicious old "Republican Guard" rose up and became a driving force within the Syrian born cult of Daesh (ISIS).

A heartless and determined enemy who are nonetheless being inadvertently aided by certain recalcitrant members of our own congress and senate who are also doing considerable harm to our country, by ironically doing almost nothing at all!

Ostensibly they are doing this to frustrate their political opponents, but nearly paralyzing our government in "a time of war", almost rises to the level of treason! According to the United States Constitution, Article VIII,

Section III "[anyone]...giving them [our enemies] aid and comfort within the United States (or elsewhere) is guilty of treason."

And while I am accusing no one of actual premeditated treason, the very idea that such actions, or lack of action, might inadvertently suborn treason, is most certainly some serious "food for thought"!

Moreover, it seems especially egregious when the perpetrators were supposed to be our own leaders. In the very least, until they can be removed from office by the American people through the power of our vote, it appears that grounds for a politically independent "Grand Jury" style of investigation and possible summary dismissal from office may even be warranted! (But, even if nothing is done, history will eventually reveal the truer story of the considerable damage that was done to our poor struggling democracy.)

But, returning more directly to the rather unpleasant subject at hand, it was not too long before Syria was almost completely overrun, and shortly thereafter Daesh swarmed southward and attempted to take Erbil, the capital city of the Kurds!

The Kurdish people that are scattered across the mountains of Northern Iraq, Syria, Iran and Turkey constitute a liberal and independent Islamic island, in a seething sea of Muslim extremism.

Along with increasingly unstable Turkey, they are our among our staunchest Muslim allies in the civilized world's current struggle against radical Islamic terrorism. Which is an ancient evil that we probably should not even dignify by saying that it has anything whatsoever to do with the modern Muslim religion!

At first in many areas the defending "underdog" Kurds were taken by surprise and were driven into the nearby mountains. But they soon rallied and drove Daesh from their villages, towns and cities, only to discover the unspeakable atrocities that were being done to anyone unfortunate enough to be captured by Daesh.

But what was done to the innocent inhabitants of the quaint little ethnic village of Sinjar, who could not escape in time to save themselves, rises above Daesh's usual level of unspeakable atrocities! I will be kind enough not to describe such unspeakable atrocities here, as they are a matter of unhappy record if you should want to know more.

But as a Jew I cannot with good conscience recommend subjecting your psyche to a detailed recounting of such nightmarish horrors that, like the Holocaust, should also never be forgotten.

Nonetheless, the resulting ephemeral images, which your imagination is most likely sufficient enough to supply, can be "cognitively experienced" without subjecting your psyche in the process to the possibility of incurring PTSD (i.e. "Post-Traumatic Stress Syndrome"), like far too many of our dutiful veterans do!

Today, the Kurds are defended by a small but determined Kurdish force called the *Peshmerga*, consisting of a mere 150,000 soldiers that are thinly scattered across a thousand kilometer "Front" that separates their vulnerable liberal society from the nearby horrors of "ISIS" (Daesh). And to the best of my knowledge, they are also the only Muslim society that has an entire Battalion that is wholly comprised of, and led by, a cadre of very capable and determined Islamic "women of valor"!

And although a few American and coalition soldiers, who fought beside the Kurds in Iraq, continue on their own to stand with our Kurdish allies, it is puzzling (if not outrageous!) that (at the time of this writing) our country continues to neglect supplying them with the arms and ammunition that they so sorely need to effectively defend themselves.

A situation that seems especially egregious when we consider that they are among the distinct few, of predominately Muslim countries, that (at the time of this writing) are helping the civilized world in our ongoing battle against Daesh! In other words, to be exact in our estimation of the current situation, we also need to acknowledge these brave and steadfast people.

Who in every desperate battle for their very survival are fighting and defeating the modern world's most feared and ruthless enemy!

But ironically, they are also not yet recognized as a sovereign nation! (Information gathered from the Front Line®, courtesy of the courageous award winning National Geographic Correspondent, Neil Shea).

Controversial unmanned warfare through the use of armed "drone" aircraft has been an unfortunate, if inevitable, result of the modern world's response to the terrorist's craven and misguided "Jihadist" acts. Thus, unhappily, drone strikes have become necessary, and although they are precise, they are also not "perfect".

But neither are the air-to-ground strikes of more conventional "manned" stealth aircraft, or for that matter missiles that are fired from ships that are hundreds or even thousands of miles away from the war zone!

"We're fighting an enemy that denies the very essence of the Geneva Convention and makes no distinction between enemy combatants and innocent civilians. Now, war is ugly, innocents die, and it is up to us to try to minimize this as much as possible, even if our enemy does not. But if we had to wait until everything was perfect, we would lose our opportunity to do anything effective!" – Michael Hayden, former director of the CIA and NSA.

Horrific crimes of Inquisition style public torture, internet broadcasts of the cruel executions of their unfortunate victims, and the purposeful targeting and blowing up of innocent civilians in countries often far removed in both distance and culpability from the battle zone, have literally forced this unfortunate technological response from an otherwise relatively merciful and civilized people!

But, such is the sad reality of war, and the reason why humanity as a whole must strive to rise above it by cooperatively working to remove its root causes in the world.

One day, when I was a teenager herding cattle in the Santa Cruz Mountains, Paco, the ranch foreman, and I were talking about a brutal murder that was on almost every television and radio station at the time.

Like 'most everyone who discusses such horrific acts, all of our efforts to understand why, and how, anyone could do such horrible things to a fellow human being, soon reduced us to a sort of stunned, sad silence.

We rode on together, occasionally drifting away from each other in order to bunch up the cattle that we were driving back down from the green grass of the higher Santa Cruz Mountain meadows. Paco leaned a little to the side in his saddle, and gave me that special "Trail Boss" kind of look. Which meant I had better reign in my horse and listen carefully.

Paco reigned his high strung Appaloosa around. Then he reigned in close to my right side, leaned over in his saddle towards the easy riding Morgan that I was seated on, and said; "A long time ago, some drug lords killed a whole family in the small town where I was growing up [in Mexico]. We were all trying to make some kind of sense of the tragedy, when my *Abuela* 'Grandmother' said to me, *El mal presa a los debiles porque teme a los fuertes.* "Evil preys on the weak, because it fears the strong."

And it suddenly became crystal clear to me, that being strong is as important as being smart and kind, in our sometimes "crazy" and uncertain world! But, in spite of the spate of senseless violence in our modern world, in light of the massive environmental crises that humanity now faces, my father, who began his eventful life as a poor sharecropper's son breaking up clods of dirt with his bare feet behind a team of big Kentucky mules named Sally and Sam, would probably say; "Son, we have bigger 'Channel Cats' to catch and fry!"

"...it was a slow day, and the sun was beating down on the soldiers by the side of the road. There was a bright light, a shattering of shop windows. The bomb in the baby carriage, was wired to the radio" – The Boy in the Bubble, © 1986 by Paul Simon and Forere Mothoeloa.

XII

But, returning to the current state of our own nation, the aforementioned onerous unresolved issues, and thoughtlessly reactive practices, continue to divert huge sums of our country's money and resources. But too many of us unfortunately still presume that the continued use of unreformed and dysfunctional policies will somehow meet and solve our current problems.

However, just because a practice once seemed to make sense, does not excuse us from reassessing it when it is obviously no longer working very well! Is there any wonder why so many intelligent (and perhaps by now even comparatively better educated?) people around the world look at our powerful nation and shake their heads in disbelief!

Our concerned allies ask a good question, "America used to be great, and once even led the free world in many ways. What in the world has happened, what has gone wrong?" However, history does show that there is often a swing into chaos before a corrective change in the opposite direction occurs.

This "swing" is apparently an inescapable feature of our cognitive human life, with its roots sunk deeply into the phenomenon of our dualistic consciousness. We literally create the world that we live in, in the image of the knowledge and ignorance of our own split-minds! Whatever balance is found between extremes is therefore, sadly, only temporary.

Nonetheless it can only emerge, even if but for a little cognitive while, when we approach the nondualistic edge of our own foundational Awareness, which contains, but is not constrained, by any ostensive "edges". But if more sinister sorts of apparent change cannot be halted, in our collective experience of the universe, then can they at least be better managed? If not, then any sort of education is evidently useless, and we are all in serious trouble!

It is indeed a sobering commentary on humanity's progress that so many of the enlightened Masters and Mother Nature's gifts of psychoactive botanicals, both having done so much to raise the consciousness of humanity from our humble animal beginnings, seem to quite often suffer from a continuing superstitiously based litany of ignorant self-misunderstandings.

These often well-meaning, but generally self-defeating misunderstandings often fuel unreasonable fears and thoughtless demonizing. And they may even lead to unintended mayhem, rather than to an affirmation of life by honoring the relatively "benevolent" nature of our own consciousness.

Moreover, marijuana is hardly new to the West, although our politically independent scientific community's attitude toward it has been markedly different than that of many less enlightened governments!

"With cannabis there is no cause of fear, but possibly of laughter!" – Robert Hooke, 1635-1703.

XIII

There is a natural cycle of growth within the individual towards a more complete understanding, and the expression of a deeper awareness within the temporal experience of our own phenomenality.

This is the natural progression of our consciousness, which is sadly too often truncated by negativity as we grow older and experience more of the inevitable suffering of life.

But hopefully, wisdom may still come to us with time and reflection. A wisdom which not only helps to improve the condition of the individual and the world, but also prepares us to accept the more unpleasant eventualities in life that we cannot control.

And 'though strictly speaking, we illusory thought contrived beings may have no real control, in a relative sense we do, and we are therefore evidently not excused from trying!

Humanity simply cannot continue to frustrate this natural process, because as our species experiential history quite clearly shows, we do so at our own considerable peril! (I.e. History is "experiential," because our experience of it is also all in our heads!).

Our focus must be allowed to expand beyond the mere acquisition of security, power and wealth for ourselves, our families and our particular social group or country.

And although this is, in itself, perhaps no great revelation, it remains, nonetheless, a demonstrably important one to treasure, and to keep forever in the depths our heart, and the forefront of our minds. Because whether we want to admit it or not, humanity's continued existence apparently depends upon it!

We need to listen to what the accurately informed and our own common sense tells us, rather than to whatever the loud and insistent voices of misguided hate and fear mongers, religious fanatics, self-serving politicians and greedy corporate executives may repeatedly say!

Because their disingenuous (insincere) "guidance" generally represents either their own special interests, religious superstitions, or the "chameleon agendas" of an emerging and perhaps unstoppable corporate oligarchy, instead of the real and compelling needs of humanity at large.

However, America has proven that it still has the heart and will of a free, and at times perhaps even wise people, by electing our first black president! But not because he is (or is not) "black", but simply because he, like Franklin D. Roosevelt or Abraham Lincoln, was a presidential man of his times and, moreover, was also the right American for the job!

Indeed, many of us who are old enough to remember Dr. Martin Luther King's prophetic civil rights speech on the steps of the Lincoln Memorial in Washington D.C., and his subsequent assassination in Memphis Tennessee, wept openly and without shame when Barak Obama was inaugurated.

And 'though ever dwindling in number, those brave and decent souls who worked tirelessly for civil rights in an ignorant and violent time, and even put their very lives on the line in Selma Alabama and elsewhere in the American Southland, will never forget the cost, and neither should we!

It was a good day for America, and moreover it signified the beginning of a "brave new era" in our tarnished civil rights history. Finally, after the extensive horrors of black slavery, our country's devastating Civil War to correct the horrific problem, subsequently shameful "Jim Crow" decades and seemingly intractable racial prejudices, it seemed as if the American people were finally beginning to live up to our own precious Bill of Rights.

But it remains a delicate balance, and racial prejudice is disturbingly far from over, especially in certain areas of our otherwise great country. Which seems particularly perverse, since according to Henry Louis Gates Jr., host of the popular PBS show "Finding Your Roots", recent advances in Genetic Genealogy have revealed that despite our apparent differences in skin color, hair, and so on, we are all actually rather recently related cousins!

And if not always, then most surely now, as the patriotic (but nonetheless slaveholding) Thomas Jefferson (also a man of his times) eloquently opined, an equitable liberty always requires our "constant vigilance!"

Because ours is a liberty hard won and hard kept, by the blood of our forefathers, and sadly by that of every subsequent generation of young Americans. It is a precious liberty, special and rare in human history. And even if for these reasons alone, and even if it is in many ways still only a wishful dream, it is a liberty that we can ill afford to take for granted.

But I suddenly find myself at an utter loss for words, because our precious freedom is so much more than any author could ever adequately describe, in any sort of missive.

And when I compare it to the sad state of the people in nations that have little regard for the civil rights of their own citizenry, it becomes difficult to adequately express my gratitude for having been born in the USA!

So I will simply say, that so long as we strive to be just and inclusive in our freedom, I trust that it will always be a liberty that is worth fighting for! Nonetheless, we have currently (in a fashion) "lost our way". As a result, there are millions of angry, currently unemployed, geographically isolated, and (sadly) often under or inaccurately educated white people in the American Heartland.

But their manufacturing jobs have been lost primarily due to innovation and advances in technology, production and distribution. Moreover, the United States has become an ethnic and cultural "melting pot", and whether you like it or not, the arrow of time cannot be reversed!

Simply put, as Americans, for the good of our country, it is our civic duty to ensure that every citizen has an equal opportunity to participate and succeed in our economy. Hence, it is always worth the struggle to find our way back to a veritable "freedom for all Americans" once again, because anything less destabilizes our present, and jeopardizes our future!

Hence, as we advanced toward the final culmination of our chaotic 2016 election year in America, no doubt, history will show that we should have been more informed, and even *wary*. Because uncertain times tend to invite not only anarchy and fascism, but the destructive deceptions of demagoguery as well!

Prelude to: An Unwanted Detour

Please Consider The Following, and indeed *far too much* of this chapter, to be a brief *de rigeur* "detour" from the happier "redwood trail" that we've been walking together. Sadly, it has been brought on by the truly heinous crime of another craven "deranged" shooter, "shooting up" a bunch of innocent people in our country!

Sometimes these pathetic creatures prey on innocent children. Or even on elderly Black Americans as they sit in prayer on a church bench. But this time the slaughter took place in a clearly nonviolent "gay nightclub", in our lovely tropical state of Florida.

My dear concerned wife, who's a retired school teacher, insisted that I include the following observations, since we don't currently hear anyone else offering them to the public!

An Unwanted Detour

I Do Hope That My Unabashed Patriotism is not too unsophisticated, or possibly even offensive, for anyone's more refined sensibilities! Indeed, if so, I must humbly beg your pardon, because I'm just an uncomplicated "tree man". Who grew up (almost alone) in the great redwood wilderness, of rural northwestern America. The only living son of a humble, patriotic fireman, who was also a proud Oklahoma veteran of World War Two.

And I suppose that, perhaps just like you, I've also become a lot like someone that I deeply admire and love. In my case, this was my widowed grandma's second husband. My beloved "old fashioned", cowboy hat wearing, lariat twirling, tobacco chewing grandpa! (My bigger-than-life, Ozark "hillbilly" grańdaddy, William Oliver Ingram).

But despite the tenderness and love that he always displayed to his family and friends, my big "teddy bear" grańdaddy was a study in contradiction. Because he was once a "tough as leather" Deputy US Marshal that rode in the lawless 1800's Indian Territories of the Western District of Arkansas, for the infamous "Hanging Judge", Isaac Charles Parker!

I suppose that's probably a big reason why I have the attitude that I do. Because, just like a good many of you, my fellow "Baby Boomers", even though I'm not getting any younger, I'm still always ready to "give my all", for the sake of life, justice and freedom! But only when it's the right thing to do for the exigent circumstance at hand!

And if you're my fellow American, barring outright vehement prejudice, crime, or treason, whatever your personal views might be, this *is* America, where we are all free to hold different views, and even speak our minds out loud, if need be.

So, my fellow Americans, of all persuasions, races, heritages, and choices of partners in the mysterious processes of finding our hearts true love, the above *always* includes you too!

After all, although we do maintain a strong military "for safety's sake", we are still a peace loving nation. But always with the expectation that we are also, if need be, at a moment's notice, a "warrior nation" too!

Hence, we need to take a page from Israel's "playbook" against terrorism, and always be prepared to be a nation of citizen soldiers, as well as peacetime ones! It is, quite simply, still our *"call to duty",* as an sovereign and democratic people, to take care not only of our individual selves, our soil, and our leaders, but of each other as well!

And, while speaking in this vein, in this troubling time of senseless mass shootings, I would like to offer some humble advice to anyone that finds themselves "under fire" in an unarmed crowd. Trying to shoot the shooter, unless you're "sniper accurate", is clearly *not* the right thing to do!

When a shooter holds an assault rifle in their hands, the only *effective* thing to do, is to rush them instantly from all sides, and overpower them, to stop the ceaseless killing. Because, and I'm sorry to say this, when you're facing this kind of fire power without a bullet proof vest, especially in a panicked crowd, if you run, you'll most likely die!

Although it may seem difficult at first, *we can overcome* our fear, and vow to ourselves, and agree with each other in advance, that this is what every American is going to do, and will do, *without hesitation,* if we are ever faced with this distressing mortal dilemma.

And when we clearly understand that we are doing it out of love, not just for our great country, but for each other too, then the crowd now holds the ("Greater") power, to literally tear the cruel (hopefully "would-be") shooter "limb from limb"; because (indeed) *E Pluribus Unum!*

After all, we all are going to die someday, so why not make it count for something, when it really matters! The attitude that most American Indians still hold dear applies. That we all need to keep in mind that one day we too might need to say, *"Today, is a good day to die!"*

Craven terrorists and murderous cowards, listen up, and beware! Because, when it comes to America, you're just beginning to learn that, "It's not wise to poke a nesting Eagle!" (Because our Fort Campbell is always "chock-full" of courageous American Cowboys who, like our brave **Special Forces 595**, are always "waiting in the wings" to defend Her!)

When we are wrongfully attacked, history shows that our citizens' determination to bring justice to the perpetrators is as tireless as it is indomitable. Moreover, our history indicates that acts of bravery and valor tend to spread courage in America, faster than the common cold!

But there is a deeper, perhaps even "sacred" reason why America is called, "The land of the free", and "the home of the brave!" And even if your terrorist hearts may have become so twisted by tragedy, hurt and rage, that you can no longer see, or even understand the power of universal Love, just understand this, that you are really only serving Death!

And the living, who still love life, and thus their fellow man, will never allow Death to win what is not Death's fair due! Because every time you prod the living with another cowardly attack, you only strengthen Life's essential unity, and Her resolve *to end you*, for the Love of Being's sake!

(Piae Memoriae, Todd Beamer *et al*, the courageous "gay" Rugby player who lead the charge that stopped the 9-11 terrorists on United Airlines Flight 93. (They were likely headed for the White House!). *"Greater love hath no man than this, that he lay down his life for his fellow man!"* – John 15:13).

XIV

Like many people of my generation I had nearly given up all hope that America would ever truly again be in the hands of her own people!

To many of us who lived through the civil rights and antiwar movements and the assassinations of John F. Kennedy, Martin Luther King and Bobby Kennedy, it seemed like an impossible dream that a black American would ever be elected to the American presidency in our lifetimes.

But Americans of good will voted for the impossible dream. We voted, and then we held our breath, still hoping for a better America, still believing we could make a better world!

Then against all odds it happened, Obama had won! It was a good day for America. It was a good day for social justice, and for the world. Indeed, I am compelled to mention that Barak Obama, a man of black and white, Muslim and Christian heritage and therefore uniquely suited as a president for our times, was subsequently reelected to serve a second term (much to the chagrin of the racists and many One Percenters!).

No doubt it was to the continued betterment of our nation, and perhaps even for the world at large. If only the Democrats and Republicans in our bicameral congress and senate could have rediscovered a way to put aside their political theories, resulting enmity, and often secretly self-serving agendas, and learn to cooperate again, for the good of our nation!

However, this is neither an endorsement nor a condemnation of an American president's policies and actions, which only history will reliably reveal to be either "good or bad" for our country.

Nonetheless, if I put our more recalcitrant problems aside, I was again proud to hear the somewhat timeworn presidential invocation of "My fellow Americans", because the true American majority had again made their honorable intentions clear!

For it is not riches or power alone that make our nation great, but the brave heart and increasingly common decency of the uncommonly common, but increasingly present, *homo illuminatus* "enlightened hominid".

But this was just the beginning of a grander "sea change" that will eventually allow Nature's powerful will to survive to sweep over our entire planet. And although it may sound trite in today's jaded world, this can only transpire by nurturing the natural emergence of opportunity, equality and justice on every level of human society, everywhere around the world. We can no longer afford to ignore anyone!

And while it is also true that force may sometimes be required in our currently unstable, but nonetheless rapidly changing world, as history shows, lasting change can't be accomplished with guns, or by compulsion alone, but through education, goodwill, family planning, and reinvesting in humanity's national and collective global future. Because intemperate violence, almost inevitably, only leads to more violence!

XV

No particular form of national government is going to accomplish this, and most certainly not more international corporate greed. Moreover, continued political and religious "lip service", and disjointed philanthropy, only prolongs humanity's collective problems!

And while charity is certainly helpful, benevolence alone cannot solve problems that have been allowed to proliferate and grow to such an immense scale.

A major contributor to our national woes is the dramatic redistribution of wealth in our country since the end of World War II. It is a major problem because (as many of us now know) the majority of our nation's wealth is outrageously in the hands of approximately one percent of our population!

Which means that America is now literally at the mercy of a plutocracy that holds hostage our national wealth and power.

The problem of course (besides an appalling lack of empathy for its desperately struggling victims), is that the consolidation of wealth, power and special privilege by the few often hurts the many.

This is not because there is not enough to go around, but because so many of these privileged few Americans are serving only themselves, rather than giving back to the society that provides the means for the acquisition of their wealth and power.

Indeed, our country now has the greatest inequality of wealth, by far, of any developed nation. As our former Secretary of Labor, Robert Reich, points out, nearly 23% of our nation's income is now being taken home by the wealthy top 1%, and as an inadvertent result our societies stabilizing economic middle class is rapidly vanishing!

But statistics can be misleading, so just what does this conservatively mean? In short, the wealth of the top 400 Americans today far surpasses that of the total income of the bottom 150 million among us! But why? In 1978 the average middle class male worker earned more than 48 thousand dollars a year, while the top 1% earned 393 thousand dollars; adjusted for inflation.

But by 2010, the average middle class male worker was making less than 34 thousand dollars annually, while the top one percent was now earning more than 1 million dollars! And today, in 2017, these discrepancies have only gotten worse.

Evidently, the problem that Robert Reich is really posing to us is, "How much more inequity can our economy tolerate"!

This is not a plea for communism, or even classically robust socialism, which simply transfers the social or "common" wealth and power into the hands of the few, in a different, if not more perverse manner!

It is simply a concerned citizens call *"...from America's Redwood Forests..."* for a return to the principals upon which our country rests.

Indeed, the American Revolution was fought to put an end to just this sort of unjust and harmful nonsense, and disturbingly our inadequately regulated modern corporations are, in a very real sense, simply replacing King George III!

And I can only imagine that if our founding fathers could somehow know of our current impasse, they would probably be "rolling over in their graves," or even more likely, "rolling up their sleeves" to get to work on a practical solution for the problem!

But democracies' rarely take the penultimate path, and as Winston Churchill once wittily observed, "The United States always does the right thing, after thoroughly exhausting every other opportunity!"

Yet in all fairness, the actualization and defense of liberty and justice seems to require a constant effort against the baser instincts of humanity.[93] Nonetheless, if the "powers that be" cannot be persuaded, or forced by law to do what is best for society and nature, as John Locke observed long ago, by the organized and informed will of the people, then it certainly does not bode well for our country, or for the world! (See "John Locke", pg. 546.)

This disturbing trend is not limited to the major players on Wall Street, or to the poorly regulated and greedy cancerous sort of capitalism (perhaps better called "acquisitionalism") that is evidently infecting the entire economic system of today's world.

Because capitalism is indeed evil when it is not controlled by the more universal ethics of fair trade and fully inclusive Civil Rights. Be they ensconced in a modified system of democracy, "evolved socialism", or in some other system that is comprised of a truly representative government, equitable economy and just law!

(Piae Memoriae, Muhammed Ali, 1942–2016. A larger-than-life American "Pugilist Mahatma", and an American Muslim. Forever the undisputed "People's Champion", and now, an eternal Icon of American Civil Rights!)

XVI

Aside from our tireless labors to get our great country to live up to its own constitutionally guaranteed "Bill of Civil Rights", in a country with "free speech", conjecture about such important matters will no doubt always abound. As will the differing opinions of Americans on our everchanging current issues!

But such is the adversarial nature of the ongoing public, political and jurisdictive debate, which is an essential feature of any healthy system of democratic government.

Nonetheless, when we put our speculations about better systems of government, law, and economy aside, the overarching fact remains that humanity is always One in its essential Universal-Consciousness.

In much the same way that a diminutive African woman named "Eve" is the physical progenitor of all modern humanity; ever-dwindling white racists included! (See "Eve", pg. 531).

It is human societies conceptualizations of borders, differing languages and cultures that are arbitrary. Therefore, for any substantive change to occur there must first be a widespread awakening to a deeper level of our essential intuitive Awareness, to reveal our veritable Oneness.

This awakening needs to reach into every level of society, from the hidden oligarchy to the most destitute and illiterate of humanity. And this is indeed always possible, because revealing and enabling the courageous and compassionate heart of the common person is ever its singular and most important requirement!

Without this, whatever plans are made, or actions taken, will ultimately fail, because they will continue to be disrupted by the widespread self-interest of humanity's billions of ubiquitous unenlightened illusory egos. This may sound impossible to accomplish, because it is, if we continue to think in conventional terms, and act in conventional ways!

The world worships the power of the intellect, but Awareness is different from intelligence because it is also metacognitive and intuitional, or "apperceptive". Awareness arises spontaneously in a manner that seems to involve the entire organism of our body-mind.

Our intellect, on the other hand, must be carefully addressed and educated in a timely, sequential manner, most effectively when we are still young and our naturally inquisitive brains are still physically developing.

A deeper Awareness however can arise at any age, because it is the internal revelation and realization of deeper truths about the nature of our consciousness, life and world, which are universal in their very nature.

We all share in this essential Awareness because it is the very essence of who we are. Mother Nature is constantly tending towards the fuller revelation of this Awareness and has been about it since the advent of the Big Bang Singularity that set our entire universe into motion, around fourteen billion years ago.

Therefore, humanity needs to encourage conditions that enhance, rather than detract from, the natural progression of an emerging deeper Awareness in the general consciousness of humankind, because technology alone will most likely not suffice to save us from ourselves!

In other words, the problematical world that humanity has created is analogous to the frequent downstream flooding of the great Mississippi River's extensive shoreline and terminal Gulf Basin. This situation is caused not by any one human act or natural catastrophe. It is, instead, the result of countless tiny harmful incursions into the balance of nature!

For example, whenever we settle along the riverbank of our nation's powerful Mississippi River, we build up a tiny dike to secure our holding against its occasional flooding, which we rightly interpret as the dangerous natural power of the river.

But it is the accumulation of these small insults to the river's powerful natural processes of annual flooding that has created the overall inundating problem. Yet our great American Mississippi, like the more sinuous Egyptian Nile, renews the fertility of the surrounding land with depositions of organic and mineral materials in exactly this manner.

Not only in this, but in many other natural matters as well, Mother Nature tends to be rather "messy", but is nonetheless quite effective, if only we can find a way to cooperate with her often subtle, but sometimes overwhelming methods!

And due to the recent "Superstorms", which are no doubt beginning to be generated by global warming, New Orleans, like the world at large, despite whatever protections that science and technology might currently provide, is becoming an increasingly perilous place to live!

In point of fact, by the time of this initial writing, as if to further presage the horrors that may come, Superstorm "Sandy" had already dealt an unprecedented blow to our countries entire Eastern seaboard!

XVII

But there has never been any real mystery to this occult wisdom, for it is the fundamental property of all men. It is among humanity's oldest and perhaps most commonly ignored knowledge. Prophecy for instance, a form of this wisdom, is but the art of revealing the potential of the obvious, when for some reason it has become obscured by current events.

But it does not require an insightful spiritual prophet, but simply a somewhat prescient scientist, to predict what will happen if we don't find a more "carbon free" way to meet our ever growing energy demands, because our suffering planet's overheating problem is now quite widely understood, and simply speaking out more loudly will not change anything!

Moreover, we need to be wary of those who would downplay the seriousness of this growing problem either for their own self-serving purposes, or to "protect the public" from the many "evils of hopelessness".

Because, our ever warming planet-wide problem will require a thoughtful series of practical and effective planet-wide solutions.

Indeed, in light of the massive environmental crises that humanity now faces, my father, who began his eventful life picking cotton in "Dustbowl Era" Tahlequah Oklahoma, would probably say; *"Son, we just need to stop talk'n, 'n fret'n about it, and just get 'er done!"*

However our situation is not really hopeless! We most certainly can accomplish this grand task, we simply have not yet fully understood what it means to possess the power to assist in unfolding nature's ongoing evolutionary endeavors, thorough humanities higher visionary capabilities.

But this will likely only happen in time to avert our ever growing global disaster if we can quickly summon the necessary determination and courage to put aside our political and cultural differences, and our past and present grievances, and act together as a global community to save ourselves!

And we can only do so with a resolve that is based not upon exploitation, violence, or a wishful "supernatural" solution, but upon the certain knowledge of humanities fundamental oneness, in the life affirming force of a universe that has somehow manifested both life and mind out of an almost "insentient" matter, and this out of an apparent nothingness!

Moreover, contrary to the protestations of a great many poorly informed but well intentioned environmentalists, solar and wind power alone cannot effectively supply the world's current and ever growing power needs. (By replacing the glut of coal and oil burning electrical power plants that are currently destabilizing our planet's atmosphere with carbon dioxide).

To honestly believe so is simply another futile indulgence in ineffective "wishful thinking" that is most likely also based upon a concoction of outdated and inaccurate information! Professor Mark Z. Jacobsen, of the Stanford Energy and Environmental Department, in his recent "Wind, Water and Solar" use paper (cite: Energy Environ. Sci., 2015, 8, 2093) predicts that, even with almost total conversion, by 2050 we could conservatively supply only around 39% of our country's energy needs with WW&S power (i.e. Wind, Wave and Solar).

But we should also not too readily dismiss other newer technologies in "carbon free" energy production. For example, in the arid Arizona Desert, 100 miles Northwest of Las Vegas, there is a new and much more powerful iteration of the more traditional "low yield" sort of solar power plant, called the Crescent Dune Solar Reserve.

The plant has 10 thousand large mirrors that automatically follow the sun on its daily journey across the open desert sky and reflect and concentrate its power onto a heat collector that is located atop a 600 foot tall tower.

The searing heat is transferred into an extensive array of tempered metal pipes that are filled with constantly recirculating molten salt, which is then stored away to be used in energy production at night and during the rare cloudy desert days.

The intense heat is used to produce a flow of constantly recirculating steam, which drives an array of turbines that produce a constant 110 megawatts of electricity. Which is (a truly impressive) one fifth of the power that is produced by the average coal burning power plant!

In Iceland, experimental deep drilling is being done to utilize our earth's high internal molten magma temperature, of 1,700 degrees Fahrenheit, to produce a constant flow of 800 degree Fahrenheit steam that is powerful enough to drive a series of immense electricity producing turbines.

It is said that the average 5 kilometer deep geothermal well can produce as much daily energy as 10 average oil wells! However, this is not only an inherently risky business, but it is also evidently limited to the available sites where it could be effectively utilized.

Nonetheless, research in deep drilling geothermal energy production could help us to learn how to harness the incredible power of an erupting volcano! Which could prove very useful in the not-to-distant future, when the Super Volcano that underlies Yellowstone National Park decides that it is time to erupt again!

And while geologists still currently disagree about when this might occur, in anywhere from a few thousand to over a million years, when it finally does, as it has done on several occasions in the past, it could well cause another planet wide extinction of life, if we cannot find a way to mitigate the event!

And last, but not least, a Canadian researcher named Louis Michaud has further developed the pioneering work of the late visionary French scientist Norman Lout by developing a working prototype of an Atmospheric Vortex Engine (AVE).

An "AVE" works on the same principle that creates and drives tornadoes and hurricanes. In other words, a layer of cold air suspended above warmer air creates an area of low pressure below the cold air that generates an energetic spinning vortex.

Curiously, when the opposite occurs over a body of water, generally only harmless fog is created! Like that which rises up from the Pacific Ocean, and like a slow moving tsunami eventually spills over the Santa Cruz Mountains. To join together in a blithe, if not turbulent reunion, with the fog that is steadily washing in past the Golden Gate Bridge to fill the great briny basin of the San Francisco Bay.

Sometimes the hoary mist rises up so high that only the tallest mountaintops can be seen. From the air they look like lonely little islands, awash in a stormy, slow motion sea.

But, returning to the topic of our energetic "vortex", the vortex of spinning "AVE" air is used to turn a turbine that generates electricity. And although it is still in in its infancy, the potential exists for generating and co-generating massive amounts of "carbon free" energy. By harnessing the constant stream of wasted energy in the column of hot air that rises up in industrial chimneys around the world.

Energy efficient vortexes could also be created by utilizing captured solar heat and rising aqueous thermal currents. The amount of energy, for example, that is produced daily by storm vortexes is about a thousand times greater than all of the energy that is being produced by other means over our entire planet!

Nonetheless, if we are realistic in our assessment, even when we factor in these promising new technologies, we still fall far short of our modern worlds ever growing energy needs.

And although we may well one day be able to meet all of our energy needs with these sorts of creative alternative means, we also need to act quickly, and safe nuclear power production seems to be the best, if not the only practical method at hand!

Indeed, even without the breakthrough advent of new world-changing "moonshot" technologies, we already possess the knowhow to fix our persistent energy need problems and only need to look at France's own successful nuclear power plant solution to see how it can be realistically accomplished.

Among some of our Silicon Valley's more practical visionary enterprises, like that being currently undertaken by the "Bill Gates Foundation", are the creation of workable designs and prototypes for smaller, meltdown-proof "breeder" type nuclear/electric power plants.

These are a *nouvelle époque* of "fourth generation" sodium and thorium "intermediate-loop" TWR's "Traveling Wave Reactor(s)" that can be primed with nuclear fuel, then buried near the cities that they will supply in order to reduce the electrical power that is invariably lost during the transmission of electricity over long distance power lines.

These reactors will continually reuse their nuclear fuel for many decades until its radioactivity is finally reduced by around 80%, at which time it can be used for other technical applications, or safely stored away until it degrades to a negligible level of radiation (i.e. superheated sodium and thorium are used to produce stem to drive the TWR's electrical generators).

But of course our nearby Lawrence Livermore Laboratory (if not some other research facility, public or private) will likely soon make the breakthrough discovery that will allow us to start using clean "fusion", rather than fission powered nuclear reactors.

Along with wind and solar power, and the replacement of the internal combustion engine with an efficient electrical or hydrogen gas powered motor, we already possess the means to halt our current global warming problem, and perhaps over time, as we replant our forests and stop killing our oceans, to even reverse it!

But we will most likely not be able to accomplish this if we do not (soon) educate ourselves with reliable facts to overcome our more unreasonable fears.

For instance, unbeknownst to most people, by agreement with the former Soviet Union, by 2013 the United States had already purchased nearly

16,000 nuclear warheads to be used for fuel in our own reactors, in fact, they now supply almost 10% of the electricity that is being produced by our countries reactors!

And insofar as the runaway development of nuclear bombs is concerned, of the 37 countries that have the technology to produce them, since the first bomb was dropped on Hiroshima in 1945, only 9 have actually done so.

Thus, even though we cannot un-invent nuclear weapons, the international community can always decide to not make any more of them. And by simply enforcing our international agreements (and preventing rogue nations from building them, we can eventually turn them all into nuclear fuel, to help reduce our continued dependence on fossil fuels.

However, we also need to somehow overcome the considerable economic resistance to change that keeps our country tied to the continued self-destructive use of coal and oil as fuels to supply our ever growing energy production and transportation needs! But to do so with effect, we need to become a nation that is in the hands of its own well informed "citizen scientists".

And (G-d forbid!) that some terrorist group might one day set off a nuclear explosion in a country with nuclear weapons in their arsenal! (Or even more likely, if an explosion should occur, "by unhappy accident!") But since the possibility does exist, we also need to have effective prophylactic international agreements in place to preclude the possibility that it not inadvertently set off World War III!

Nonetheless, regardless of the reasons why, both big business and government have been slow to respond to our ever-warming global crisis. Yet at the distal edge of effective potential solutions science may have found a rather unusual way to help manage our ever growing problem, as we methodically ween ourselves from requiring fossil fuels as our primary energy source. Ironically, we have done this by observing the harmful effects that Super Volcanoes have on our earth's atmosphere!

When Mt Toba, a "Super Volcano" in Sumatra, erupted 75,000 years ago, thousands of kilometers of volcanic ash were forcefully ejected upward into the sky, in a massive fiery plume that reached over 30 miles high, all the way up into our earth's stratosphere!

From there it spread out until it covered the entire planet, shrouding it from the sun in a dense fog of volcanic dust. Soon our earth's temperature began to drop, and chilly, winter-like conditions set in over the whole planet.

But most of the volcanic dust settled back to earth in just a few years. And the earth's climate would have returned back to normal again, but for the accompanying release of massive amounts of gaseous sulfuric compounds into the stratosphere. There they slowly spread out, along with the volcanic dust, until they enveloped the whole planet in a dusty twilight sky.

Then the real problems began! Because when the sulfur compounds started to interact with the water molecules in our stratosphere, sulfuric acid began to form. This generated conglomerate sulfuric compounds that began to act like a sunscreen, and the entire planet was soon plunged into winter.

First all the plants died from the cold and lack of sunlight. Along with them, the insects and cold blooded creatures died off. Then the starving warm blooded animals began to perish too, and the eruption of Mt. Toba became a life-defeating "extinction event"!

But, we have also learned from the progression of these catastrophic events that the geoengineering of our planet's climate to offset global warming is possible. This can be accomplished by releasing sulfate aerosols from high flying inflatables and aircraft into our earth's stratosphere to create a planet cooling "sunscreen".

In other words, our situation is far from hopeless, because we already know what we need to do. Nonetheless, we also need to organize as a global community, in order to develop a logistical plan (of global proportions). But doing so may well prove to be the most difficult part of our international task!

Does it not then seem both reasonable and desirable to honor the processes that assist in the fuller emergence of an immortal consciousness into our all-too-short and limited experiences of life? But the kind of change that is needed must also emerge as the will to change our currently limited perspectives of who and what we truly are.

However, while we certainly need to understand the lessons of the past, and pass on our successes to the coming generations of humanity, we have nevertheless reached a rather dangerous impasse!

The concept of living responsibly, which has been rightly called our "social contract" with succeeding generations, has perhaps never been of greater importance. But what many here in Silicon Valley's present day "Consciousness Renaissance" atmosphere now understand, the consciousness upon which all concepts rest, is now of equal if not even greater importance than it ever was, if we are to reach the next level of humanity's naturally unfolding destiny!

The founding fathers of our young countries recent past surely never envisioned many of the challenges that our modern world presents to our current welfare, and humanities future survival!

In the 1960s the Grateful Dead rock and roll band euphemistically observed in song, to paraphrase; *"When ya' can't go back and ya' can't stand still, if the thunder don't get ya' then the lightnin' will!"*

Perhaps this is really a statement about the human condition in general, but it can also be broadly applied to our current problems. In other words, while preserving their vital fundamentals, many of the models of the past need to either be upgraded, or perhaps even abandoned, in order for modern governments to more effectively represent our higher nature; although "deeper" Nature may in fact be more to the point!

But besides what's already been said, when I am asked, "Ok, I've heard enough 'doom and gloom', what can we do other than watch the world slip away into chaos and destruction?" My answer is almost always the same, "Be vigilant, personally investigate, and diligently inform yourself with accurate information, and only then can you effectively exercise whatever power that you may still have!"

In other words stay informed, stay in touch with your elected officials, and **vote.** The "computer revolution" gives the astute average person an uncommon advantage in this regard!

Listen to those of our leaders and doyens "of good record" who yet retain their human decency, independence of thought, and personal integrity. And have some faith that Mother Nature is essentially life affirming and even "on our side", if only we are on "Hers"!

Exercise your economic power in the market place to effect positive changes in the ever growing power of the corporations that nonetheless must cater to the buying public in order to prosper and survive. Support and maintain our workers unions, which are still essential to maintain the fair balance of economic power that preserves our middle class. (Which every chairman of our beleaguered Federal Reserve Bank, from the heroic Paul Volker to the failed "Free Market" Ayn Rand advocate Warren Greenspan, have affirmed is critical to a stable American economy).

But do not use violence, otherwise you will lose whatever advantages that you may currently have, because if you are perceived as a physical threat, you will most likely be "crushed"! (*Hoc pro* אחי, Raymond J.).

Do not attempt to abandon the working systems that may simply require reformation, but work to influence them positively and change them from within. But beware the seductive and misleading control tactics of the deeply entrenched emerging global oligarchy.

Support public education, science and technology because they are currently nature's best instruments to hasten the inevitable growth in the evolution of our consciousness, and the preservation of our planet!

Finally, our current American "Revolution" must begin as a grass roots "evolution" in consciousness, the practical first application of which is the open sharing of information through enhanced global communications. Indeed, in a very real sense, when taken altogether, these events are actually the widespread effects of everyone's emerging essential Awareness!

"Pray" deeply, and meditate regularly, with great faith in the Divine nature of our own immortal Awareness, and learn all that you can about the true nature of your own consciousness and the great, if not "sacred", mystery that is life.

But always remember, that your prayers are never finished, until you act on them, without intention! And finally, understand to the "depth of your very bones" that positive change is not only possible, but that you too are not only an important, but even a necessary part of it!

And last, but certainly not least, beware the sometimes disingenuous advice and self-serving manipulations of the intransigent primitive monetary institutions and recalcitrant political powers and persons that are still selfishly trying to dominate and rule our planet, but are instead inadvertently nudging it towards a sure destruction.

"The best defense against bullshit is vigilance, so if you smell somethin', say somethin!" Which seems like sound "expletive" advice, indeed! From the last "Daily Show" that was hosted by the incomparable Jon Stewart!

XVIII

But returning to our initial subject (marijuana), there is a significant difference between inexpert dabbling in the spiritual uses of ethnobotanical psychoactive substances and their more traditional use to facilitate desired states of consciousness that can accelerate the enduring biochemical changes that often accompany the state of mind known as Enlightenment.

However, it is also important to understand that our supposed "Enlightenment" is really just another (eventually passing) state of mind, because it is only known by our own ideational person, who is also an object of consciousness. And of course, whatever appears within our consciousness is also our object, so we (in Essence) cannot possibly be it either!

It is also certainly mere wishful thinking to believe that Enlightenment can be achieved by taking a pill, or by simply smoking, snuffing, or ingesting a psychoactive botanical! Their recreational use, especially among those who are still physically and mentally developing, can interfere with their education, wellbeing, and eventually perhaps, even with their later livelihood!

For earnest adults, in the beginning, psychoactive drugs may or may not be of assistance in opening our inner doors of perception. But they can only take us so far, and soon will likely become just another foreign obstacle.

Indeed, perhaps it is best not to get involved with them at all because we need to exercise good judgment in life, and drugs tend to unduly narrow our focus and cloud our reason. But habit can also become a "drug". Continuing to use force, for instance, rather than reason, to fix the world's problems only makes them worse!

The will to action, which the Yogi's call "Rajas", must be guided by informed compassion rather than poorly informed reaction and mere religious, political, social or financial self-interest. The results of doing otherwise are the all-too-familiar turmoil, violence and instability of today's rapidly changing world.

But sadly, so long as we remain Self-ignorant this is almost invariably what transpires, because humanities illusory and intermediary "animalistic" ego is still ignorantly in charge.

Nevertheless, the choice of cooperating with, or surrendering to, Mother Nature's plan for a more evolved universe, or not, is always ours to make; albeit perhaps in illusion. But of course, everything is always in the unseen "hands" of the consciousness manifesting Divine!

XIX

"Historically speaking", the "fully Awakened" rarely discuss such things, choosing rather to help the individual who comes before them to learn straightaway about the true nature of their own consciousness. This is the traditional mark of an accomplished, fully Awakened "Master", whose words shoot like arrows past our intellects and our illusory egos.

They sometimes strike directly into the very heart of our delusions, and then they have the power to "kill" our habitual over identification with our ideational person(s) and universe(s). As they reawaken the sleeping apperception of an occult, but omnipresent, and inherently sentient Reality.

But the great Masters of the past who taught the "negative way" to enlightenment, lie buried in the past. We now live in a different and much more complicated time that requires that we take positive action together as a species to ensure our planet's survival. And we will likely succeed only if our actions are based upon a deeper and less selfish, if not grander, "Self-understanding".

However, even if we don't yet know it, we have always been a "sleeping giant"! This is because whatever is experienced within our consciousness is actually impersonal Universal-Consciousness experiencing *it*-Self. The Enlightened and the fully Awakened clearly see this, which gives them the edge in communicating their insight.

But for the rest of us it is difficult to see that there is only our experience. There is no "object", and no "subject", because both are simply an illusion. And in their absence, relative space and time are also meaningless and effectively nonexistent.

Nonetheless, our inner universe of convincing illusions is also the play of Universal-Consciousness! Yet words cannot contain this perception and our divided mind cannot fully understand it. Because it must be directly apperceived from a perspective that lies outside of, before, and literally "in between" our dualistic consciousness!

It is good to keep this caveat in mind, while not dwelling overmuch upon it, because if we do, the dwelling itself rapidly becomes just another illusory object. Which then demands the involvement of our illusory subject/objects (i.e. "egos"). Thus binding our ideational subjects and objects conceptually together in an even tighter embrace!

Such is the essential nature of any honest self-inquiry into our so-called "worldly bondage", and the resulting apparent bondage of our planet to violent human confrontations and perhaps even the eventual annihilation of all life. Which is itself just another illusion, but one that we can no longer afford to ignore!

It seems difficult indeed to see that our own true Self is what we truly seek, and that humanities continued widespread Self-ignorance (and ignorance of our true place within Nature) may well even lie at the very heart of the world's current woes! But this difficulty too is only a persistent illusion and even the determined "Seeker" eventually grows tired from the effort required to inadvertently keep it alive by constantly pursuing it!

And when it is eventually surrendered, we are, simply put, whatever then remains. But if we only act on our own behalf, in the not-too-distant future, whatever remains intact within the Web of Nature, may not be worth either saving, or surrendering!

At this juncture many Jews begin to recall the wise words of the 11th century Sephardic astronomer, philosopher and Kabbalistic Master, Mūsa ibn Mayūm (Moses Miamonides), which I am happy to share with you at the end of this rather sobering portion of our shared wilderness adventure; ים אני רק בשבילי מה אני... *Im ani rak beshvēli, mah ani...* "If I am only for myself, what am I...".

An observation that is no doubt well understood by my fellow former probationary Zen Monk, Jerry Brown (i.e. our current California Governor). Because, evidently, the fate of this lovely land of yours and mine, lies in no one else's hands but our very own!

"This land is your land and this land is my land, from California to the New York islands, from the redwood forests to the Gulf Stream waters, this land was made for you and me... ". Lyrics from, "This Land Is Your Land" by Woodie Guthrie, recorded in 1944.

Indian Tobacco

In Another Time And Place, that in my mind today, seems "long ago" and "far away", when I was a teenager riding out on horseback to round up some stray cattle, I lit a cigarette that I had "borrowed" from my mother's purse. "Paco", the ranch foreman, slowed his horse and reigned in beside me.

He didn't say anything for a while and we rode silently side by side until I'd finally smoked about half of the purloined cigarette. I fully expected him to lecture me about the perils of smoking, as my parents or teachers probably would have done, but he did not. Instead, Paco explained that tobacco is a plant that is especially sacred to the American Indian.

He told me some of the Pomo Indian legends about tobacco, the reverence that it is accorded in tribal ceremonies and how it can heal our body and spirit, or even kill us if it is ignorantly misused! "Tobacco", he explained in length, "is strong medicine!"

Within a few days, after pondering the ancient "Wisdom Words" of his slowly dying tribe, I stopped smoking. Somehow the understanding took hold inside of me that the casual use of sacred *"Tōbah-Chō"* is most certainly "not a good thing"! Smoking tobacco without understanding the significance of the plant, and its proper use, seemed like an insult to American Indians, and I never smoked again!

I

The great Advaita Sage Nisargadatta Maharaj, who sadly passed away from throat cancer in nineteen-eighty-one in the same year that I got together with my lovely wife, advised people who are seeking Self-Awareness not to get overly involved in what he called "social work".

He also taught that quitting your job and retreating from society in pursuit of spirituality is not only fairly useless, but will likely only cause grief for you, your friends, and your family!

Moreover, he also advised learning all that we can of the special knowledge involved, while earnestly and unceasingly applying it in the processes of self-inquiry and meditation, but to otherwise; *"Pursue your work and lead your life as usual!"*

It seems fairly clear that he was advising us to give the world its due as we quietly cultivate our own understanding. Which seems like sound advice indeed! But we also seem to have no choice but to pursue our work (albeit in an illusory world!).

And insofar as the world's "spiritual" needs are concerned, just like any of the other problems that we inevitably encounter in our experiential world, it is easy to point out problems and complain about them! But other than what we have already discussed, what can be done?

Perhaps teaching our children to understand and respect ethnobotanicals after educating ourselves about their traditional spiritual and medicinal uses, is a good place to start.[(94)]

It also seems absurd to keep wasting so much time and money on prohibition, harsh penalties and often life destroying prison sentences as part of an impressive, but in many ways misdirected drug war.

"Impressive" because of its cost, scope, violence and the sheer number of people who are arrested in connection with drugs. Who are then hurried by questionable "plea bargaining", or "three strikes" judgements through an already overburdened criminal justice system. (Only to languish in overcrowded, and often now "privatized" prisons that try to make a profit from the human misery of incarceration!).

And "misdirected" because the so-called "War on Drugs" seems to have actually increased the street value of the drugs that it has striven to keep off our streets, while not appreciably reducing their availability.

II

Apparently the "War on Drugs" is having the same unintended reverse effect that Prohibition in America, really the "War on Alcohol", created during the nineteen-twenties. Like then, there is now so much profit to be had, that gangs and cartels spring up as quickly as grass after a wildfire. But just like a Hydra, as soon as one head is removed, another soon crops up to take its place!

And those who survive generally become even richer and more powerful, once their competition is finally eliminated, which often happens as a strategic "tip off" to law enforcement. In today's world, drug crime organizations have become so wealthy, powerful, and influential that unfortunately they, rather than law enforcement, with all due respect to the inimitable Elliot Ness, are often now almost "untouchable"!

Because the same simple economics of supply and demand that drives national economies also regulates underground economies, prohibition, harsher penalties and clandestine, or even devastating Special Forces "military style" attacks on the supply side of the underground drug economy have not worked out very well.

If anything, the problem has been exacerbated by the use of reactive force, rather than a reasoned and informed response to the root causes of the drug trade.

This is not a critique of law enforcements "boots on the ground" method, but of the overall effectiveness of the mindset that is behind the methods that are currently being employed. The reasons for its shortcomings are relatively simple, even if the scope and operative aspects of trying to manage the international drug trade are not.

Because drugs are generally small, not very fragile and are often quite malleable, they are also easy to disguise, and to smuggle. Furthermore, most people can either be bought or threatened into cooperation. Tactics which wealthy and ruthless drug lords are apparently quite expert at!

Bribery, blackmail and intimidation are very effective tools in enlisting the right people to hide and facilitate the smuggling of drugs. If something goes wrong, they are either murdered or simply "disappear" and a new entryway is created elsewhere.

In a country as large as the United States, our shipping ports, airports and borders alone present so many opportunities that we have to be satisfied with whatever drug shipments that our already overburdened network of law enforcement can intercept!

But it is easy to be misled by the reports of drug seizures in the tens and even hundreds of pounds, or even tons, until you consider that only a small fraction of the illegal drugs that enter our country are ever discovered!

It is also easy to be further mislead by the large numbers of "small-time" drug dealers who are subsequently arrested as they move their cartel smuggled drugs on our streets. For here too, only a small number of these overabundant retail drug dealers are ever arrested, and the smart ones often profit from the arrest of other small-time drug dealers by taking over their competitors' territories and customers.

The illegal drug business is particularly "cutthroat" on this level and strategic tips to local authorities to eliminate the competition are not uncommon. Also, but not as common, are the sensational drug-trade battles between rival cartels and organized gangs.

III

Blood-spattered, bullet-riddled, burned, and mutilated bodies make sensational press, and the fear generated lends public support to the War on Drugs. Sadly, on occasion, some brave and self-sacrificing undercover narcotics agent is discovered and murdered. This makes a big ugly media splash and fuels the fires of this ill-advised, so-called "war" that took another law enforcers life in the line of duty.

Their bravery and ultimate sacrifice are never in question. But if we want the whole truth, we must also ask their families, fellow officers and friends, whether their self-sacrifice is worth the sometimes meager and often transient suppressive results that it may have obtained, since they are the ones now forever deprived of their consortium (i.e. overall companionship).

Sadly however, it is not unreasonable to assume that the bloodletting on all sides of this intractable "war" will continue unabated unless a more thoughtful long term method is applied. It also seems reasonable to say that these fallen heroes, dead drug dealers, overdose fatalities and unfortunate prisoners are all victims, including their families, in this ugly faux "war" that no one is really winning!

We can only hope that the greedy emerging global oligarchy (like the dangerous incipient decay that can secretly lurk inside a tree's trunk) is not covertly involved. Because they are apparently quite capable of functioning "above the law", by trading in an ever mutating variety of "black pools".

Which are the current monetary "instruments of obfuscation" that are being purposefully created by the mathematical and techno-geniuses who are being carefully courted, and often hired at great expense by many investment firms from prestigious universities like Silicon Valley's peerless Stanford University and Boston's inimitable "MIT".

Without perpetually updating our progressively irrelevant investigative methods and technologies, we really have no way of knowing all of the financial connections that may exist within an increasingly inscrutable and ever faster array of computerized financial instruments of trade.

Therefore, without being an "alarmist", we could even reasonably ask if this could be another clever way of clandestine "money laundering"? How could we possibly know if even our own underfunded, evidently mismanaged, and understandably often beleaguered, F.T.C. (Federal Trade Commission) is unclear, if not totally unaware, about what is actually going on! And while generating ever longer "prime numbers" may protect us from computer hackers and cyber warfare, unfortunately it cannot protect us from our own greed!

The real fuel of this ostensible "Drug War" is not the drugs, or even the people who sell or use them, it is ignorance, poverty and the strong human will to survive and prosper. Because the human will is easily corrupted when avarice and survival fear enter the picture.

But is this not the case in any sort of war? This is why the wise old Hiroshima suburbs Roshi, who sat lamenting upon a hill of smoking rubble that was once his Zendo "Meditation Temple", sorrowfully exclaimed to a Nippon News journalist; "What indecent foolishness is this thing, that even decent men sometimes proudly call war!"

In the end, perhaps the best advice has already been given by many of our own experts, wisely given and yet unwisely dismissed! They have repeatedly advised that we arm our children with accurate, non-hyperbolic information about drugs. And they have also advised the decriminalization of "soft" drugs like marijuana.

Through better education and early intervention (marijuana detection kits are in most of our local Drugstores now), the demand side of the illegal drug trade's economic loop will eventually decrease. And by incrementally decriminalizing drugs, the outrageously high price of illicit street drugs will also drop.

No longer able to reap such huge rewards for the considerable trouble of drug growing, manufacturing, and trafficking, organized crime will eventually be forced to turn to other more lucrative pursuits. Rather than farming cocoa in South America for cocaine, coffee or bananas may do!

IV

The Taliban were a horrible scourge to the Afghani people, and they literally tore apart not only their country, but also their ability to fairly govern themselves by forcing them to live under a repressive and twisted "Medieval" sort of fundamentalist Muslim regime.

But, ironically, because the use of opium is prohibited by the Islamic religion, the foreign occupying Taliban power prevented the Afghani farmers from growing their relatively small crop of opium poppies.

However, after we soundly defeated the cruel occupying Taliban and drove most of them from the country, the farmers began to grow their poppies again in order to feed their poor starving families! And with no one to control them, they were soon growing and selling ever greater amounts of opium, already processed for quick conversion into heroin.

But unfortunately, because of widespread governmental bribery and the lack of sufficient resources and manpower, the Afghani government is doing little to stem the rising tide of heroin that is steadily flowing out of their country.

Thus, rather than allowing traditionally unregulated poppy cultivation in Afghanistan, which only produces more illegal heroin to be sold by terrorists, legally empowered farmers could as well sell their (now taxable) opium for legitimate medical use and the financial welfare of their war impoverished communities.

And they could also be compensated in the same fashion that American farmers are for their produce, in order to ensure a stable moderate medical market for their opium products, and the ability of the Afghani farmers to sustain themselves in lean times. Nonetheless, in a very real sense, especially in the beginning, we will probably "get what we pay for"!

(Versus a probable rampant "plague" of heroin addiction, with its accompany astronomical personal, medical and insurance costs, it would simply be a wise investment in the health and welfare of our nation!).

Indeed, we do this in America not only to stabilize our own economy, but as a wise investment in the survival of the hard working people and agricultural businesses that feed our nation.

When we invaded Afghanistan to free it from the Taliban, we also became responsible for ameliorating any substantive negative effects that pursuing our enemy into Afghanistan has had upon the country.

Moreover, if we continue to do nothing, the impoverished Afghani farmers will simply return to the practice of growing their lucrative illegal opium poppy crops, and the Western world may soon become inundated with an almost uncontrollable flood of heroin!

And to make matters only worse, the Taliban regularly steal any excess profit from the farmers, whom they now allow to grow poppies, in order to finance their terrorist strikes on the West!

But an extensive period of oversight and training for alternate and sustainable ways for Afghani farmers to supplement their income, and the creation of the financial instruments that are necessary to establish and maintain international markets for their products, will likely need to be provided, and supervised.

And we need look no further than the already existing "World Bank", which specializes in just such enterprises, for a practical solution. In the long run this, or something much like it, provides a better and much less costly solution than continuing to use ever more escalating violence!

Indeed, what large "pyramidically" organized American crime syndicates, like the infamous hierarchical "Cosa Nostra", did after the repeal of alcohol's prohibition is rather telling. The big crime syndicates turned, for example, either to gambling, or to some other morally "gray-area" enterprises, which while perhaps not socially ideal were nonetheless now taxable.

But it may also be the real reason why the terrorists and drug cartels secretly don't want the profitable illegal drugs of our times to be more effectively managed! To understand the intransigent nature of these massive problems, like a financial detective, we need only "follow the money".

Yet their solution is evidently just as difficult as the problem itself is and, as ever, "the devil is in the details"! Moreover, insofar as our planets eventual destruction from global warming is concerned, drug lords, "God Fathers" and even gangsters have families too, and troubled times sometimes creates "strange bedfellows"...

V

Our experts repeatedly tell us that drug abuse is essentially a health and socioeconomic problem that will most likely remain unmanageable if it is treated merely as a criminal one. It is simply an exercise in common sense to understand that drug related crimes involving theft and violence, like any other such crime that involves our persons and property, requires immediate legal action!

But recreational drug use is apparently a more subtle problem that is not only not so obvious to our more "common sense" mind, but that our countries best trained medical minds have continually maintained is best dealt with through education and expert medical intervention.

Not surprisingly, as it turns out upon closer inspection, those who turn to drugs to make themselves feel better need psychological counseling and psychopharmacological intervention. Rather than an unceremonious trip to a dehumanizing prison that will probably soon make them even more disturbed than when they went in!

Because the most incorrigible criminals are the unstable sociopaths and psychopaths who will likely commit crimes with or without drugs.

People who use drugs make a choice to do so, so the firm line between "victim" and "perpetrator" is blurred and there is little evidence that this choice is positively impacted by unduly harsh and uncompromising law enforcement.

But while petty theft and muggings to support drug addiction must always be vigorously pursued and prosecuted, mere arrests and incarceration are an ineffective means to manage the continued "abuse" of non-prescribed and prescription drugs. But what does seem to have a promising long term effect are education, medical intervention and the creation of economic opportunity and fair paying jobs.

Simply put, socioeconomic hopelessness, and even our often well intentioned mismanagement, are the major reasons why gangs, drug abuse, drug arrests and drug related deaths in America happen mostly in our ghettos. Indeed, when we witness the rare televised interviews of the members of many of the more successful street level consortiums, it becomes quickly obvious that they are generally the intellectual equals, in terms of native intelligence, of the people who are interviewing them.

In the end, we are often left with the sad realization that but for the crime that they are circumstantially compelled to commit in order to survive, feed their families, and attempt to prosper, that this is exactly the kind of people with the courage, wit and drive that our country so badly requires in its current time of need!

For many of the same reasons, the countries that produce and smuggle drugs into America are mostly comparatively poor countries. Thus a more circumspect look at our countries current drug problem reveals that it can be better managed by attacking its vital socioeconomic roots. Rather than turning our inner cities into heavily policed ghettos, which soon become interminable "battle zones", or by trying to build a "fortress America" to keep the drugs out.

Because clinging to false truths, whatever their origin or our investment in them, will only continue to exacerbate the problem!

The veridical truth, is that the ever growing stream of "blue-sky" (wishful thinking) tax assets that is being continually invested in our current so-called "Drug War" does not generate economic prosperity. Simply because it does not help to "grow our economy" in the same way that peace time, profit based investing in American businesses does.

And while legitimate "wars" are sometimes necessary for the common defense of our nation, pursuing them when better ideas and less costly alternatives are available, only adds to the world's human misery, as it increases our already staggering national debt!

VI

At a time of economic uncertainty, job loss, health care and environmental crisis, ever looming inflation, governmental gridlock and plummeting property values in many parts of our nation, we need to explore more economically feasible methods of resolving our multifaceted national problems.

Our better informed experts have been advising, for some time now, that poverty and the lack of economic opportunity, rather than financial greed alone, are the basic causal factors that fuel illegal drug trafficking around the world. Perhaps it is time for us to incorporate their advice in crafting a more effective and sustainable method to help manage this runaway problem?

Sometimes it is unwise to mix metaphors even if it creates arguments and catchy phrases that we believe might support our point of view and seem to make it more convincing.

Inexact truths may win an argument or an election, but generally lead to ineffective solutions in our everyday world. There is a big difference for instance, between law enforcement and "war" and the confusion that is being generated is costing our nation dearly![(95)]

But great leaders, like great scientists, geniuses, or even great nations, sometimes make well-meaning but nonetheless "great" mistakes! However, this is how humanity learns and hopefully progresses from learning about whatever went wrong.

This seems so obvious that I won't belabor the point with more facts and explanations. Indeed, even the writing of this relatively dark missive is somewhat unpleasant and deciding to include it in an otherwise science and metaphysically themed book has not been an easy decision.

Moreover, writing about possible solutions to our nation's problems without appearing to be either "preachy", judgmental, condescending, or simply "opinionated", without alienating anyone, is a rather dubious task! But as Americans we all share in a portion of any blame, just as we are all also burdened with finding and implementing solutions to our countries problems. And evidently, just this sort of participation is a good citizen's burden in any successful democracy!

Nonetheless, regardless of wealth, station or opportunity, there is certainly a lot of self-medicating going on to ease the impact of ignorance on the world. Evidently not putting higher truth into action, wherever and whenever it is needed, only contributes to our global misery.

Therefore I will take a chance that you will indulge my dabbling about in troublesome matters that are apparently neither scientific nor metaphysical, at least in their outward appearance, but whose world-wide urgency nonetheless demands our concerned citizens national discussion! For as Socrates long ago correctly observed, *"You cannot teach a starving man philosophy!"*

And until we are able to extricate our poor beleaguered country from the mire of the recalcitrant problems that it is currently deeply mired in, it will remain difficult to get on with the crucial business of working with other nations to preserve the world's economic system and to save our poor suffering planet! But we will never be able to accomplish this, until we include the compassion of the human heart in our visionary plans to do so!

"The very essence of our democracy is rooted in the worth of the individual that life has meaning that transcends any manmade system. And love is the greatest force on earth, far more enduring than hatred or the unnatural divisions of mankind."

J. Edgar Hoover

The Sky is Falling!

"The Sky Is Falling!" cried Chicken Little, and everyone in the barnyard suddenly became terrified! (An excerpt borrowed from, "The Remarkable Story of Chicken Little" by John Greene Chandler, published in 1840).

In this charming child's tale lies a rather sobering adult message. When we don't take the time to investigate the source and validity of anyone's alarming claims, often with the best of intentions, we can easily make the same foolish mistake as poor Chicken Little did!

But such is sadly the case with the ignorant refusal of a few well intentioned parents to inoculate their children against clearly deadly communicable diseases!

This is an especially bad choice, where diseases like small pox and polio are concerned, since they have been nearly eradicated by the same life-saving processes that these parents are ignorantly denying to their own children, and thereby selfishly endangering everyone else's.

Such irresponsible choices place these ignorant, but well intentioned, parents in the same category as that infamous harbinger of deadly disease once known as "Typhoid Mary"!

Indeed, beyond the defense of any individual's personal civil rights, when children's guardians place them and others children in harm's way, societies only responsible choice is to compel their compliance by force of law.

Democracies are understandably sensitive to the rights of the individuals that comprise their citizenry, but nonetheless often difficult decisions must occasionally be made, for the greater good of the democratic society at large.

In like vein, the vitriol and boycott against GMO's (genetically modified organisms) harms not only children, but adults as well! In fact this is an especially silly position, because it is often based upon a profound ignorance of either the science involved, or the antiquated *halachic* (i.e. temporally related ritualistic) portions of Jewish scriptures that non-Jewish religious fundamentalists often base it upon: *"Thou shall not mate different kinds of animals, plant your fields with different kinds of seed, or wear garments made of different materials."* – Vayikra (Leviticus) 19:19.

Because, when we are finally able to correct the incipient problem of an emerging agronomic monopoly (which tends to "crop up" with any entirely new agricultural invention), GMO's may well offer humanities best chance of repairing our devastated tropical forests, as well as our species eventual colonization of other planets, moons and planetoids.

But they also offer us the science based hope, that we will be able to continue to feed our displaced, distressed, and most likely *ever-growing* world population! As global warming steadily pushes our planet's trees, plants and crops to the edge of their adaptability.

And while I have never met a tree trimmer who is not also an environmentalist, every arborist, consulting arborist and tree researcher that I personally know also understand the compelling need, for the greater good of all humanity, to continue the genetic research for solutions and practical applications in these venues.

Arborist researchers, like Dr. Joe McBride of UC Davis, for example, have recently begun to compile important practical data on the specific urban and suburban adaptability of key tree species to global climate change. Firstly to identify species that are doing well in already hot areas that can be planted in areas that will soon be reaching the same temperatures.

Next we will be researching what can be done when the limits of native tree species adaptability in the temperate Taiga and Tropical Rainforests is finally reached. In order for them, and ultimately even the human race, to survive and hopefully even to thrive, in the warmer and more carbon dioxide rich atmosphere of our not-too-distant future.

But while not all of an arborists research is exhilarating, it is often extremely important for the sustainability of our current civilization. Such as arborist Dr. Nelda Matheny's (of "Hortscience") research into the safe and effective use of recycled and reclaimed water during California's current period of "ENSO" sustained drought (i.e. global climatic events that are linked to climatological anomalies).

But one of the most poignant examples of the problem, which is being exacerbated by the current unedited glut of natural science misinformation and excess of inexact environmentalist hyperbole, is that of the many well-meaning environmentalists that inadvertently helped to halt our countries nuclear power production of electricity, when we finally stood poised on the brink of building a new generation of safe and non-polluting "breeder reactors".

Ironically, they were playing right into the hands of the same oil and coal industries that in the course of a few decades have placed our entire planet in jeopardy (with global warming). From the almost inconceivable amounts of invisible carbon dioxide that is being dumped, in ever greater amounts with each passing day, into our planet's already overburdened atmosphere!

As the time worn, but nonetheless still true, adage goes, "What we don't know, can hurt us!"

The Riddle of the Sphinx

"What walks on four legs in the morning, two during the day, and three in the evening, before the setting of the sun?"

~ Sprach die Sphinx ~

ALMOST NOTHING COMES UPON US WITH SUCH UNEXPECTED FINALITY as our own approaching old age. Everyone knows of its inevitability, yet we are almost never prepared for the advent of its eventual restrictions. Yet the measured aging of our bodily vehicles is not without its benefits.

The strong lines that moor our awareness to the flesh are (barring the onset of a serious disease) "gently" loosened, as nature prepares to eventually cast them free. But, despite its many charms, our apparently dualistic life of ever varying individual consciousness is at times a burden that is difficult to bear!

This is why, during our "halcyon" youth, and our "efforting" adult years, we generally lose ourselves in activity. Eventually however, we are forced to slow down, in our latter years. It becomes a time to deepen our understanding as we begin to "take stock" of our lives. We are given the opportunity to prepare for the inevitable decline and physical death that awaits all of our psychosomatic vehicles.

If we are lucky, we might even be able to step out of our unconscious involvement with the ongoing stream of our individualized consciousness, which holds us ever-so-tightly in the conceptual grip of our objectified person(s) and universe(s).

These are our purportedly "Golden Years", but not simply because we may be fortunate enough to afford a rest from our labors and enjoy the supposedly "good things" in life. Indeed, it serves us much better to realize the transient nature of our life while we are still young and are able to begin to deepen our understanding, even before advancing age finally forces it upon us!

The answer to the riddle of the Sphinx is "humankind". An infant crawls around on its hands and knees, while an adult walks upright on their two legs, and the elderly among us must often lean upon a staff as they negotiate their latter days. While the Truth that sets us free from the illusion of our mortality, is the imperceptible "Staff" of our own purely immortal and Self-sentient Awareness.

The Shaman's Way

Our Life Force Constantly "Talks" to our sentient "Staff of Life". Moreover, there are no impenetrable mysteries in nature, only ignorance in our understanding of *Dai Shen-Zen's* "Great Nature's" ways. Every living thing is in constant inner communion with every other living thing.

But "person" living things talk and think of things experienced in their (only apparently) individual minds and we soon forget how to listen to (and to "speak" with and through) the living web of impersonal Consciousness that connects all things together in nature.

We have sadly forgotten how to fully appreciate or knowingly participate in the "dance of life" that sustains the great tribe of the living through the endless parallel dimensions of Universal-Consciousness that underlie the tiny corner of the multiverse in which many still erroneously believe we alone exist, as solitary physical beings.

But even the "Life Force", that brings forth life from within apparently insentient matter, is finally comprised of what the naturalist Shamans of tribal humanity still understand to be the "Great Spirit" (i.e. the Supreme) that lies beyond the bounds of the multiverse, and yet is always present in even the most humble of things. Therefore, when something is taken from nature, something else is always returned.

In our sensorially detectible universe this appears as the physicist's "Conservation of Energy" law. Knowingly participating in this process is the ᏣᎳᎩ *TsaLaGi* "Cherokee" Indian way, and according to my late Keetoowah uncle Leonard, it is the doorway to the nearly forgotten "Shaman's Way" as well!

Yet having already forgotten so very much after our apparent "fall from grace", life soon appears to be our very own private "struggle for survival" amongst other desperate creatures that are simply doing the same, and such teachings begin to seem quite frivolous indeed!

But appearances can also be deceiving. So what might really be going on, on a deeper level, which could shed more light on our planet's rather curious situation of apparently conflicted life and mind?

On The Battlefield of Split-Mind

ON THE BATTLEFIELD OF *manas* "mind" the only way to win the "war" of Self-awakening is to let the Force of Life *it*-Self fight our losing battle for us! Evidently, the powerful faith of maintaining a lucid abidance in this, our ever witnessing Self, prior to our dualistic consciousness, is the only objectively unobstructed Way (or "Tao") to do so. Such is the imperceptible virtual "battle" of maintaining an enduring faith in our deeper, truer "self-sentient" Selves.

Thus, over time, water washes even the hardest of stone back into the silicone molecules of its constituent sand. But even molecules and atoms eventually reduce back into the quantum "Wavicle Sea" that emerges from within our universe's immeasurable pre-primeval Singularity of not "not-nothingness".

Indeed, it is only from this indefinable beginning that it, and all things, eventually emerge as "objects" of consciousness within our binary minds, thus bringing the universe more fully into "existence". But even our dualistic subject-object mind is not really ours, because it always belongs exclusively to the powerful כוח של חיים *Koach shel Chayim* "Force of Life" that brought it forth from deep within our universe's initial primeval Singularity.

This is the occult self-sentient Force which finally brings the entire universe into "existence" as a manifestation within our evolved consciousnesses. And eventually, this now "universal" Force, is what "watches" our naturally evolved ideational subject-object consciousnesses dissolve back into the not “not-nothingness” from whence it once rather mysteriously came!

In this we presently have very little choice, when the body that supports it finally wears out and stops producing the ephemeral consciousness “material” that we witness as our evidently “subject/object” selves.

However, this is clearly beginning to change (see “Jñaneshwar”, on pg. 547), but until that time it remains difficult to grasp that we are actually imbedded in an only evidently “surrounding” universe. An only apparently “exterior” universe that is in veridical truth also comprised of virtual “interior” space-time bounded experiences! Which invites us to question how such a quizzical state of affairs could have possibly even begun to transpire!

Supersymmetry Breaking

A “Perfect” Primeval Singularity is by definition “unbreakable”. Moreover, apparently by characterization, and in fact, our universe’s initially “perfect” Primeval Singularity was also, most likely, of an “impersonal” Awareness (i.e. יהוה, الله, G-d, etc.) comprised! This is because our universe’s perfectly balanced, and thus unbreakable, infinite pure potential can only be expressed *virtually*.

With all due respect to the intrepid discoverers of the “Higgs Field” and the “Higgs Boson” (and especially to the brilliant Nobel Prize winning Dr. Higgs!), what purely physical method could possibly break the perfectly balanced and infinite inertia of our universe’s all-inclusive primeval Singularity?

Especially when we must admit that the laws of physics (as we currently understand them) could not yet even possibly exist! But nonetheless “something” must have happened, because we are somehow currently having this discussion!

This leaves us with the curious necessity of trying to find a reliable way to include consciousness in the quantum equation of the fundamental mechanics of our universe. Which is borne out in the necessary interaction of our observation with quantum wavicles in order for them to “exist”.

Thus the quandary of multiple universes being "sparked" into existence by our every experience is apparently simultaneously created, both locally and non-locally. And it is done so within an endless quantum "sea" that is comprised of countless "Consciousness Force" initiated perturbation wavicles or strings of "vibratory" energy that apparently "float" (much like the surface of tension constrained water) upon the sub-quantum Singularity of Universal-Consciousness that our entire multiverse eventually emerges from. (See "Higgs Field", pg. 543).

Yet if this is indeed so, then what could the likely locus of interaction of our essential Awareness with the extended and evolved universe of our simultaneously "inner/outer" experiences possibly be!

This is a contrary question to be sure! But, one that with a little diligent investigation, we have discovered, has already been asked and answered, many times, in the secret revelatory language and occult metaphysical traditions of the world's great religions. But only if you hold the right keys!...

For example: ... התו ובהו וחושך על פני תהום ורוח אלוהים מרחפת על פני הסימ
והארץ היתה *Ve-ha-ahretz hytah tōhu vavōhu, vechōshech al pney tehōm, veruach Elohēm merah-cha-feht al pney hamayim*... "And the [expanding] 'Singularity' was [still] void and without form, and darkness was on the face of its formlessness, and [then the Oneness of] the Supreme [that was latent] within the [still formless] Universal-Consciousness began to expand [the Universal-Consciousness] into the incipient diversity of the sub-quantum sea...". A Kabbalistic "Midrash" of Bereshit (Genesis) 1:2.

It does not take a genius, to understand that sacred wisdom teachings such as these, should not only be updated by including the discoveries of modern science in their definition of meaning, but that they should also be preserved for the sake of all humanity!

And moreover, that the time has finally arrived, to throw open the doors of the ancient Mystery Schools, and to share their sacred teachings with all of humanity. To help us to adjust to the currently unfolding Singularity of Consciousness as it increasingly impacts almost every facet of our lives!

For these sorts of deeper Truths do not belong to any one religion. They are by their very nature, "universal". Because they are a reflection of the Divine Consciousness that glimmers in the "heart" of the subjective eye of every single one of us! (If you still doubt the veracity of this statement, you need to check your cynicism at the door. And take another *closer look* – into the eyes of 'most any newborn 'babe!).

Only in this manner, can the lamp of this "uncommon-sort-of" עמוקה אמת *Emet Amōkah* "Deeper Truth" be passed on from דור לדור *dōr le-dōr* "generation, to generation".

Thus, if we are fortunate, during the course of our busy modern lives, we may one day "stumble" upon these ancient שלום *Shalōm* "Peace" imbuing wisdom teachings. But when we do, we may soon discover that we've just become the latest custodian in a long line of inadvertent guardians of their ancient Sacred Flame!

(Pea Memoria, Sir Isaac Newton, *illuminate homine nove!*).

Our Whole Mind

THE THREE DIMENSIONAL "in and out" sides of things meets beyond our common experience of four dimensional space-time, in the "fifth dimension" of our whole Minds. But this is not to say that the totality of our Superverse does not extend far beyond what any sort of imaginative dimensional physics or even exacting mathematical interpretation may (currently) explain.

Hence the probabilities and possibilities of our awakening consciousness nearly defies the reasonable imagination! We need only to observe the tremendous impact that even a recently emerged biological consciousness has already had upon our own little corner of the galaxy. (In an amazingly miniscule amount of time, when compared to our fourteen billion year old cosmos!) If Consciousness itself is not Divine, then whatever else could it possibly be?

Indeed, even if most people don't yet know it, our individual consciousness is much like a "dam" that contains the entire universe in infinite spacetimeless potential, just waiting for our experience to reveal its infinite presence, one spatial experience at a time!

But if this is indeed the universal arrangement of things, then what of the many teachers who claim that they can somehow engineer our "Awakening", when we are already not only conscious, but also quite necessarily aware of being so!

Samadhi

THE YOGIS HAVE DEVISED quite a trick, because *syncopē*, as it turns out, turns off our mind, leaving only our essential substrate of Awareness in its wake. But when our *syncopē* ends, so does our experience of Samadhi!

Moreover, simply repeating this exercise could in no way yield a different outcome, although it could most certainly lead to some sort of brain damage that might appear to be a durable sort of "spiritual" awakening!

(I.e. *Syncopē* is the loss of consciousness due to a drop in blood pressure and a resulting *hypoxia* "reduction of oxygen" to our brains. Which is also sometimes caused by the simple act of standing up "too fast"!).

Hence, even the epiphany of a thoughtless "nirvana" is apparently temporally bound. And evidently, since consciousness is necessary to any understanding, our understanding of consciousness is clearly necessary to any usefulness that such a temporary nirvana might provide.

Otherwise, when we first experience a night of deep dreamless sleep, we would all awaken fully "Enlightened", and moreover we would remain persistently Awakened as a result!

Evidently, ignorance is no of friend of either "Enlightenment" or Awakening, although it sadly often does serve to "line the pockets" of both disingenuous and well-meaning "spiritual" teachers alike. Ergo, as the worldly wise (often from costly personal experience) almost always observe in practice, *caveat emptor* "buyer beware"! (See, "Samadhi", pg. 572).

Consciousness and the Law

Because The Universe Is, in effect, "Impersonally" personal, our consciousness carries with it the onus (obligation) of its own social, as well as "higher" responsibilities. And evidently, since there also exists no "personal god" that can be appealed to as the final arbiter of the law, apparently the best that we can do is to make a reasoned decision that is based upon whatever "facts" may be at hand.

And since no other basis can be said to reliably exist, the best that we can do is to pursue a reasoned debate amongst factually based (differing) opinions, upon which a non-biased "best judgement" can be made.

Otherwise anyone with a self-serving agenda could claim to speak for a god that is unable to speak for himself. Indeed, the historical Catholic inquisition (and currently, the violent fundamentalist Islamic cult of "Isis") is a grand example of what can go horribly wrong with this sort of delusional "Theocratic Law"!

Moreover, when we firmly believe that we are merely a body, or even a body with an unlikely "personal soul", we often feel that we lack certain things, which leads to acquiring them one way or another. Sometimes even when it requires stealing them (like a less aware בהימות *behemōt* "animal") from others!

We Are What We Lack

We Try To Surround Ourselves with whatever we feel that we are personally lacking. But we only lack whatever we allow ourselves to feel that we do not yet already have. Moreover, this only sounds like so much "doubletalk" until we are finally ready to dismiss it too! Because whenever we are ready, we can always sink back into our-"Selves", and rediscover that the entire universe is already arrayed "before our feet"!

However, as we vigorously pursue what we believe we lack in physical and mental capacity, as well as enhanced health and longevity, we are also entering a new age of human possibilities.

A new "Singularity" that is being brought ever-more-rapidly on by the impact of Consciousness, as it continues to emerge from within our universe, as life and mind, from within energy and so-called "matter" (i.e. matter is really of energy comprised, $E=MC^2$).

But it is an unpredictable "new age" that our traditional morality, ethics and contemporary law (including our religious, political and even economic systems) are still rather ill-equipped to deal with!

Nonetheless, because the entire universe is already arrayed "before our feet", and evidently also within the grasp of the Consciousness by which it is known, just whose "feet" are we really talking about here anyway!?

The Footprint

Tracking Skills Are Often Essential to our continued survival in an unfamiliar part of the wilderness. And even within the relatively unknown wilderness of our expiring humanoid brain, there exists the latent footprint of our immortality, just waiting to be seen!

For several hours after our "deaths", apparently the electromagnetically memorialized, biochemically initiated, dendrite distributed information that contains the algorithmic-like codes of our brain, remains electronically accessible.

This is the electromagnetic "footprint", or perhaps more precisely the internalized cerebral "fingerprint", of our experiential persons and universes that is unique to each person's mind. Conventional computer technology is capable of storing and operating such a system. Evidently, information retrieval is the only substantive problem.

But since we are working with a non-living brain system, with a properly crafted legal "imprimatur" that is based upon principles of properly informed and carefully reasoned law, the doors of science and technology can be thrown fully open, to achieve its speedy resolution!

Nonetheless, big science projects inevitably require the rare vision, faith and courage to invest "big money" without the certainty of an immediate return on the money that's been invested.

And in respect to the advancement of human knowledge, although many people don't know it today, the important contribution of our past president Bill Clinton to science and technology, with the approval of the Hubble Telescope and Human Genome projects, cannot be over emphasized!

Since president John F. Kennedy's initiation of NASA's successful Apollo moon mission, this is arguably one of the greatest political contributions to science and technology that has ever been undertaken, because it has opened the doors to the exploration of the (strangely conglomerate) "inside/outside" universe in which we human beings apparently coexist!

Indeed, as a consequence we have already discovered around eight thousand likely habitable earth-like planets and we have also unraveled small-but-significant portions of our constituent human genome that are already contributing to the cure of several of humanities most recalcitrant maladies!

But what will most likely transpire, regarding the reawakening of our individual consciousnesses, and whatever they may contain of our "persons and universes", also reanimates the moral quandary of human experimentation.

Nonetheless, the ages old inertia of life seeking (its own) immortality may well soon swiftly surmount it! No doubt a new event is beginning to rapidly emerge; the fusion of computer intelligence with the biological "algorithms" of our symbolic human awareness.

But the emergent hybrid of consciousness will be an unpredictable amalgam of man and machine, with predictable, and possibly even horrific, dysfunctions along the way to its full functionality. Whether the effort is morally sound is a serious question that needs a sufficient answer; and hopefully, before we proceed too much further into the quagmire of its non-resolution!

Yet when this issue is finally resolved, without a doubt, near immortality may, in our not-too-distant future, be "Coming soon to a computer store near you!".

But what might the nature of such an amalgamated mind be? The most obvious difference is also the most obviously significant, it will be "dis"-embodied. Indeed, being disembodied will allow the human mind to exist in perpetuity, stored within the electromagnetically operated "cybernetic" machinery of a computer's information storage and retrieval systems.

When this A.I. hybrid stage is accomplished two "sci-tech" directions will concurrently emerge; a disembodied consciousness, and a re-embodied one.

However, somatic cultivation and harvesting will again create some serious moral and ethical questions, which are hopefully not insoluble conundrums! And again, for the aforementioned reasons, "reanimation" is also on our sci-tech horizon. But most likely this will be accomplished by using living bodies, either human, non-human, or an amalgam of both, and no-doubt artificial genetic manipulation as well.

Indeed, the possibilities of genetic engineering are in themselves astounding (not the least of which may be the genetic enhancement of our future astronauts that will enable them to endure the long term exposure to radiation and the rigors and privations that are inevitably associated with any sort of sustained space travel).

The confluence of genetic and cybernetic technologies will soon define the future of humanity. In short, the current singularity is that of "mind", rather than of "matter", as mind seeks to consolidate its control over the latter.

However, since mind can also evidently exist outside of life (i.e. a "living" body), is mind nothing more than a conglomeration of thought, and "thoughtless" information? In other words, at what point (and how) does our subjective person emerge.

The principal avenues are clearly those of a likely intrinsic subatomic awareness (i.e. an "incipient sentience") and a subsequently emerging awareness (i.e. "embodied sentience"). As a result, a non-individual, non-local, non-conscious (pure) Awareness seems to be the percipient state of the universe, and noticeably so at the quantum level.

And although we are incapable of describing such a fundamental awareness, evidently it is also the source of the sentience by which it (i.e. fundamental Awareness) is known reflectively in our consciousness, and by which anything at all is known by the super-subjective "I" within our dualistic consciousnesses!

A superlative "Awareness" is thus the percipient of the universe, and in a practical sense, since any sort of experience is apparently precluded without it, the universe may be present in some fashion, but cannot be said to "exist" without it!

Without awareness, there is no cognition, somatic, cybernetic or otherwise. And without cognition, apparently there is only oblivion.

Since substance is as clearly necessary to mind as mind is to substance, both are most likely aspects of some essential "somethingness". Most likely this somethingness substance is the pure "Awareness" (for want of a better word) without which neither substance nor mind can exist.

This unalloyed fundamental Awareness, which could even be said to be the "footprint" of G-d, is apparent throughout our cosmos (and even more evidently so on the quantum scale!).

And if this thesis is correct, the Divine is presently beginning to more noticeably follow its own cosmic "footprints" back to an upcoming event of truly universal proportions, that of an emerging universal sentience!

Evidently, the universe (actually the "multiverse") is beginning to "wake up" and the most exciting current events of science and technology are beginning to sound the alarm that is finally making it happen.

(Dedicated to Silicon Valleys newest "infant" AIs; "Deep Blue", "Bob", and "Alice" too!)

Luminous Trees!

Some Rather Unsuspected and illuminating surprises in tree research are almost literally "right around the street corner"! Although it sounds like highly unlikely science fiction, incredibly the streets and sidewalks of our nation may soon be illuminated at night not by the brightly glaring orangeish light of an endless array of overhanging silvery metal street lights, but by the softly glowing green leaves of living "bioluminescent" trees!

This could be accomplished by splicing the luminescing genes from bioluminescent marine bacteria into the *dendrochloroplasts* (chlorophyll cells) of tree leaves. Thus creating a new brand of softly glowing street trees, that together with "starlight friendly" LED (light emitting diode) street lamps might even allow us to once again discern the nightly wonder of a star filled sky!

Indeed this has already been done with an otherwise nondescript species of houseplant, which now glows somewhat eerily in the dark, with the gentle greenish light of a bacteria's bioluminescence!

(Whoever said that science is not fun was probably not a big fan of Halloween either, both of which can be frightening to the superstitious and the poorly informed!)

But even though green glowing trees do not appear in Nature, they do seem quite beautiful and pleasing to the human eye. Which invites the inquiring mind to wonder why. And this innately *human* curiosity has led to the unlikely use of genetic bioluminescence to help us unravel the many mysteries of the human brain!

Yet beyond brain structure and function, by what method can we empirically reveal how the mind achieves either sentience, or qualia? The current answer to this question leaves us with little choice but to approach the problem in a different way. Which we can dialectically do, when we ask a pointed question like; *"What is the true nature of beauty?"*

Beauty

Everything In Our Experience of the universe exists only in opposition to its apparent opposite. Light to darkness, positive to negative, and ignorance to understanding, for example. Yet all opposites are resolved in the consciousness by which they are known. This is to say that everything within consciousness, is of consciousness comprised. But consciousness as we understand it to be, is only present in the living.

If life is a spontaneous, but statistically remote, possibility that has emerged from a chaotic burst of energy, then that which favors its existence should also be present in its makeup. In simpler "creature" life forms, this is obviously the case; mouths are for eating, stomachs for digesting, and anuses are for elimination. And although we human life forms share these same essentials, we also seem to have something more within the makeup of our consciousness, even beyond the phenomenon of our sentience.

Because even our elaborated sentience is present in rudimentary form, in even the simplest of life forms, and can be easily explained along with almost everything else in our consciousness, by Darwin's illustrious "survival of the fittest" theory. However the presence of more indefinite things like "love" and "beauty" are not so easily explained, as are those elements of intellect and emotion that clearly favor life's survival.

Take sharks for example, some species of sharks are more intelligent than others, yet the "stupid" sharks seem to survive just as well as the smart ones, and so both are present in our seas in roughly the same forms as they were millions of years ago. If survival is life's "goal", at least among sharks, the emergence of greater intelligence as an outcome of survival pressure seems to be, in the very least, an incomplete explaination!

We are surrounded, for example, by single celled life forms. They inhabit the land, water and air that we have all evolved in, in essentially the same forms as when life was in its earliest stages. They inhabit the same special niches on planet earth that evidently encouraged their emergence, and have allowed their continued survival.

Since life emerged some four billion years ago, despite the occurances of several planet wide extinction events in our earth's history, water, land and air have not much changed, and neither have the single celled forms which continue to inhabit them.

Yet some have arranged themselves into cooperative multicellular groups, with highly organized duties. That have even given rise to creatures the size of dinosaurs, whales, and elephants! And we know that this is genetically so, because every living thing grows from a "seed", which contains a foundationally identical set of instructions, to grow into whatever multicellular forms that inhabit the earth today.

Thus "Eve" was not really a woman, but a single celled "seed," which eventually contained the instructions for the appearance of her intelligent human form. Hence, the seminal question is, why is this so, since single celled forms are clearly the most robust survivors!

The only cogent explaination seems to be some sort of process of genetic "mutation". The problem here, of course, is the random nature of genetic mutability.

Suppose, for instance, that a simple genetic code is randomly mutated by structural failure, brought on by some sort of cosmic radiation, or perchance by a systemic anomaly arising during mitosis (i.e. genetic "code-splitting"). Because life's simple genetic code represents the basic structures and operations necessary to its survival, nearly every mutation would be fatal!

Therefore a very, very narrow window of probability exists, for the eventual emergence of humanity, from within an insentient "matter" that is ostensibly composed of simple, strictly "physical" energy! But evidently, this narrow statistical window is the only one open to evolution; so how might such an unlikely occurrence as complex humanities eventual appearance ever come to pass?

Logically, based upon the nearly inestimable trillions upon trillions of single cell life forms that have come to pass on earth over the past four billion years, the sheer random "force of number" could conceivably be cited as the essential causative factor of earthly humanities eventual appearance.

Moreover, we have recently discovered that thousands of earthlike planets with the right amounts of carbon, hydrogen, nitrogen, oxygen and liquid water are orbiting life-giving stars in the propitious "Goldilocks Zone" necessary for life as we know it to evolve. Hence, life has probably seeded itself all over our universe on millions of "life-permitting" planets, moons, comets and asteroids that are scattered amongst the stars!

But even broadening the statistical base of probability for the emergence of life, by including the many possibilities of extraterrestrial life forms in the mix, does not necessarily dismiss the problems that are inherent in our current understanding of life's appearance, let alone the eventual appearance of human, or human-like life forms with higher intelligence!

Nonetheless, here we are, endeavoring to understand ourselves, an act that apparently emerges as a factor of our sentience operating within our subject-object consciousness, and bounded by our experience of space and time. But even here, we now understand that our dimension of experience is limited by the range of our sensory organs, including what could be called our "sixth sense" of cognition.

Indeed, our entire detectable universe is likely only the miniscule portion that we are currently able to sense of an inestimable "Superverse" that extends, perhaps endlessly, in sub-quantumly interconnected, but nonetheless dimensionally contiguous (I.e. "parallel") universes. In other words, the statistical base for these inestimably remote occurances (such as our own appearance within the universe) is actually perhaps in the rather large "googolplex" range!

Yet, even by the near infinite extension of possibilities that naturally emerges from a spacetimeless, and quantumly energetic post

Primeval Singularity *quark-gluon* "soup", it seems clear that the "arrow of time", which allows the extension of objects in space, is guided by the most probable outcome of the events that comprise our entire Superverse's manifestation!

For this reason, the probabilities themselves constrain its constituent universes, and most certainly the appearance of higher life forms within them, to the range of possibilities that are available within their total range of space-time extension.

Therefore, we are limited in our own universe to the most likely possibilities that are available within its overall space-time extension. From its inception to its (likely) inevitable conclusion (although some astrophysicists opine that it must go on "forever"!).

When we search for order within the space-time extension of our own universe, we are tempted to believe that it is guided by the intellectual rules which we create in order to more accurately understand its mysterious internal workings.

But what is really going on, has nothing to do with the supposed "laws" of physics, which are simply our observations of probabilities extension. Since whatever is going to happen, could only happen in the most probable manner of its own energetic expression, at any particular moment in our universes overall, ever-expanding existence!

And of course "universes" without end may well be there, but they may as well not be, if they are not "known" to exist! Therefore the emergence of consciousness, love, and the perception of beauty are most likely not simple anomalies. Instead they are simply the space-time extended properties of our early unfolding universes emerging physics.

As a result, a mathematically conditional, because statistically predictable, "cause-and-effect" relationship can be only partially substantiated by the facts! Thus, an outwardly appearing "reactive" sort of non-conditional "evolution" is, no-doubt, the truer vehicle of their eventual appearance in our subject-object consciousnesses.

Evidently, it matters not whether we believe in the requirement of consciousness to the appearance of (perhaps endless) universes in the Superverse that contains them, since the facts are apparently just as self-evident as they are necessary to its eventual appearance!

The Chiron Conundrum

From The Subjective Perspective left is left, and right is right, which creates the conundrum of rendering other subjects into chirons of ourselves. But so strong is our attachment to our subjective position that we make the conscious effort to alter our own perception of the obvious.

This is to say that our "Supersubjective" perspective is inherently sentient and as such it is always the percipient Subject (i.e. "Divine" substrate) that observes its own subjective selves (i.e. "our" egos) observing its other subjective selves (i.e. "their" egos). Thus creating the subjective conundrum of directional ambiguity.

For example, if you were in a spaceship that was travelling through a dark and starless expanse of space, without some sort of independent on-board navigation equipment, it would be impossible to accurately determine any relative direction.

In like fashion our dualistic minds rely upon a purely conceptual perspective of "left" and "right" to reinstate the requisite dualism of an apparent "subject" and "object(s)" that the hominid hippocampus evidently requires in order to navigate within an experiential universe of its own psychosomatic creation!

Nonetheless, while these deeper sorts of considerations are in themselves admittedly quite compelling, how can we utilize them to better understand, and perhaps even more fully reveal, the penultimate Source of our attachment to the line of higher sentience by which they are known?

The Line of Awareness

EVERY ARBORIST AND TREE TRIMMER KNOWS that it is wise to "tug hard" on a climbing line before ascending into a tree, to test the firmness of its purchase. But when the line appears to tug back, it also seems wise to investigate why!

Indeed, when the line of awareness in our "Tree of Life" (by which we ascend into consciousness) is tugged upon when we meditate, the "line" not only tugs back, but it becomes quickly obvious that it's point of attachment lies somewhere prior to, or "above", our everyday subject-object consciousness.

And because the line of our awareness is apparently attached to some point that's far above our biological Tree of Life, this seems to defy both practical experience and ordinary reason!

At this point we "tree people" generally open up our problem solving tool box, and in this case by choosing the sharper tools of a deeper sort of intuitively involved *(prajnaic)* reasoning, an in-query for reliable answers naturally ensues.

So too begins a relentless "tug of war" on the line of Universal-Consciousness, which holds our experience of person and universe together. On one end pulls the familiar temporal experiences of our person(s) and universe. And on the other, spacetimeless impersonal Awareness.

In between, lies the strong cord of pure Consciousness Force. Which seems to somehow anchor our individual consciousness to the endless pre-Singularity spacetimeless "heavens" of pure potential.

This is the essential "Elsewhere" from which the Primeval Singularity first emerged, and that still clandestinely, through a quantumly elaborated Universal-Consciousness, contains the entire Superverse!

But paradoxically, the occult anchor point of our consciousness can only be revealed by exposing the hidden connection of our own "branch and twig" of life, to the grander Tree of Life that contains and continues Life's incredible space-time journey in our own little corner of the unfolding universe.

Eventually, there comes a time when we begin to understand that the mystery of our own consciousness can only be solved by climbing up the down hanging rope, to see its point of attachment first hand!

But when we do, we also soon discover that we are not really climbing "up" at all, but actually "inward", toward the immortal spacetimeless "Elsewhere" source of both our detectable universe and our mind's own mysterious sentience!

Mountains

"Why Does G-d Choose To Suffer?" This is a good question, which cannot be accurately asked until we understand that we are the Supreme. But even with our "enlightenment" a substantial gap seems to remain between our new understanding and the Awareness by which it is known.

When the top of a tall mountain is finally reached the view is extraordinary, but it is impossible to climb any higher. From here every path leads back down the mountain. But the memory of the adventure remains, until our tired body is no longer able to bear the Divine heights.

How tenuous (and at times even painful!) but magnificent, is the climb up and down G-d's mountain!

Santa Cruz Mountains
Northern California, USA

Chapter 7
Sempervirens "Ever-Living"

Twilight Magic

As I Lay Here Today, in my usual foggy morning "Turiya" state of mind. I watch myself remembering and forgetting. But I cannot tell if I am awake, or perhaps inside a dream. Nonetheless, there seems to be a mysterious rhythm going on somewhere "outside", above my head, *pit-pat, pit-pat!* (I.e. *Turiya* is a Sanskrit term for "lucid dreaming").

The sound of "fogwater" dripping from some sempervirens needles high above me, *pit-pat, pit-pat...* Remembering and forgetting, there is a sound, *pit-pat*, on the rainfly just above my head. *Pit-pat!* And it awakens me.

Tentatively I open one eye, *pit-pat!* I can see the sound source, *pit-pat!* of water dripping, large shiny drops, *pit-pat!* They dimple the gray nylon shell of our tent as they strike, *pit,* and are tossed back up into the air, *pat!*

When the downward dimple springs back up like a trampoline, a rainbow spray of droplets "magically" erupts into the twilight morning air with a bright flash. Then a glowing halo forms around the droplets gleaming centers, as they splash back down onto our tent's gray nylon shell. *Pit-pat, pit-pat!* Then perhaps a hundred more!

These large glistening "fogdrops" form slowly on the endless dark-green sea of rainforest needles, wetted by last night's dewy fog. They coalesce, fatten and eventually fall, wetting the forest floor as they strike our tent's gray dome far below.

Opening my other eye to take in the colorful show, my upward view is suddenly obscured by a misty white cloud. But my "breath cloud" quickly dissipates and the overlapping circular patterns of the glowing fogdrops rapidly reappears.

"O' how marvelously they do glow!", as they are illuminated by the early morning shafts of sunlight. With a veritable rainbow of colors, that radiate outward in brilliant ever shifting patterns and transparent shades of yellow, orange and green, purple, red and blue!

And quite soon their kaleidoscopic glow almost magically transforms our humble tent's prosaic gray shell into the similitude of a watery, stained glass *como-una-ventana (*"window-like") ceiling. "How beautiful!" I am moved to muse.

Then snuggling back down into my warm goose down sleeping bag, to the rhythm of the watery tympani (drumming) above, I close my eyes and soon drift off into a dream. And at first in my lucid dream, I seem to know that I am merely dreaming. Yet in knowing that, I begin to wonder why and how I could possibly be awake inside a dream!

But finally growing tired, from all the effort of remembering, I soon forget that it is just a dream. And when I finally open my eyes, yawn and stretch, then slowly climb out of my warm down sleeping bag to greet another day, I can no longer (with very much conviction) say, which universe is real, and which is really just another dreamy play!

Magicus Naturalis

THE "PHENOMENAL" UNIVERSE, this is to say, our illusory "personal" experience of the universal manifestation, is almost literally *magicus naturalis* "natural magic". In other words, the universe that we experience is almost literally a clever stage act of divine legerdemain. And impersonal Universal-Consciousness is the secret "stage" upon which the Divine "Magician" plays out her clever tricks.

She does so within the living cognitive field of our perpetually incomplete, and thus ever shifting, illusory dualistic perspectives of individual consciousness. Hence, this ephemeral cognitive material is the actual gossamer "thought stuff" of which our entire experiential universe is comprised!

Mother Nature perpetually seeks to restore an abiding balance to the endlessly flowing stream of our ever opposing subject-object experiences. But as things are at present She only succeeds in remaining functionally conscious through us, by compartmentalizing our endless experiences into the transitional conglomerate realms of our past memories, present imaginations, and future anticipations.

Because even our supposed "present" is really just another fantastic imaginary mental construction. A portion of the Universal "Manifestation" is thus preserved in our dualistic consciousness'. But only during the inexorable course of our little experiential lifetime whiles!

The incomparable East Indian Advaita Sage Nisargadatta Maharaj once quizzically noted, "In Consciousness anything is possible!"

This is the equivalent of saying that just as in our ephemeral sleeping dreams, which are also the product of a consciousness wherein almost anything seems possible, so it must be in the "waking dreams" of our more "universal" Consciousness. (In which we are the dream character by which Universal-Consciousness experiences the apparent limitations of our more persistent waking dreams!).

The waking dreams of Universal-Consciousness, which we experience as our individual and collective realities, are thus a trick of the "Divine Magician". For just as the *"Ladino"* speaking Kabbalists of Spain, the "Sephardim", once correctly maintained; *La Vida es sueño,* "Life is but a dream!"

Our puissant experiences of person(s) and universe(s) are thus a "slight-of-hand" illusion that is occasioned by constantly shifting perspectives of spatiotemporal scale. Which is subtly performed by a Universal-Consciousness that performs its "magic", hidden away from our view in an implied fourth dimension.

There it is imbued with the simulacrum (likeness) of a reality within three dimensions, by the "smoke" of ideational time and the experiential "mirror" of our mind's representational interior spaces. The phenomenal universe of our experience is thus a "magic act" of space-time illusion that is created by the Divine with symbolic space-time "smoke" and memory "mirrors"!

I

But, how is this Divine legerdemain accomplished? What we actually experience is our own impersonal Universal-Consciousness, as it emanates from within the unmanifest noumenal Supreme (i.e. "Elsewhere"). Currently "It" (i.e. our Supreme source) views *it*-Self three dimensionally. Apparently in the only way that is presently allowed by our species rather primitive present stage of Self-conscious evolution!

And of course I am not simply referring to Darwin's classical theory of "Natural Selection". Which is most certainly demonstrably correct, insofar as it goes! But instead I am referring to the deeper impersonal Consciousness that is in fact the כוח החיים *Koach ha-Chayim* "Force of Life".

A Force that not only allows Life's intricate processes of evolution to function, but that also in no way limits the astounding present possibilities of its expression in our dualistic consciousnesses, or the future probabilities of its eventual fuller expression. Which most certainly brings into question any theory of strictly insentient "Natural Selection"! (That masks an unlikely "spontaneous generation" of life!)

Our familiar ideational "persons and universe(s)" are thus in fact evolutionarily extended self-objectifications that are accomplished within Universal-Consciousness by creating the imaginary objects of our individual consciousness'. Which are then ideationally separated within Universal-Consciousness.

This is likely accomplished by an artful action of focus that creates different perspectives, fundamentally based upon apparent symbolic differences in position and scale, which we experience in three dimensions, from a four-dimensional space-time point of view.

Our experiential universe is thus manifested by the force of Universal-Consciousness operating within a four-dimensional range, by the careful sequential arrangement of mental "quantum". "Quanta", *ut hoc loco* "as here used", are the smallest possible aspects of our tridimensional consciousness. These creative aspects of perspective are also known as "mental modifications", or v*ritti* in Sanskrit.

Thus, our mind's basic oppositional symbols of three-dimensional consciousness are the resulting "subject/objects" and "object-objects" that appear within a four-dimensional space-time field of impersonal Universal-Consciousness. These are, in turn, all witnessed by the fundamental presence of our primeval Supreme Awareness, from at least a "fifth" dimension! This is because whatever appears in our field of experience must be viewed from a superior dimension.[(96)]

But strictly speaking, as strange as it sounds, even experiential dimensions have no independently substantial existence, since like everything else, they also "exist" in our consciousness! This is to say that we cannot possibly be anything that appears "before" us. We cannot, for example, ever be the "apple" object that we may be looking at, simply because we are always its experiencing subject.

But what then must the quanta of our experience be arranged into, in order for them to perform their intended functions? Certainly they must first be arranged into the psychosomatic instrument of their own perception, our familiar "body-mind". Because both our bodies and our minds must first be manifested, in order for the universe to become "present" in any living sense within our individual consciousness'!

II

From the subject/object (i.e. the "ego's") perspective of our psychosomatic instrument, our personal memory also contains the apparent "history" of our collective group memory. Our collective memory and its symbolic representation in written language, and in the more exacting language of mathematics, also contains the discoveries of science.

If these discoveries arrived at by the empirical scientific method and rigorous logic of predictive mathematics and physics are correct, then there certainly seems to be an impersonally Conscious "evolutionary" process of some sort going on!

Moreover, as the "Periodic Table of the Elements" and the spectrographic analysis of starlight reveals to the unbiased and astute scientific eye, it is also rather intricately involved in the eventual appearance of our biological instruments of perception from within the "dust" of stars!

However, before the appearance of our human body-mind instruments (of Life's continuance and universal perception) the manifestation of the cosmos within consciousness could not be said to exist *per se.*

Indeed, any sort of apparent existences, like space, time, energetic perturbations, quantum sized "strings" of vibratory energy, and any other possibly "objective" thing, must have their basis in the conceptualizations of our consciously experiencing minds.

Because without our minds time does not "exist", and without time space cannot exist. Just as it does within the relatively motionless and spacetimeless interior of a cosmic "Black Hole"!

In the absence of space-time, only spacelessness and timelessness prevail. This is apparently analogous to the physical condition that is brought about by the extreme gravity in a Black Hole.

But if you exclude the idea of a discrete phenomenon of gravity, in the absence of motion, space-time and phenomenal objects, it seems most likely that this is somewhat like the noumenal, primeval state of our observable universe.

In other words, the objects of our dualistic perception seem to warp our experiential space, which then appears to function much like a separate, physical force of gravity.

"Time" is thus our serial observation of this process acting upon the sequentially appearing phenomenal objects of our individual consciousnesses. Indeed, some theoretical astrophysicists call this peculiar condition of spacetimelessness "Elsewhere".

It is evidently what prevails outside of, and prior to, the appearance of our universe and the currently "invisible" forces (like gravity) which also seem to govern it, be it "experiential" or otherwise! But what then might this spacetimelessness be?

Spacetimelessness

In The Absence Of Gravity, energy, matter, consciousness and every other possible kind of imaginable phenomenon, what could possibly remain? It seems most likely that the nature of this primeval spacetimelessness is a pure equanimity of absolute Force and Supreme Awareness, in other words, an "Awareness-Force".

Otherwise, how could something, or any "thing", possibly emerge as an all-inclusive "Primeval Singularity", which was then somehow forcefully obliged to go Bang!

Remarkably, this primeval spacetimelessness was able to arrange, and even continues to arrange, for a nascent condition of purely potential energy to somehow appear *in eo* "in [experiential] fact" as if it were comprised of a wholly independent energy, matter and space-time.

Moreover, that it then manifests as our apparent "persons", by which it appears to exist in just such a fashion, indicates that its essential nature is likely a boundless Consciousness-Energy potential of pure Awareness-Force! And because it is evidently prior even to our consciousness, energy, space and time, this essential Awareness-Force is also essentially "transcendent".

But because it also manifests its pure potential into the energy, space, time, life and consciousness of our universe, it evidently must do so by extending itself into an immanent Consciousness-Energy that is capable of generating the subquantum energy "string" perturbations that eventually give rise to our universe.

Moreover, since it is the *Parabrahman* "Supreme" from which an apparently infinite potential of Universal-Consciousness energy somehow subsequently emerges, and then gives birth to what we experience as our person(s) and universe(s), it is also likely (in its "unextended" state) in a state of perfect balance. Or in affective terms, it is the awakened Yogi's *Ananda* "Bliss" of persistent transcendental *Salam* (*i.e. Sōlh, Shalōm,* or "Peace").

I

From a state of immanent (indwelling) energy potential, as Universal-Consciousness, the Parabrahman differentiates and then evolves, or perhaps (in a sense) "devolves" and then evolves, through several stages. It is important to remember that this takes place within pure Universal-Consciousness, prior to the emergence of any sort of phenomenal subject-object consciousness, and therefore it also transpires within an eternally present "Moment".

This grand Moment is the eternal first Moment of the manifesting universe, as well as our own veridical locus of Awareness. (The brilliant contemporary metaphysical author, Ekhart Tolle, refers quite eloquently to this eternally present Moment in his insightful little book, *The Power of Now.*)

But it does appear as "historical" within our linear tridimensional consciousness. In other words, the real Moment is timeless and spaceless and therefore only capable of containing something phenomenal in potential!

However, currently we can only understand the apparent evolution of the universe within the context and reach of our mind's available three-dimensions of psycho-sensory perception. Which operate relative to an unseen, but intuited, fourth dimension of inferred space-time that is most likely being witnessed from a fifth dimension.

From this perspective we see that first energy appeared, then matter, and then the life principal (or "Force"), followed by mind, all operating within a field of apparently ever expanding and ever changing space-time! Space-time is thus not a "substance" per se, in that it must simultaneously remain the same, while it is constantly changing.

Water, for instance, undergoes subtle but substantive changes in its molecular structure as it assumes the various forms of solid, liquid or gas. But it nonetheless always remains, in its basic molecular essence, simply "water".

However, in a more distant geological time, a preponderance of our lovely planet's collective waters actually began their earthly existence as (comet delivered) atoms of oxygen and carbon, swirling about in a sea of seething lava!

Similarly, while space may appear to contract or expand, it remains substantially vacuous and unchanged! What we experience as empty space, however, is actually an all-pervasive supersubjective aspect of Universal-Consciousness.

Which allows the daily cyclical flow of apparent "object-objects" ("vrittis"), and unlikely "subject/objects" (egos) to holographically appear. And seem to either stand still, or move independently about within our minds.

Hence, only within our familiar dualistic consciousnesses does the transcendent Supreme Witness become apparently immanent and is thereby enabled to experience the symbolic (i.e. "algorithmic") volumes and contents of our outer and inner spaces, through Its apparent "us".

What appears to be "outer" space however, is actually only the imaginary projection of an experiential event, which transpires internally within the biological machinations of our dualistic, subject-object consciousness'. Hence, our sense of time is irrevocably linked as a matter of perspective to the concept of space.

This is so since our perception is a phenomenal event that requires the measurement of three dimensional spaces into a fourth dimension of Universal-Consciousness that is essentially invisible to our cognition, but not to its probable fifth-dimensional "Supreme" Witness. "In" side and "out" side space are thus irrevocably interdependent!

This can be thought of in much the same way that Einstein perceived the universal force of gravity. Because the appearance of an object evidently "warps" the space around it, the larger and/or more "massive" or dense that the object is, the more it warps the space which must always appear to be around it! In other words, there exists a hidden connection within our fundamental Universal-Consciousness, between apparent objects and the space that they seem to occupy.

This certainly challenges the question of an actual "inside" or "outside" of our universe. Which is not to say that the observationally based laws of physics cannot function quite well if this subtle distinction is disregarded entirely! If they did not, the universe would not be able to include the dualistic minds by which it is known to exist.

II

The subtle connection between the seen and the unseen is most likely our Universal-Consciousness, which ever underlies the phenomenal appearance of space! Space-time must therefore appear constantly to expand, as the apparent objects of its manifestation within Universal-Consciousness appear causally to interact and change within its continuum.

The difference between the supposed phenomenal objects of our consciousness and the presumed objects outside of our consciousness is also fundamentally uncertain, not only because both are probably comprised of Universal-Consciousness energy, but also because of the hidden perceptual link between space and time, which originates within it.

Therefore impersonal Universal-Consciousness, like the astrophysicist's "Dark Energy" and "Dark Matter", is invisible. But we can know that it exists by the "footprints" that it leaves in our consciousness (i.e. "dark" simply means that at this time – whatever it is – it still remains unknown).

Thus time can be said to be the serial displacement of space by our next apparently present perceptual moment. However this causation exists purely as a matter of perspective, and it is also therefore inferred. It is the relative phenomenon of a mental manifestation, which transpires only within our consciousness.

In precise point of fact, nothing has really transpired on the basis of an independently operating causality at all! The entire universe of our experience, the only universe that can be said to "exist", thus appears to be expanding ever outward from every inferred point of perspective within it, as if each were the center of the entire manifesting universe!

This is because everything transpires within our minds, more essential, Universal-Consciousness, and nothing is either really ever happening or changing anywhere "else", except as a matter of perspective, as it does within a dream.

Each potential point of personal perspective is thus occupied by impersonal Universal-Consciousness and so appears to be the center of the cosmos, relative to every other possible point of our apparently "personal" perspective(s). And indeed, whatever rules or laws of manifestation may be carefully inferred from the information currently at hand, they seem to be in support of the mysterious legerdemain of Universal-Consciousness!

However, it will only remain mysterious until we reveal its truer nature and our place within it, directly through the root of our own sentience, by our own "Enlightenment".

But through the ever advancing scientific method, and the stream of resulting technological innovations, with better instrumentation such as the soon-to-come "neutrino" powered micro- and tele-scopes, we may one day reveal its fuller universal nature in a far less subjective way (see "Neutrinos", pg. 556)!

In the meantime, we can infer that it is like a clever stage act, a mirage, or an illusion, which brings the universe into the cognitive realm of our subject-object experiences. Whatever we experience is thus a virtually objective space-time manifestation that invariably occurs within the "inside/outside-in-ness" of our individual consciousnesses' manifestation.

Therefore whatever we experience is actually a phenomenon that occurs within the universe's substrate of Universal-Consciousness. Incredible as it might seem, when we think that it is we alone who are looking "out" at the universe, it is really the universe that is looking "in" at *it*-Self!

Moreover, "It" (the Universe) simultaneously appears to volitionally move through our familiar, but faux, "in between" world(s), through its truncated imaginary "us". Thus incredibly (all-the-while), the Universe experiences it-Self as if Its imaginary little "we" were nonetheless an independently veridical "us"!

In other words, we prosaic "people" entities are just as necessary to the existence and evolution of our evident universe, as its probable Supreme source is. And Universal-Consciousness is the potential "stuff", and invisible force, of the primeval and absolute Supreme, from which everything is wondrously articulated into our implied realities of apparent existence!

But if this is so, then just how might this artful articulation possibly be accomplished?

Strings

The Universe Is Carefully Orchestrated into existence with a beautifully mathematical, even "musical" symmetry. It is, on the quantum cusp of its energetic expression, a wonderfully articulated, perhaps even Divinely composed "symphony", being played upon a myriad of invisible "stringed instruments"! Indeed, at the very basis of our apparently ever expanding universe lie the "strings" of Consciousness-Energy *quanta*, of which it is evidently altogether comprised.[(97)]

From this perspective, the essential character of all "quanta" is that of perfect equanimity. In other words, the perfect primeval Singularity is in essence an absolute "Peace"; *Salam, Sōlh,* or *Shalōm.* Therefore in essence, "Salam, Sōlh, or Shalōm", all refer to the "Abiding Peace" of the Supreme (i.e. the pre-Singularity "Elsewhere"). From which all quanta emerge, and then articulate into the inscrutable quantum "sea" that underlies all subatomic diversification.

Quanta are thus a subtle and sublime expression of the sub-quantum impersonal Universal-Consciousness (*Elohim*). Which lies upon the very cusp of undifferentiated primeval Awareness.

From here, they can be said to articulate into their various expressions, of apparent matter and energy. Suspended, as it were, in "internal" and/or "external" space-time, depending upon your point of view!

To clarify, we can transcend the apparent limitations of our illusory individual consciousness, first in the realizations of pure logic and then by direct experience. Because from apparent beginning, to evident end, everything most likely happens only within, and to, our universe's impersonal "sub-quantum" Universal-Consciousness.

And it is from the (cosmically puissant) energetic presence of this Universal-Consciousness that endlessly differing iterations of our universe are quantumly empowered to "split-off" and form into our ever growing multiverse, as we experience them within our dualistic consciousnesses.

Thus we are indeed "co-creators" with the Divine! Furthermore, the subtle Universal-Consciousness (which also determines the manifestation of the universe in our life and mind) is limitless, yet essentially changeless, and without any "first", and therefore even subsequent "material" cause. Therefore our entire universe is ultimately without causation, history, fate, or even purpose!

I

Evolution can therefore be modeled as a recursive, spiral motion of impersonal Universal-Consciousness, "closed" in respect to the continuity of its subtle Universal-Consciousness substance, yet "open" to its Force of apparent change, in that it takes place in the spacetimelessness of an eternally "present" moment.

At least this is as far as our unassisted human mind can presently reach, when we try to conceptually understand and embrace the universe of our common experience. Yet, when we assign the concept of number, and then count the resulting transitory "present" moments of apparently passing space-time, it has taken many billions of years to reach our seemingly ever-present moment! But any conceived "moment" is really just an illusory space-time moment, ideationally comprised of our cognitive experiences.

However, our entire lifetime experience actually takes place within an eternally present "Moment". Which is an enduring sentient moment that is itself a conglomerate ideational "event" that is comprised of our cerebrally collated cognitive experience(s).

But all of our experiences are in themselves nothing more substantial than our brain's ongoing "sixth sense" interpretation of its own serial stream of discrete "code-like" sensory events (the sixth of which is our brain's interpretive faculty of cognition).

All of which is going on in our cognition automatically, like the blood that is constantly being circulated in our body's veins by our continuously beating hearts! Thus we, our *experiences*, the apparent *places* in which we seem to be having them, and the various *moments* in which we seem to be having them are all (most likely) only aspects of a Universal-Consciousness!

If this seems at all difficult to grasp, the simple truth of the matter is that all of our supposed pasts, presents, and futures – and all that they could ever contain – exist only within Universal-Consciousness and therefore in a space-timeless and eternally present, but only experientially tangible, "Moment".

Thus, when examining the Universal Manifestation from different perspectives (as we evidently always do!), it is important to keep this caveat in mind. But it is also important to understand that everything, including the Universal-Consciousness Force of which the entire universe is apparently comprised, eventually resolves back into the Supreme "Elsewhere" from which it once ostensibly emerged!

Now, as the awakened Sages suggest, keeping in mind that everything is most likely Universal-Consciousness, we can proceed with our discussion. The "quanta", or in Sanskrit *vrittis,* of which the universe of our experience is comprised, are arranged into the Universe's psychosomatic instrument of three-dimensional perception.

This is our familiar, sometimes-called in metaphysics, "body-mind". Which is certainly necessary to our every experience, even if our experiences are actually only illusory or incomplete in themselves. However even this rather fundamental fact generally escapes our attention, because of our unexamined experience of it!

II

Our existential dilemma really begins with the sudden appearance in our mind(s) of the illusory subject/object ego(s) by which the objects of our consciousness are conceptually known. Because our "ego" is actually an unfeasible subject that is also simultaneously required to be an object!

Nonetheless, the objects of our consciousness are experienced as if they constitute an internally experienced, but actually independent external universe. Indeed, after its phenomenal appearance, our body-mind is automatically informed by our five senses.

But the often unacknowledged sixth sense is our *cognition*. Which arranges and interprets our brain's representative sensory codes (of the other five senses), and is thus functionally limited to the perception of three dimensions.

This sensory information is received subliminally as a virtual stream of *vrittis*. These are then differentially expanded by our sixth sense of cognition into the various objects of our apparently individual consciousness. In other words, the electrochemical "sensorial" impulses of our bodies and brains are subliminally received on the higher end of their Consciousness-Energy spectrum.

They are then expanded within our individual consciousness by creatively changing the focus of the independent Awareness, which is witnessing their "expansion" within the endless potential of Universal-Consciousness, within an inferential fourth dimension. And it is a curious fact of our perception that any given dimension can only be fully cognized, or viewed, from a superior dimension!

Hence, we can perceive space as time only from the perspective of an essential Awareness that witnesses it percipiently from at least a fifth dimension of Superspace!

However time remains as invisible as space, because the cognitive aspect of our subjective instrument of experience is itself comprised of a Universal-Consciousness, which also comprises the imagined objects of our dualistic experience.

Therefore Universal-Consciousness is much like the invisible lens of a telescope or a microscope that is being adjusted by the mechanical instruments, both on, and within its "frame" (i.e. by our senses and brain). While the independent Awareness of the Supreme gazes upon our phenomenal universe through "it".

This dynamic aspect of our universe is called the *Saguna Brahman* "Manifest Divine" in ancient Sanskrit, *Allaha* in ancient Aramaic, and *Elohim* in ancient Hebrew.

The malleable experiential "lens" of Universal-Consciousness (i.e. our body-mind) is evidently subliminally focused on the quantum level, in a differential manner, by the sixth sense of our biological instrument of individual consciousness, in order to bring a "rapid-fire" stream of conceptual objects into cognitive focus.

In this way, the qualia of our compelling space-time involved three dimensional experiences of (person and universe), are manifested within our binary subject-object consciousness. Which then takes on the realistic appearance of exquisitely formed holographic projections!

III

Our cognition is in effect our superlative or "sixth" body-mind sense, which organizes the ongoing stream of our brain's neural input into the apparent objects of our consciousness. But these imaginary mental objects are not inherently conscious.

Our everyday consciousness is thus brought about by the Universal-Consciousness, of which they (our "mental objects") are comprised, which is indeed always conscious, albeit in a fashion prior to dualistic consciousness as we generally understand it!

Sentience enters our dualistic consciousness because of the essential presence of the Supreme Awareness, by which everything in our universe is known to exist. In classical Buddhism this Awareness is known as our *Bodhi* "Buddha Nature".

And although this term is generally conflated with *Nirvana* (or *Nirguna*) "without form", its more exact meaning derives from the ineffable Supreme source of our entire manifest universe!

But there is really no separation of It (the monadic Supreme) into discrete parts, which then must be somehow connected by the presence of something other than what they are! Because like everything else in nature, each organized part actually transitions into every other part, like a grand cosmic puzzle.

Moreover, they are not static in their illusive individuality, but also act much like the transition zones that eventually connect a trees outermost bark to its innermost heartwood.

Therefore everything is essentially always the sum of "One", only masquerading in our apparently individual consciousness as innumerable binary combinations of space-time separated experiences! And thus the presence of space-time in our cognition evidently has the relative value of *Zed* (zero) + 10^{-43}, which allows the diversification of our experiential universe into virtually infinite combinations (see "Planck Time", on pg. 564).

But if any part of this interdependency is disrupted, the whole heuristic edifice crashes and our entire experiential universe abruptly ends! Just as a tree dies if its cambium is cut through all the way around its trunk.

Except that trees are much tougher than we are and so take much longer for the Life Force, vital breath, and food "essence", of which all living things are evidently comprised and sustained, to then disassociate and return back to our essential substrate of primeval Universal-Consciousness. (Which is then consciously present only to other living things, as the experiential materials from which they too seem to emerge).

The main function of all of this focusing of Awareness through our "split-mind", is to expand our experiential quanta of vrittis sufficiently in the fourth-dimension of Universal-Consciousness, which we experience as space-time in three dimensions, in order for us to cognize them.

This is accomplished by focusing our attention upon each conglomeration of experiential quanta, over enough duration for them to "expand" ideationally within our consciousness, and then to be noticed by us. The cognition of symbolic volume as well as experiential surfaces is thus quite necessary to our every experience of phenomenal qualia!

Our cognition is principally composed of our intellect, emotions, imagination, and memories. These four cognitive aspects complete the arrangement of the organized quanta that appear within the reflection of Universal-Consciousness. Which we sense as the space-time that sustains the experiences within our split-mind as qualia (see pg. 512).

Hence the apparent "darkness" that we see when we close our eyes is actually the "light" of the occult Awareness by which it (the apparent "darkness" that we see) appears within our consciousness. And in the subtly sentient "dark light" of this Awareness, our evident persons and universe(s) are experientially created.

They are generated by the automatic, subliminal, psychosomatic processes that are constantly occurring therein, and are subsequently "reflected" back into our foundational sub-quantum Universal-Consciousness, where they become sentiently "known" through the auspices of our essential Awareness.

Our "person and universe" are therefore only conceptual, or ideational in nature, and are thus essentially the phenomenon of our transient individual consciousness. Even our necessary experience of "absence" is thus the conceptual "presence" of (us) its experiencer!

But, while we can divide the universe's system of sentient binary consciousness into its main parts in order to better understand how it functions, it is important to keep in mind that we are in fact a psycho-somatic system that necessarily incorporates all three (i.e. Awareness, Universal-Consciousness and individual consciousness) in order to function!

Thus the thoughts of which the concepts of our person and universe are composed, are in fact the equivalent of phenomenal "ghosts" appearing within the biological machinery of our brains.These sentient processes of Awareness transpire through the interdependent presence and actions of our fundamentally unreal individual consciousnesses, appearing within its substrate of impersonal Universal-Consciousness.

In other words, we experience the "consciousness-objects" (i.e. thoughts) of our cognition through an inferentially personal consciousness that operates like a reflection showing up within, or upon, the impersonal Universal-Consciousness that is the actual basis of the very quantum of which our soma and psyche are comprised.

Our every purely contingent experience is thus finally known only within the actual intemporality of our Universal-Consciousness, by the sentient reflection of our actual *Nisarga* "Supreme Nature" that is being constantly "cast" therein!

IV

But of course this is a "knowing" of an entirely different order, since it is not only spacetimeless, but also "subject-objectless" in nature. It is therefore an "all-inclusive" knowing, without any sort of separation, and must happen from within an Awareness that unites rather than divides.

Indeed, our every experience is actually not of our own making. Because it is perceived indirectly by the Supremes' virtually "transcendent", yet within our binary consciousness "immanent", Awareness. Which thus imparts a strong feeling of sentience to its phenomenal aspect as an apparent "us".

This happens without any effort on our part. In other words the Divine is constantly arranging to have the apparent experience of living "your" life as you! The simulated tridimensional volume of our mental objects are thus experienced within an inferred four dimensional space of actual Universal-Consciousness. It is cognized by the ideational measurement of a three-dimensionally "surfaced" four-dimensional volume, and it is "measured" serially (into temporality) by its duration.

We experience this as the continuing conceptual framework of an apparently ever "moving" four-dimensional time. All of which transpires in an imaginary manner, the only substantive portions of which, are the unseen biological machinery that generates the experience, and the "secret" underlying Universal-Consciousness of which everything in our universe is evidently comprised!

Each conceptual "frame" can therefore be measured or experienced relative to another, all of which then seem to have an existence within and over a veridical time. Without this implication of duration, nothing could be cognized!

But to be distinguished as distinct, which is also necessary for cognition, the objects of our consciousness must also be conceptually separated from each other by the inferential measurement of spatial distance.

The phenomenon of travel in space over time, between apparently separated objective places, is represented as such within our collective individual consciousness'. Despite the fact that it interdependently exists outside of our individual consciousness, most likely within a cosmic substrate of Universal-Consciousness, as a malleable space-time.

This is a curious situation likely caused by the current structure and limitations of our evolved bodies and minds.

This creates conundra, like the fullness of emptiness and the emptiness of fullness, limitless beginnings and endless endings, evidently self-organizing chaos, "fractal" surfaces, bilocation, quantum entanglement, the possibility of F.T.L. (faster-than-light) travel through wormholes, and even so-called "Grandfather Paradoxes"!

Moreover, it allows the existence of an infinite multiverse that we know almost nothing about! (See "Time Travel"; "The Grandfather Paradox", in the fifth paragraph, on pg. 591).

But these sorts of conundra only exist as such due to our present ignorance about the fuller nature and role of consciousness in our universe, and in the multiverse at large, as it exists almost entirely beyond our currently quite limited knowledge!

Thus, all of our habitually redundant identification with objectification is evidently extended and known by the Supreme through its expansion as Universal-Consciousness acting as a body and brain, which we incorrectly believe is exclusively our own!

In "reality" (if such can even be said to exist at all!), every interdependent thing is most likely happening only within and to Universal-Consciousness, in much the same way that a dream is manifested within our only apparently seperate "individual" consciousness.

As a result, even our supposed everyday experience of "reality" is thus quite unreal. Because it is all done within Universal-Consciousness, by the subtle "smoke and mirrors" of our individual dreamtime consciousness' experiential cognitive machinery.

In other words, the clearly ideational "person and universe" of our common experience(s), are actually just the subtle artifacts of our illusory individualistic perceptions. Thus, our seemingly substantial universe is largely just *an illusion!* (*Hoc pro* אחי, Lenny J.)

V

"The future moves around the present and is cancelled out by the past"; said the illumined Advaita Sage, Nisargadatta Maharaj. And although he was a man without formal education, he spoke with a percipient authority from within his profoundly impersonal "personal" experience of the universe, having fully awakened from imprisonment within his own body-mind's recurring waking dream.

What the Awakened Sage describes is his direct experience that the "future" exists only in our anticipation of it, and is experienced along with our purported "present", only in the evident "past".

In other words, the imagined past exists only in our memory. Which also contains the ideational similitude of a future that will never actually be! Therefore both also never really exist, except as concepts within our individual consciousness'.

Because, evidently, almost as soon as we begin to become aware of our own existence, we almost instantly become "self-deluded" into believing that it is all "Real"! (I.e. By our habitual identification with the automatic subliminal operations of our cognitively implied individual consciousness').

But ours is actually a rather insubstantial little consciousness. Which is likely reflected by an underlying Universal-Consciousness. Ultimately from its original Source, that lies within the provenance of our ever witnessing supreme Awareness.

So, it seems quite logical to ask; "Is our body-mind and experiential universe actually even real"? To which, evidently, the only cogent answer can be, yes *and* no! Since it is only the unchanging reflection of one absolute substance/Source. Which illumines our somas and enables the shared mutual experiences of our individual consciousnesses, and the powerful ongoing illusions of our largely mechanical psyches.

Therefore even our common experience of the evident "present", as we most generally relate to it, is never really experienced by (an) "us" at all! This is because veridical reality forever lies within the essential Awareness of the persistent pre-singularity Supreme. Which is thus always experiencing "our" lives, through the vehicle that we wrongly assume to be exclusively our own (individual) body and mind.

Evidently, we are actually the absolute Supreme, the "footprints" of which science is just beginning to reveal and quantify. And thus we are not merely the constant cognitive stream of ephemeral and transitory figments of worldly experience that only appears to be our "persons" and "personal experiences" of life!

Hence, Buddhas are never actually born and only appear when they fully awaken to this fact! It is good to keep this caveat in mind, while not dwelling over-much upon it. Otherwise, even this occult knowledge could easily become just another entangling net of concepts!

"Memory is the scribe of the soul." – Aristotle.

A Radical Departure!

The Very Idea That We Are Not An Incorporeal Soul, our body, or even the content of our minds, is a rather bold assertion! And moreover, it also seems to depart radically from our common knowledge and experience, and thus certainly begs an explanation before it can be easily accepted!

By the time that our sensorial machinery has actually processed its information into a cognizable thought form, it has already secretly slipped away into the "past". But what of our evident present? Surely it exists somewhere! Yet if the present does exist, it could only exist in the somewhere of an "Elsewhere", since it is clearly not accessible to our tridimensional psyches and somas.

This is to say that all that we can ever really know is the contents of our concepts, which take both space and time to construct and even more spacetime to cognize!

Ergo, we are, quite simply put, always just a cognitive step "behind" whatever we can sense or think! Therefore we either have to admit that our purported "reality" is itself not real, that it is simply another concept or collection of concepts, or that it ever lies in a transfinite, or perhaps even a "spacetimeless" moment before our ability to sense it dualistically.

At least for now, the only real authorities, on the rather slippery "esoteric" subject of our consciousness, are still the awakened Sages and Masters of the past. And they generally concur that the actual "present moment" exists as a spacetimeless "über Reality" of Universal-Consciousness.

Our familiar subject-object consciousness thus appears to be the creative art of a projected, or emerging Universal-Consciousness and its percipient source of witnessing Supreme awareness, operating within, and as, our body-minds.

Thus, our body-minds cooperatively (and subliminally) "create" our experiential objects of consciousness. These consciousness objects of thought then appear as apparent "reflections" within our essentially spacetimeless Universal-Consciousness.

But in order for our individual consciousness' to accomplish this astonishing fete, they must also do so to an independently witnessing Supreme Awareness.

Simply because all experience, as we currently understand the phenomenon, requires both an observed object and a witnessing subject! This invariably must also include the phenomenal objective ego "subjects" by which anything objective is known.

Therefore, due to our present limitations, we experience everything tridimensionally, as a seemingly actual subject surrounded by real objects extended in, and separated by, space and time. However, everything is apparently the force and elaborated energy of Universal-Consciousness. Which works in the most marvelous of ways, to allow the monadic noumenal Supreme to experience itself in potential diversity (i.e. through us, as Its "avatars").

But from our currently limited perspective of understanding, evidently it can only do this in three dimensions, as ever changing and inter-reliant space-time perspectives. (Thereby creating a convincingly real phenomenal manifestation, which we then mistake as our very own!)

Thus, it is important to understand that everything is essentially, always, the absolute Supreme. Which is most likely then an *actual* "natural", rather than a *suppositional* "supernatural" Supreme. A percipient "First Cause" of the Primeval Singularity.

(Which will eventually be more fully revealed by science, or perhaps even continue to be inadvertently "disproved"!) And not the phantom deity of humanity's unenlightened past, which continues to menace our species progress and our planet's very survival!

In other words, a fundamental "Supreme" is not only the probable Source of our universe, but also its supreme Witness. And the Force of Universal-Consciousness is its extension into our present quantum universe of virtual wave and particle.

We experience this as a virtual composite of quantum material that possesses a potential mass and energy, which acquires its fuller "actualized mass" and energy on a three-dimensional scale, by the subtle consciousness force of our experience of it, within a (not-strictly) "ideational" four dimensional space-time.

Moreover, our space-time universe is simultaneously bounded, constrained, and impelled by the presence of the rest of the endless multiverse that it is connected to. This appears to our dimensionally challenged consciousness to be an invisible and mysterious Dark Energy and Dark Matter that nonetheless profoundly effects our visible universe!

We have evolved within this elaborated universe as biological instruments with the capacity to cognize the virtually limitless potential of Universal-Consciousness within a four dimensional space-time continuum, by utilizing a phenomenal three dimensional subject-object perspective.

Whether or not an imaginary exoteric "god" created our illusory experiential universe is thus quite beyond the obvious point of its grander Consciously universal existence and our natural compulsion to understand it.

This is why the Guatama Buddha replied, with the quizzically self-contradictory statement; "There is no difference between Samsara and Nirvana", when he was asked by his most learned disciple to expound on the differences between the two.[(98)]

On the other hand, if we are only aware of a present, future and past, which are not only partial, but in effect an illusion, and we are in fact just another part of it, we cannot even reliably say whether or not we even truly exist at all!

If "being born" means the awakening of our consciousness as an illusory body-mind within an illusory dualistic universe, which we then take to be real, then having actually never been born, we are essentially little better than a zombie or a walking corpse, until we realize the truer nature of our experience.

But, having never been born, then we can also never truly die! Apparently what does eventually "die", as we begin to wake up, is our habitual identification with our illusory persons and universe(s).

This is what the *Avatar* "incarnation of god" Sri Krishna likely meant in the epic Hindu tale of the *Bhagavad-Gita* "Song of the Divine", when he advised the stories protagonist (Arjuna) that; *"Whatever exists will never not exist and whatever does not exist never will, but their meeting place is seen by those who see things as they truly are"*.[99]

This clearly states the oneness of Samsara and Nirvana, but also says that there is a grand dream of sorts going on, a dream in which we participate, but which is not merely our own dream.

Personally we are only dream characters, until we wake up to our true impersonal status as the Universal-Consciousness of which our durable waking dream is actually comprised!

But ultimately, this too fades entirely away. As it eventually dissipates back into the primal spacetimeless Awareness. Which is the universal substrate of *Sat* "That", by which our only seemingly "personal" persistent "Waking Dream" is known.

This is our Superverses absolute and supreme Awareness (i.e. יהוה *Yahweh*, الله *Allah*, Deüs *G-d*, ∞ "*Elsewhere*", etc.). From which our fundamentally spacetimeless "Monadic" 非人格 *no arimasen* "impersonal" (i.e. a Japanese Soto Zen term) Divine consciousness, eventually emerges into a state of experientially individual diversification. Which it does by way of our ideational "personal" dualistic consciousnesses.

Indeed, evidently only a largely misunderstood, but nonetheless thoroughly absolute and necessary G-d, could manifest such an astounding experience, in such an incredibly multidimensional universe!

The Essential Understanding

"All Of This Sounds Interesting", you may be thinking, "but is there not a simpler way to understand? What is the most essential knowledge?" I must confess that I do agree and will try to render everything down to what I understand that a living Buddha might say are the essentials.

However, it is essential *a priori* "in advance" that we define our terms. First we must do away with the plethora of metaphysical distinctions regarding consciousness and agree to use the term "Mind" to represent them.

Next we need to know just what "understanding" means. In the automatic processes of our cognition, understanding necessarily follows upon perception. In other words, if anything is not at first perceived, it cannot subsequently be known or "understood" to exist.

But here we refer to knowing the specific and particular existence of something, by which it becomes "real" to us, and thus not merely to an act of our sensorial perception and psychomechanical cognition.

Our sensorial perception is thus, in each and every portion of its cognitive process, existent within our essential Universal-Consciousness. Hence, we are also not referring to the mere physical processes of neuronal activity in our brains.

This is to say that "form" exists contiguously with "function", primarily as a matter of our interpretation of perspective and scale, and that both are defined and modified by the Force of Universal-Consciousness, of which they are comprised, and within which they exist.

Moreover, "form" in our universe is not merely made up of some physical material, it is the very quantum stuff of which the universe itself is comprised. Indeed, it is none other than the Self-same, self-Aware energy of Universal-Consciousness.

Which appears to have arranged itself, seemingly by tedious "natural selection", through a durable molecular recording process of genetic evolution, into our present day human bodies and brains.

Our brains and bodies, without question, together underlie the dualistic appearance and functions of our seemingly individual consciousness, within the phenomenon of our individual minds.

But they are no more solely responsible for the appearance of consciousness within our minds, than they are for their own projected symbolic appearances within our (only apparently "individual") experiential universes!

However, the "understanding" that we are specifically referring to is an uncommon understanding. Its instrument is primarily our intuition rather than our intellect, although clear and certain intellectual understanding most often does precede it.

And because our intuition is the primary vehicle of this, our uncommon "deeper" understanding, perhaps we can together agree to call it "apperception" to distinguish it from perception, and our more common dualistic mode of intellectual understanding?

Now, having defined the more essential terms in our discussion, our communication becomes much easier (for greater clarification, see "Apperception" pgs. 504, and 517).

The essential understanding seems to be that everything exists within mind. This is to say that everything "exists" only within mind, and can have no other real independent existence whatsoever! But this is not to say that nothing is there. Just that it is not fully so, until it is *experienced* into existence (see "Microtubules", on pg. 550).

If you fully understand this, then the further apperception inevitably follows that our mind, as such, does not exist independently at all! This is because our "mind" is but a symbolic representation of what we are, as well as whatever we experience, and therefore it cannot be perceived as an actual object, which is separated from that which perceives it.

Yet, to "exist" our minds are apparently as dependent upon our bodies, as our bodies are dependent upon our minds. "Co-dependents" of consciousness, they have no separate presence, and can exist only experientially, due to the necessary auspices (assistance) of an independently witnessing (and purely self-aware) Presence.

Our mind is thus most easily understood to be a deeper expression of the curiously personal descriptive noun, "I". Indeed, when our mind rests in itself as "I" it is whole, or "whole Mind". And when whole Mind turns outward to objectify (manifest dualistically), it appears to divide us into subject and object, and is now the illusion of our "split-mind".

Therefore when I become an apparently personal subject, whole Mind splits into an ideational "I" subject/object, and its locative object "that", or person object, "you". In other words, the basic illusory dichotomy of so-called "self and other" is then established!

But un-objectified, I always remain "I" in my essential whole Mind, and apparently you too remain "I" within your whole Mind, since we are both "functionally" sentient in quite the same way! Therefore whatever I can say of you, you can also say of me, since we are both evidently also implied "objects" in our split-minds, of the same essentially impersonal, yet nonetheless apparently personal, "I".

In fact, this *is* the "essential understanding"! However, this essential understanding alone is probably not enough, for we must also come to understand how whole Mind, by objectifying to know *it*-Self, becomes our split-mind.

In explaining, by simplifying, we will also be reviewing, which is apparently necessary to do quite often, and from as many critical angles as possible, until a deeper understanding of our own consciousness finally begins to sprout and grow within us.

Evidently, we must do this in order to answer every doubt and misunderstanding that might arise, until our Self-understanding has matured and at last become self-sustaining and complete.

Therefore, all that we can probably accomplish here, is to define and encourage this process through understanding. Because only when every doubt and misunderstanding has at last been removed, can our minds become truly "ripe", or ready for the deeper intuitive understanding of nature's true apperception to dawn.

But even this evidently more enlightened state of mind is most likely not-quite-yet either a completely "Self-Realized", or a fully "Awakened" one, because our transcendent Awakening is neither a process nor a mental state.

It is, quite simply, the reawakening of the deeper lying, foundational presence of our more authentic, primeval Awareness. Which ever abides prior to the appearance of our individual consciousness and all of our split-mind experiences, which are subsequently contained therein!

This is to say that it is not the actual complete metanoesis that is the shifting of our locus of identity away from our apparently split "subject-object" mind and back to our actual whole Mind. It is not yet the reintegration of our self-identity back into our essential "I".

But the achievement of this contingent apperception, whether it happens all at once or gradually, is apparently the necessary precursor of any full or true Awakening. For this reason many Sages consider this sort of apperception to be synonymous with the state of so-called "Enlightenment", or in traditional Rinzai Zen parlance *Kensho,* a "cutting through" (of illusion).

This is nonetheless still just a rarefied intuitional mental state and it will pass with the eventual death and deterioration of our body-minds. So now, having become an *Illuminatus* "Enlightened One" by finally reaching the top of our subject/object's experiential "Tree of Life", how can we possibly climb even higher, before we eventually lose our existential grip and fall?

A Red Tide

SOME WOULD SAY, and perhaps not without good cause, that our space-time bounded life is much like a turbulent "Red Tide." A sanguineous tide that first rises up as a vital living "wave" but eventually must fall back upon itself in hunger or exhaustion, as it indifferently feeds upon itself, in the necessary pursuit of its own survival.

But every lifeform began as One, eventually spreading out like "rubicund" corporeal water into different space-time places.

Only over the long and slow passing of considerable space and time, did Life become discernable as many (only outwardly) different, and thus concomitantly "individual" forms. With the ability to discern their own apparently different existences, within ostensibly different, but nonetheless locatable spatiotemporal positions (i.e. "places").

But eventually, Life did assume the appearance of being many effectively "illusory" individuals, located within many different forms, but nevertheless emerging from a common root.

Thus, only by extending itself over considerable space and time, did Life eventually assume the genetically interrelated architecture of different living forms that we now understand to be our planet's extensive biological "Tree of Life".

What most researchers, and even a few brilliant Biologists, don't generally consider, is that Life has somehow found a way to "invent" symbolic space and time, in order to do so! To "real"-ize this essential feat of Self-deception, the "Force of Life" (or "Life-Force" if you prefer) had to somehow durably "create" the illusion that it actually existed as discrete individuals.

To accomplish this illusion, Life needed to somehow progressively generate durable three dimensional forms, within the evidently "invisible" space-time illusion that it had finally accomplished through a long process of trial and error.

In order to do so, "Mother" Nature has artfully assembled "living" organic structures. Mostly using the "stardust" atoms of carbon, hydrogen, oxygen and nitrogen, *(COHN).* Which were violently birthed within the hearts of exploding stars.

Novel "organ"-isms, with the ability to experience themselves in different symbolic, but nonetheless changeable four-dimensional spatiotemporal positions, as they went about the important business of surviving, reproducing, and thriving.

Which Nature could only do, through a carefully preserved, yet evidently malleable and impressionable, and constantly growing consciousness. A consciousness that is also being phenomenally symbolized in various durable ways. Within what appears on our scale of observation to be "solid" matter!

Thus began Life's interminable effort to "survive" long enough to progressively advance a durable-yet-malleable genetic design! But the essential identity principle that resides within the Force of Life, and within all of its space-time extended forms, remains the same as it was in its genesis. And so, even though our bodies age, and our minds must eventually change, our veridical natural Self-identity remains ever unchanged.

Instantly, and almost always, as we steadily from our infancy grow, we also come to "know", and when we do "know", we also instantly know that we *are!* (This is to say, that we only then seem to "exist" within space and time.) Thus the Force, Form and essential Awareness of Life are, and evidently must be, present throughout its presence in our comparatively insentient universe.

And so, the universe comes into and passes out of being like the ebb and flow of an ocean's tide. Existing, as it evidently does, in such a speculative fashion, all that can be accurately said, of our ongoing experiences of an evident "universe" and "person(s)", is that they are our mind's interpretation of a biological event that is spontaneously occurring within our brain!

This is the "Red Tide" of universal Life in motion, generating its experiences through a phenomenal cognition, which only seems to be ours alone. But we cannot be whatever appears before us in our cognition of it. Since it is always our object, and we are always its subject, and a subject cannot ever be its own object!

Moreover, this "pretty much" includes *everything*, except for our essential Selves, upon which our every experience depends. Which is the presence of the impersonal "Sentience" of the Supreme, by which we know that we and others exist. And can thereby experience the apparent existences of our own and others experiences.

The sum of which comprises the existence of our entire universe! Both within our essential wholeness as the Supreme, and within its extension as our only evidently "divided" minds! And if this is so, then in "Essence", the entire human race is nothing (more-or-less) than the Supreme/Elsewhere, latent within Universal Consciousness, exploring its own manifestation!

ما یک قطره نو اقیانوس، اقیانوس در یک قطره!

"You are not the Ocean. You are the entire Ocean, in a drop!" – Rumi

Cosmic Bits and Universal Bytes

VRITTIS are the cosmic "bits" and universal "bytes" of a ubiquitous quantum Manifestation.

Continuity and the Tao

THERE IS A IS CONTINUITY of Awareness in our universe that links the phenomenal and the noumenal with our physical vehicles of (Its) individual experience. Some have called this continuity the "Tao".

The Dual Play

WHATEVER WE BELIEVE is "our" experience, is actually consciousness cognizing itself, as though through a character in a dream.

Thus, from evident "top" to actual "bottom", everything in our universe is the dual play of consciousness, appearing to itself as "us", within its ongoing cognitive play.

But we cannot be either this consciousness or its play, for the simple reason that we *witness* the endless universal performance of both consciousness and play!

Likewise

EACH OF US IS NOUMENAL to the others phenomenality and likewise to ourselves, but only through the presence of the Life Force within our body-mind, which is taken in with the food that we consume. This *Chitti* "Consciousness-Force" is an organizational "element", the *Sattva* principle of our identity. Hence, what many believe to be "enlightenment" is only the cognizance of this principle within the Universal-Consciousness by which it is known.

However, this ostensible "enlightenment" is also clearly transitory. It does not abide beyond the existence of our own physical bodies because it is simply a greater cognizance of our own internal sense of "beingness".

Thus, poor misunderstood "enlightenment" is simply a more correct "Cosmic Perspective", in which we can all share while we are still embodied. However, it is also quite likely a גשר *Gesher* "Bridge", between the mind of man and the spacetimeless, pure Awareness of the *Parabrahman* "G-d the Supreme". But with the final dissolution of our body, our "Gesher of Understanding" vanishes too.

This is the seminal meaning of the "timeless", and once well known, Japanese Rinzai Zen Kōan about a man who climbed to the top of a "fifty-foot-pole". Which (in contemporary terms) first asks the rather inscrutable question, "How can a person possibly climb higher than the cerebral 'executive top', of the operating system of the human branch of the 'Tree of Life' that he is currently in?"

Next, it reminds us of the inexorable "Law of Entropy". Which the Buddhists call the *Dharma* "Truth", of impermanence. Because, although the person sitting on top of the fifty-foot-pole may well be able to see for a great distance, perhaps even much further than the people who are on the ground below, nonetheless, eventually they must either fall to their death, or slowly descend from their erstwhile "lofty" position!

But in contra distinction, when our "Gesher of Understanding" finally vanishes, then only our reawakened aspect remains, being no different than the supreme "Parabrahman" (יהוה *Yahweh,* الله *Allah,* ∞ *Elsewhere,…*), even before and after every possible universe in the multiverse has come and gone.

Nonetheless, in the end, we are mostly, if not wholly, merely consciousness events that are spun into a temporary sort of "I Am" beingness, from the ephemeral "cotton-candy-floss" of our own consciousness.

We dwell inside an ideational mind that is essentially comprised of "mist and whimsy". Indeed, it even seems unlikely that our individual experiential "universes" even exist, as we commonly experience them. And in the end, only the Supreme ever abides! Thus, we are merely a rather miniscule portion, of a rather "Grand Conundrum", indeed!

No Difference!

I Can See No Difference! This is to strongly reiterate that Consciousness is "consciousness", no matter the method or material of its expression! There is an unbroken line of (its) ascension through an endless field of potential ("possibilities"). From the appearance of the first Primeval Singularity, to this very moment of Universal-Consciousnesses appearance as our evidently individual consciousnesses.

Therefore I can see no reason not to propose an imaginative hypothesis, which is itself enabled by consciousness, that what we are currently witnessing, perhaps first here in "Silicon Valley", is Consciousness (with a capital "C") starting to take control of the quantum material of which it is made.

Since there is clearly a technological solution for this to transpire (either biological, mechanical, or more likely an amalgam of the two), it is only a matter of sufficient space and time before Universal-Consciousness, operating as our "individual" consciousnesses, will gain control of itself on the quantum level.

When this transpires, consciousness will eventually become universally aware of itself on the membrane level of its constituent strings (for a quick refresher, see "String Theory", pg. 581).

At this moment the universe will become a sentient, single über Mind. Space-time and energy constraints will become irrelevant, and the universe will be taken back up into itself, as it were, into the Elsewhere from which it emerged.

But we could as well say that our pursuits are simply broadening and deepening our horizons. Nonetheless, the final results are quite the same! So, why not call it like it is? Our universe, through our apparently individualized consciousness', is simply beginning to take a greater control of itself on the quantum level. And it is doing this through the ever growing presence of its essential Sentience, within our rapidly growing dualistic hominid minds.

In light of this likely fact, the question of when and how consciousness emerges, becomes a non-question, in that it is always either "here or there", or not "here or there". Except that whatever exists can quantumly be both "here" and "there" at the same time! As only apparently different expressions of the same quantum substrate.

Just as a spoon and a fork are both made of the same metal and are always present as such. But of course in our three dimensional realm of cognition the same spoon (or fork) cannot occupy two different spaces at the same time! Physicist's aptly call this quantum phenomenon of bilocation "simultaneity".

Yet, it also seems to be an unavoidable fact, that whatever is either "here or there" in our usual three dimensional experience of four dimensional space and time, can never be both quantumly "present" and "not present" in the same space, and at the same time!

Because whenever Universal-Consciousness appears or disappears, its appearance as either "somethingness" or "nothingness", within our dualistic individual consciousness', continues to depend upon the presence of its polar opposite, to either be, or not to be, there in space and time.

Whole Mind: Split-Mind

"I" Remains Eternally "I"! In other words, as long as our body lives, whole Mind always abides. Moreover, our whole Mind is more the real, while our "split-mind" is doubtlessly unreal! Because our split-mind is merely the phenomenal effect of our whole Mind splitting into the perspective illusions of our familiar subjects and objects.

Indeed, the entire phenomenal manifestation of our person and universe, within our split-mind, is an illusion being generated by redundant cerebral processes. This results in the reflection, or eventual transference of identity from our true essential Witness in whole Mind, to our necessary (yet patently impossible!) subject/object ego in split-mind. Which is actually an object (ego) that only appears to be an independent subject within our phenomenally split minds.

The ego subject in our split-minds, however, is only one half of the projection of our whole Mind into an illusory objectivity. The other half is the cavalcade of apparent objects that are experienced as such by their own subjective aspect, which is our evident ego. But this is most difficult to see, because our illusory ego even has its own redundant false self, the *persona*, or "formal face" that we present to other egos!

Thus both our subjects and objects are actually opposite poles, or complementaries, of the same faux objectivity that is being ideationally projected, as an inference of our whole Mind.

Our whole Mind is therefore always here, *(t)*here and every *(w)*here"! Indeed, here, there, and every possible מקום "where" are actually always within an unobtainable "here and now!" Hence, they are also actually never precisely "here", *nor* exactly "there", within our experience of "being there", in the here and now!

Entertaining mnemonics (memory aids) aside, simply put, we exist within our whole Mind's phenomenal *subjective* experience of its own *objective* portion. Which is always witnessed here in the eternally present moment of its own experiencing!

I

This is also to say that our evident persons and universe(s) exist as events within the experience of the subject portion of our phenomenally split-minds. However, both our subject and object portions are essentially, always whole Mind!

They are, quite simply, the complementary aspects of our whole Mind. Which has involuntarily bifurcated and morphed itself into them. Thus generating the compelling experiences of our persons and universe.

But our ostensive "persons" and our (individual) "universes" are merely concepts of thought appearing within our split-minds, and therefore have only, or have only there, a relative existence. Nothing, or no real "thing", therefore has any actual independent existence, either as a subject, or as an object. Furthermore, any existence whatsoever is only a transitory illusion within the subjective portion of our ideationally split-minds!

This illusion is created by the ubiquitous reality of our whole Mind. In other words, our whole Mind is also the real "ity" or "ans", *ut hic* ("as used here"), of the presence of *it*-Self within its every, only apparently dualistic aspect. The dualistic illusion of space-time is thus the projected (ideational) field of the whole Mind in which our dualistic cognition operates.

Therefore, with all due respect to the inimitable Sigmund Freud, it is only because of whole Mind's presence within the consciousness of our body-mind's objectival "ego" subject in split-mind, that our phenomenal subject not only knows, but also "knows that it knows", and thus appears to be (and acts as if it truly were) objectively sentient!

Because, while a "cigar" in our nighttime dream world is sometimes just a cigar, the ego "subject" in our daytime dream world is always just an ego object. And moreover, both are merely the illusory products of a passing dream!

However, because our split-mind's subject/ego is actually a purely phenomenal object, it can only *know* of its own existence objectively, and be *Aware* of its "knowing" noumenally (non-objectively)!

Thus our everyday dualistic consciousness, of person and universe, appears to be a closed system of subject and object. And our objective persons and universe(s) only seem real, because they exist "to" and within, the purely sentient presence of the eternal Witness, which is our ever-present whole Mind.

Yet, even this essential knowledge must be expanded as a thought form, an "ity" or an "ans", within our subjective portion of split-mind, in order to be intellectually understood!

But intellectual understanding, like any other sort of perception or cognition, is still essentially only an objectification within our split-minds.

Just as when we stand between two mirrors, our dualistic consciousness appears to be a closed system that seems to go infinitely on and on when we gaze into it!

Therefore at this point in our inquiry we are in near agreement with the French philosopher Vincent Descombes's modern "Philosophy of Mind", which attributes every mental event solely to some underlying neurological or cerebral activity. However, consciousness cannot operate in a vacuum! It is also dependent upon an independent witness, space and time.

II

As apparently individual subjects in split-mind, we are able to experience the similitude of reality within our thoughts only because our essential witnessing Self, through "our" whole Mind, is the actual Reality by which we know that we exist.

Nonetheless, our essential witnessing aspect is invisible to us, because it is "That" by which we experience a dualistic reality within three dimensions. And this is not at all difficult to understand when we consider that it is impossible, for example, to look into our own eyes without the aid of a mirror.

But the "mirror" in this case is like the faux reflective surface of an ever illusive "Negev (desert) Mirage", because it is actually our own consciousness, which contains both our ego subjects and "its" (our) phenomenal objects.

The sensorial information we receive is expanded within three dimensions, which gives it apparent volume. It is then imbued with a virtual reality of existence within an ideational time. Because we witness what we experience with our transcendent Self which is embedded within the actual spacetimelessness of our undivided whole Minds.

This seems to happen sequentially, because one thought form must be "dissolved", like a pillar of salt suddenly collapsing into the Dead Sea, before another can rise up to be experienced and understood within the desolate dualistic-desert of our split-minds.

Those few who have by-all-reports "awakened fully", and thus perceive this process more directly, describe it in the most curious of ways! Accordingly, the inscrutability of their unique נבי *Navi* "prophet-like" perspective emerges, no doubt, from their impersonal and more universal point of view, which only presently seems to be far beyond our own!

Thus, even a nonapparent "nothing", always remains "something". Just as vacuity is never truly empty and even absolute zero can never be reached! And since this is another aspect of our essential understanding, and although it is a rather dubious task, I will do my best to make it intellectually understandable too. This is where experiential space-time enters the picture, in order to render the universe visible within our three-dimensionally limited consciousnesses.

III

An important point that is often missed, is that whatever appears to be present within three dimensions, will forever remain within three dimensions! In other words, whatever we experience is experienced from a "superior" dimension. To further clarify, we cannot see the experiencing medium of our own impersonal Universal-Consciousness when it is acting as our whole Mind.

We cannot see it, because it is what is doing our whole Mind's seeing, into the virtual universe that is being spontaneously generated in our dualistically "split" minds, on a subliminal level, by the "biomechanical" instrumentation that is housed within our body-minds. Indeed, it is precisely our whole Mind's seeing which creates, or makes seemingly real, that which we experience as our individual persons.

Only thereby are we able to "perspectivally" share our, only apparently, "individual" experiential universe(s)! Which, like an invisible stream of ever-flowing water, are thus merely autonomous phenomenon.

Consequently, our seemingly "unquestionable" experience of a four dimensionally bounded, space-time defined "reality", is not unlike a cascading waterfall that is being constantly refreshed by our split-minds four streams of ideationally extending dimensions. Which are comprised of the same everchanging Universal-Consciousness! …see, Wu Roshi's "Waterfall" allegory, on pg. 213.

"A river flowed out of Eden to water the garden, and from there it parted, and became four streams." – Bereshit (Genesis) 2:10.

IV

However, a puissant clue to the presence of our whole, or often inexactly called "higher" Mind, is that of our qualitatively different experience of space-time. This experience, as phenomenally postulated by Albert Einstein, was also rather ingeniously called by him the "fourth dimension" of *space-time*.

Simply put, space and time are interdependent aspects of our cognition that actually transpire within our whole Minds, but operate within the mechanism of our split-minds.

But empirical science has not yet taken the next step, beyond mere "Instrumentalism", of connecting our cognitive tool of implied four-dimensional perception with the necessary fifth-dimensional perspective of our foundational Awareness, to which it must appear to "exist", in order to be understood (to exist).

This is simply because it can only be viewed, as we have seen, from a superior dimension. Just as three-dimensional objects can only be seen in their full dimension, from a fourth dimension! (See "Instrumentalism" pg. 546, and "Pragmatism" pg. 566).

Is it not patently obvious however, that it is only within the relatively higher dimension of our whole Mind, that the objects of consciousness, which represent conceptually our persons and universe, are extended and separated in space over time? Is it not also obvious that our need to do so depends upon the limitations of our faculties, and could only happen sequentially as a result?

In other words, our mind's objects of consciousness are dependent upon an awareness of the *act* of their cognition, to exist meaningfully! So, whatever else could our whole Minds reliably be, if not an aspect of some sort of "universal" Consciousness?

Everything that appears before an experiencer is thus a cognitive act of his consciousness, and exists only within this act. But the awareness of cognition is intrinsic in our whole Mind. Which by focusing through the mirrored "body lens" of its subjective ego/object aspect in our split-mind, brings our perceived tridimensional objects into a relative ideational existence, within our dualistic consciousness.

But the focus of our whole Mind's awareness is invisible to our three-dimensionally constrained split-minds, because it is necessarily interpreted as the empty space between objects and their extension in times duration, both of which can only be known by the inference of comparison.

The relativity of this comparison is the measurement within our consciousness of an object's existence over time as its "duration", and the distance of time between consciously existing objects as its (only apparently physical) "space". Space and time are thus cognitively inseparable and interdependent aspects of one another.

For instance, the greater the duration and distance, the "older" the object and "farther away" that it appears to be. We assume that the object of our experience is farther away because it appears to be smaller and less distinct, and it takes a seemingly long duration for us reach it, and for it to expand in full detail within our three-dimensional cognition of it.

Consequently we cannot even say, with any real force or form of certainty, whether this actually happens through an experiential change in perspective and scale, or an actual "physical" space-time journey!

V

But of course, the reverse also appears to be true, except that the "arrow" of time is thus cognitively bound to travel in only one direction.

(Except insofar as "antimatter" is concerned, as Werner Heisenberg so aptly pointed out, where time apparently moves backwards in relationship to our positive matter universe!)

Yet in actuality, it is from what we interpret as the space between objects, and the time that appears to measure it, that the objects of our consciousness have their beginning and their end.

And as Einstein (also) predicted, this is the reason why we go backwards in time whenever we exceed the speed of light. Simply put, this is because there is nowhere else for us larger three dimensional beings to go when (or if) we are finally able to travel at (or past) FTL speeds within the spacetime bounds of our four dimensional universe!

But in order for us to quantumly "leap" into another parallel universe like a quantum sized electron does, from one orbital level of "electron cloud" potential to another, it would likely require the same level of almost inestimable energy that caused the multiverses to endlessly split off from one other in the first place!

If we one day do succeed in "quantum leaping" into another universe, it will simply have to be done in some other fashion. (See "Frank Tippler", in the fifth paragraph on pg. 590).

Moreover, just because an object seems to be farther away, does not necessarily mean that it is! Since every ideational point within an inferred space actually touches every other point within the consciousness which cognizes it.

Just as it does in the energetic quantum "sea" which spacetimelessly connects our entire universe together in pure subatomic potential, until we experience it. And it is by experiencing it that we make it "real".

In other words, what we measure in our split-minds is neither space nor time, because we are really only measuring a projection of our own consciousness!

This incessant mental "measuring" is actually our dualistic method of cognition within four dimensions. Empirical science generally infers that this is a phenomenon of causality operating within an evidently *independent* field of space-time. This, it assumes, is a strictly physical generation and therefore subject to the surmised laws of an ensuing entropy. The necessary involvement of consciousness in the processes of the universe that we commonly experience is thus conveniently "sidestepped"!

From a strictly "Instrumentalist" standpoint this makes sense, if you are only interested in the traditional hypothesis and applications of a science and technology that is becoming rapidly outdated as we look more deeply into the internal structure of our universe.

But not in the realistic understanding of what is actually going on in our veridically singular yet peculiarly "plural" appearing universe. As our consciousness subliminally interacts with it on the quantum and subquantum scales.

Nikola Tesla the visionary "Inventor" quickly realized, for example, that it is not necessary to understand fully the phenomenon of electricity, in order for it to become technically useful. However, Nikola Tesla, the inquisitive "Scientist", also knew that understanding electricity more fully would, no doubt, one day make it even more so!

And so, he simply never stopped trying to expand his knowledge beyond the boundaries of his current understanding. Despite the protestations and derisive "Crackpot!" accusations of his scientific contemporaries. But today, 'most all of us have come to understand that our modern world stands on the shoulders of strong and determined people, with just this kind of "grit"!

My uncle Sherman was a man of small stature, with "rattlesnake fast" hands. And although he was by no means a scientist, he nonetheless had the eyes of an artist, and the heart of a lion. I suppose, this was also why he never seemed to notice any difference between himself and any other man.

"What's on the outside", he once said to me, "doesn't really matter much at all Paul (that's what everybody use to call me, before I changed my name). It's what's on the *inside* of a man that counts, when the shit really hits the fan!"

My uncle Sherman was plain spoken, and perhaps at times even "crude", by today's more refined standards. But I still remember his response to a big jilted young California Cowboy who was out looking for trouble at a Modesto livestock auction.

When the drunk young Rodeo cowboy (who still had his number pinned to the back of his shirt!) tried to provoke my uncle by making fun of his Southern drawl, my uncle Sherman just looked up and smiled a sort-of "sympathetic" little smile, puffed out his chest, and proudly said, "Yup, *Oklahoma* crude!"

Then we just turned and walked away, to buy some new calves for the ranch. We never talked about it again, but I also never forgot what I learned that day, because I knew what my uncle could have done to that foolish, unsuspecting young man!

Because, my uncle was not just one of my childhood heroes, he was also a *bone fide* World War Two hero that later became a contender for the coveted title of "US Army Welterweight Boxing Campion", before his tour of duty was finally over.

(I was always very proud of my uncle Sherman, who came to California with my father from Oklahoma during the Dust Bowl, and picked fruit to earn money to send back to my blind grandma and his younger siblings after grandpa died.) And I continued to "look up" to him for the rest of his life, even though I eventually grew so tall that I almost towered over him! And even though I became a wrestler (*got zol ophetin* "G-d forbid!") instead of a boxer in school, he always treated me like the son that he never had!

And just between you and me, I'd give almost anything, just to hear him sing another lonesome old "Hank Williams" song as we ride along together in his beat up old "Ranch Green", Ford F10, pickup truck. With the windows down, under an alfalfa perfumed, star filled, wide open country sky.

But, I must again beg your indulgence at this point. Because, for some reason, when I'm all alone in the forest primeval, occasionally the people that I have loved, but can no longer talk with, seem to "drop by" for an unexpected visit!

"Whenever they's a fight so hungry people can eat, I'll be there. Whenever they's a cop beatin' up a guy, I'll be there..." – The Grapes of Wrath, by John Stienbeck, 1939.

VI

However, as we return more directly to our enquiry, it becomes rather quickly evident that the processes of a deeper "impersonal" sort of Consciousness is apparently involved in every aspect of our "universe". Which we clearly cognize into a conceptual existence.

Thus, more simply put, since it is a *Universal* Consciousness that most likely underlies the subatomic string "stuff" of which the universe "at large", including our illusory individual consciousness', is comprised, it simply makes no sense to ignore the inevitable!

It then follows that what we experience as "time" must also happen sequentially in order to provide a thought image in our memories, to compare to another thought image in our imagined present, with the most probable next one in an anticipated future moment.

This is because the only way that we can cognize our universe in three dimensions, is by the comparison of mental objects within an unseen, but implied, fourth dimension.

Further, whatever appears to us within a fourth dimension is necessarily being viewed from a fifth, which of course does not go on *ad infinitum* "forever", except in potential. Since even so-called "dimensions" are, in themselves, also the interactive instruments of our perception (see "Infinite Regression", pg. 545 and "Objectivist", pg. 559).

This is not to say that realities of a different order are not present, just that insofar as our dimensionally challenged cognition is presently concerned, they exist only in potential until they are experienced within our consciousness'.

But once they are cognized into existence, they don't just simply go away, because we can no longer recognize their presence! They remain existent within the multidimensional underlying lattice structure of endless parallel universes, that spacetimelessly underlies the humanly visible three dimensional "veneer", of our ever expanding Superverse.

In this fashion, the entire Superverse, including our own little lives, is quite simply "Eternal!" This is to say, that if we possessed the right type of instrumentation and technique (predicated by a profoundly deep scientific understanding of space, time, and the greater Multiverse that we are all a part of) that there is no substantive reason (other than "time and ignorance"), that I should not be able to actually hear my "living" Uncle Sherman, sing an old Hank Williams song, in his rich baritone voice, at least one more time!

Because, as incredible as it may sound, somewhere within the endless interdimensional lattice-like "maze" of parallel iterations of our universe (which altogether comprises our conglomerate "Multiverse"), he is actually still driving his old Ford pickup truck, along that same old dusty country road!

And just as I happily recall from my childhood, it still meanders gently out from his fragrant, hay filled barn. Through a field of aromatic, freshly harvested alfalfa hay. Under the endless expanse of a wide open country night sky, full of bright childhood stars. While I sit there quietly by his side, listening to him sing that same old lonesome Hank Williams song, with that same old soulful, deep southern drawl!

And moreover, although it's scientifically sound, quite contrary to our more common sense, *he always will be!* In other words, and rather ironically I do believe, it is not "religion", but "science", that may one day actually be able to assist us in reconnecting with our (only evidently) lost pasts, and far too often, even with our long lost roots!

However, this can only happen, if we can manage to consolidate a sort-of future "alliance" with the slowly emerging "Strong AI" of the incipiently unfolding "Consciousness Singularity". Which is, unseen by most, beginning to subtly occur all around us.

But it is not really even that subtle here in Silicon Valley! Where our cars are beginning to drive themselves, drones are beginning to deliver airborne packages to our doorstep, and even our everyday appliances and phones are becoming noticeably more intelligent by the day!

Hence, our individual experiences are born of our own, only apparently independent, so-called "realities". Indeed, even the scientific observations that are borne of accurate and exhaustive theorization only brings the probable objects of theory into sharper focus within our subject-object consciousness, albeit generally in a very exacting manner!

But their true perceiver is the pure Awareness, which effectively stands alone, and is necessarily always prior to our every conscious experience. However, anything that stands apart from our consciousness is also apparently null. It seems to be void, even though we may know that it must exist by providing rigorous mathematical proofs of its existence!

For example, anything that may exist in what we think of as a "superior" or "alternate" dimension is invisible, yet its existence may be logically inferred within our dimensionally challenged consciousness from observable evidence, and repeatable, revelatory experimentation.

But most certainly, simply understanding this intellectually is not enough in itself, and since it also cannot be ignored, we must somehow become experts in a more direct fashion in the area of our own consciousness!

This is only accomplished by diving into and learning to "swim", in a more aware Taoist-like "Way", within the dark inner subquantum "waters" of the Universal-Consciousness of which our entire universe is evidently comprised. *"Ask and it will be given to you; seek and you will find; knock and the door will be opened to you."* –Mathew 7:7.

Indeed, upon closer inspection even the ancient mystical symbol of the fish has very little, if anything, to do with any subsequent (superstitious) religious *apotheosis,* which is actually the "elevation" of a *concept* – and not a person – into the heuristic similitude of a factually transcendent divinity.

It also has nothing to do with the Scientific Instrumentalist's theoretically conceived, yet nonetheless only hypothetically "strictly physical", intellectual "loaves and fishes". But it does have everything to do with learning to re-identify with the veritable point of our fundamental Witness' perspective!

This metanoetical "point" is the enlightened impersonal perspective of our whole Mind, as it freely "swims" within the dark and endless subquantum "sea" of Universal-Consciousness that underlies our entire Superverse.

But until our supposed "Enlightenment" finally dawns, or our empirical science eventually deepens, how do we accept the illusion of permanent "constant change"? When it is so pervasively implied!

Especially when our senses, including our "sixth sense" of cognition, tell us that it leads us eventually to the complete annihilation of our body, mind and identity, within "nonexistence". Which the Guatama Buddha quite aptly called the "Great Mystery of Impermanence". Therefore, we need to better understand what change actually implies!

Change

Change Is Difficult To Accept Only If We Cling Too Tightly to the endless stream of concepts that comprise our conglomerate body-minds. And as a result, soon begin to believe that change is not only real, but unidirectional, and unmalleable as well!

It is also frightening to realize one day, that we are not really in charge! This is the unspoken or unacknowledged reason why most people don't generally care to look too closely at the origin, or fate, of their own consciousness.

Nonetheless, by carefully observing our minds, we may eventually come to see that everything within the realm of our experience, including our illusory ego selves, is comprised solely of our mind's dreamlike algorithmic "codes" of ever changing thoughts.

But insofar as we are generally concerned, although unnoticed, our seemingly unending "thread" of consciousness is what comprises the entire universe of our experience. And when this thread is finally cut by our so-called "death", our person and universe simply cease to be!

This dreamlike thread of consciousness is what actually comprises our ostensible life. Which starts with our first complete thought, the notion that we *are,* because our existence really begins with the birth of this thought, and our subsequent experiential life eventually ends with its cessation.

But of course, insofar as the universe that we somatically sense is concerned, something is definitely "out there", but it likely exists in the form of a pure, universally consciousness "Force". However we cannot conceive it, because it begins where our cognition ends, and our perception of the universe is still currently limited to our split-mind's realm of comparative phenomenality.

The smallest and grandest forms that we can perceive are thus limited by our perceptive and cognitive abilities. Extending our sensorial range with instrumentation alone does not directly change this, but increasing the amount and quality of our understanding and our consciousness undoubtedly will!

Thus, in a manner of speaking, we are perpetually walking on the "clouds" of our own thoughts and concepts, comprised in essence of the Universal-Consciousness that is the mysterious supercosmic מעוז צור *maōz tzur* "Rock of Ages" hinted at from Toranic antiquity. This is the true "bedrock" of our astounding universe, which is only directly encountered when our perceptual thought clouds finally clear!

I

This is the Truth that our illusory ego fears. And this is also just about all that we are currently capable of knowing about our internal experiences of person and universe! Therefore, vivid or compelling experiences tend to draw us to them, like iron filings to a magnet.

And if we cannot extricate the basic principle of our identity from them, we risk becoming hopelessly identified with, and involved in, the recurring loop of a dream reality that transpires both automatically, and subliminally, within our binary individual consciousness.

This is because the universe and our persons only appear to change as the symbols of our consciousness that communicate them do, and thus our experiences always return to us as the veritable virtuosity of their own virtualness. Thus we always seem to be at the experiential "center" of all space-time bounded things.

Such is the experience of most people. But you may also discover that our mind does quite all well all on its own, generating experiences and becoming involved in them as their apparent subject, because such is its natural function!

But this illusory compound subject/object "person", besides being an impossible conundrum, cannot really be who we are, because it is being viewed from a superior dimension. Indeed, when our essential Awareness disassociates itself sufficiently from nature's evolved body-mind consciousness, everything is observed to occur spontaneously![(100)] There is absolutely no need, or room, for any obvious or discrete acts of supposed individual volition.

The perception of our purported free will's power is then seen to be a mere delusion, itself merely a portion of whatever is transpiring automatically within a consciousness that is quantumly interconnected with our entire expanding universe.[(101)] But of course, from the perspective of our dualistically identified minds, our free will appears to be absolutely necessary in order to create the desired effects within the apparent causality of unfolding events.

However, this "causality" is only the chimera of our divided minds. Everything in the universe goes on automatically, operating and happening as it will, with no need of any willful causation on our part.

Because everything simply happens, as it must, with no discrete events involved to cause any supposed "change", in whatever appears to be transpiring, within our constrained individual consciousnesses!

II

But this is also not the obviously flawed concept of "Fatalism", since our universe actually unfolds as a whole, in the same way that a clandestine arboreal flower emerges, by arranging (through its growth) the new and lovely parts that eventually make up its flowery wholeness.

The genetic potential of an entire tree's evolved design, for example, is dynamically preserved in its seed, from sprout, to mature tree, to flower, and then back to seed again.

The symbolically recorded potential of each unfolding genetic, epigenetic, and cosmically involved aspect fills a processional space in apparent time. With each processional space transitioning into the next, as only apparently temporal divisions of either the whole "unfolding" tree, or of our whole unfolding universe.

Thus attempting to interrupt and thereby fully understand the tree's "hidden-but-emerging" flower, or our expanding universe, at different stages in the organic process of its natural unfolding, is merely a reflection of our illusory individual consciousnesses presently limited "subject vs object" perspective. It may help us to understand better the biological processes of a tree's secret flowering, or the hidden fundamental functioning of our universe, but it is nonetheless an artifact of our divided human consciousness.

This is apparent, for example, in the ideational lines that attempt to differentiate one place from another on a globe. They are nowhere to be found upon the face of the actual earth, which they nonetheless attempt to divide!

Such is the *chutzpah* "hubris" of the human ego, which invariably overestimates the seemingly great subjective power of our split-minds, to completely understand everything by division and comparison. But this is just as silly as believing that we can actually block out the entire sun by obscuring its visual appearance with our thumb!

Yet until the understanding of our own volitional limitations begins to dawn upon us, our faux ego entity remains quite enamored with the illusion of its individual (divisionary) powers of abstraction, and its imagined (combinatory) control over the inferential events which it seems to see going on all around it.

And this continues to go on and on, until it finally occurs to us, through a more direct experience, that everything is actually transpiring within our consciousness, and then our habitual self-delusion (of a persistent dualistic reality) slowly begins to unwind!

In other words, our naturally bifurcated "pendulum" of experience is constantly swinging to and fro, between an "apogee" of symbolic expression and a "perigee" of non-expression, as it seeks to find an enduring balance. But actual stasis, or balance, is impossible within our illusion of individual consciousness, which depends upon comparison for cognition and the experience therein of symbolic change.

We only entertain our biologically based experiences of being apparent individuals with space-time disconnected "free wills", because of our respective illusory, dualistic subject-object perspectives. This is because our familiar perspectives of person and universe are wholly comprised of thoughts, which are themselves constructs that are wholly dependent upon the constantly streaming motion of their own symbolic manifestation!

And all of this "thinking-of-thoughts" spontaneously occurs in a consciousness that is ultimately derived from our interpretation of a veritable "cascade" of various chemoelectric processes. Which go on automatically, in our universally connected brains and bodies (i.e. via a universal quantum substrate).

But however compelling, there is really no need ever to become involved, or overly involved, because we are never the object/subject, which we invariably observe as if it were operating "before" us.

In truth, we are much more akin to the pure Presence by which our thoughts are known in consciousness to exist. Therefore while we are never really devoid of Awareness, we are in essence always devoid of (individual) consciousness, and especially so within our own universally shared primeval Awareness!

III

Our individual consciousnesses are thus dependent upon our experiential bodies, which are also comprised of concepts, and in the absence of thought, even the compelling illusion of our evidently "individual" consciousness is not yet formed. And of course, when our quantum string comprised energy bodies are not yet present, neither is our subject-object consciousness!Yet, our essential sub, or pre-quantum Awareness is necessarily always there, pervading the entire universe by allowing its projected potential identities to emerge, by way of the universal interdependence of identity with Consciousness (i.e. through the universal "principle of identity").

All of which manifested during the first Planck moments after the Big Bang, through our universe's quantum string substrate of evidently "Self-extending" fields. Which are fundamentally comprised of a subquantum force of (for want of a better term) a "Universal-Consciousness".

Thus within, and *as* "us", this fundamental principle is expressing itself in the phenomenality of our illusory individual consciousness (see "Planck time", pg. 564).

Learning to sink back into our true identity of underlying Universal-Consciousness is thus the deeper goal of meditation and self-inquiry.

Other efforts may be good for our somas and psyches but are at best only spiritual entertainment, and at worst delusional exercises that only tend to bind us even tighter to the beguiling dualistic illusions and delusions of our ideational persons and universe.

In other words, just "whom" do we think that we are trying to either "liberate", or to "enlighten"?

But since the Supreme Awareness that we already are is clearly always free, if we wish to realign our experientially split identity with our veridical unitary Source (which is also the limitless source of the multiverse that our entire universe is only an infinitesimal part of), the question of "What to do?" naturally arises.

And apparently all that we can really "do" is to learn all that we can about the truth of our own consciousness, and its deeper, truer relationship to the amazing subquantumnly Aware universe that we all share!

Eventually the understanding that we are not simply our mind's symbolic body, or even the dazzling *Zohar* inner "Luminescence" of our individual consciousness, needs to be as certain as the pain that automatically arises when we are unexpectedly bitten by a rattlesnake or stung by a scorpion in the wilderness! No one needs to tell us about our pain.

We do not need to read a book, consult with a guru, psychic, shaman, rabbi, imam, or priest; or make any particular effort whatsoever. A scorpion has just stung us, and we immediately experience the pain!

Evidently we need to have this inescapable level of understanding, and when we do, our identification either suddenly, or eventually, shifts back into the universe's essential witnessing sub-quantum Source, which of course, is our source too!

Thus our illusory participation in the ever entangling delusions of being an objectival "person" that is housed in a strictly "physical" body, which is independently operating within a veridically external "material universe", is not only impossible, but is also never truly required at all within our Consciously unfolding universe!

And while our brief adventure together in the wilderness is finally coming to its inevitable end, it is both exciting and humbling to realize that our species grander adventure of Self-discovery and universal mastery amongst the stars has barely just begun!

Spiritual Prescriptions

In A World Comprised Of Experienced Potential, we decide what is real, and which is just an illusion. But in the end, or when we finally realize that we have always been fully present to experience life, there is really no need for fear, anger, sadness or regrets. How could it possibly be otherwise?

Indeed, nature's way is to endure until the end and then to (finally) cease all at once, releasing (without compulsion) any sort of "corporeal" identification, without regret, or recrimination. Non-identification with our body-mind and the acceptance of nature's way is what all the masters prescribe.

Yet who amongst them can directly say "whom" is really doing the prescribing? On this, only the soliloquy of a profound silence suffices!

While others may unwittingly prescribe absurd or nonsensical notions of "objective subjects" and "dualistic monisms".

But instead of infecting others with another epidemic cycle of proliferating faux "spiritual" memetics, which are generally rather carelessly comprised of ignorance and confusion, perhaps they should instead wordlessly face the "terrifying silence" of their own inner stillness.

Until the profoundly silent superlative answer (that we are all in essence always the impersonal Supreme) finally sinks in!

However, when it finally does, the unsettling quandary of our supreme "Buddha Nature's" actual Oneness in diversity must then be accepted, and its full impact on our faux individual identities must then be contended with!

"G-d, grant me the serenity to accept the things I cannot change, courage to change the things I can, and [the] wisdom to know the difference." – Reinhold Niebur (1892-1971).

Thoughts *'n* "Things"

Thoughts Are Clearly "Things" Too, but evidently, it is also quite impossible to clearly distinguish between the two! (i.e. between "thoughts *n'* things"). So it seems best to simply refer to these events as "objects" of thought. And of these, are all apparent "things" comprised. As a result, there must also be a "thoughtless" and independent observer (of these thought comprised events).

Indeed, "who" else could we subjectively be? But by gaining the understanding that we are formless as well as thoughtless, now we must also begin to seriously wonder how we could possibly be whatever appears before us?

Therefore we (in Essence) can be neither of "thing" nor of "thought" comprised! Hence, whatever appears before us is much like a passing illusion. But perhaps, and even more, our observed persons and universe(s) are like a passing dream, from which we are destined, no doubt, to sooner or later awake!

Standing On the Razor's Edge

Sooner Or Later, "once on the Way", we will eventually find ourselves poised in inner space, like a flying fish suddenly suspended in the air before the drying light of a brightly luminescent inner "Sun".

Standing all alone now, we will soon discover that we are precariously perched upon the slim "razors edge", of a fuller metanoetical understanding, of our everyday lives. But it is a precarious place indeed, no longer being sure of just "whom", or even of "what", we are truly comprised!

Like a flying fish, we are first in the wet sea, then back in the bright light of the drying Sun. Back in and back out – again and again – 'till we are finally entirely done in! (What-with-all of this moving back and forth between phenomenal space-time and noumenal spacetimelessness!).

Until we realize that it is all just a grand phenomenal dream, spontaneously going on, above, as it were, the "reflective surface" of a never-ending, and "Cosmically Conscious" subquantum sea! Nonetheless, wherever we seem to find ourselves, when we look around, it is quite impossible not to notice the profound suffering of the world!

Rivers of Good Will

Perhaps There Exist "Rivers Of Good Will" that eventually create serendipitous events? For they do seem to happen with a much greater frequency than can be metrically (causally) explained.

There are certainly "rivers" of discontent, war, and misfortune, which are apparently balanced by the opposite condition(s). Hence there evidently exists no way to follow one or the other for long, except perhaps delusionally!

Perhaps a better inquiry is to discover the origin of the primary Subject that is the meeting place of both. That "place" is the Awareness of consciousness, and whatever thoughts that consciousness may contain. We may never fully understand "how", but if this is really "where", then just "when" could this possibly ever happen?

No Time – For Anything!

Is "Today" Really Forever? Apparently, or so the venerable masters of old seem to want us to believe! But is there any verity to this claim? Upon closer examination, does "the middle" still hold?

At some point the universe – for whatever reason – or perhaps none at all, suddenly began. Even if we suppose that it has always "just been there", this does not explain the inevitable breaking point on the curve of causality, as it effects universal force and form.

This is the indefinable "somewhere", where any experiential thing somehow becomes its opposite; from subject into object, and back again. Ergo, the middle "does not hold", because the proposition is unavoidably unsound!

Evidently, only "nothing" can be proven to exist, by "something" that clearly does not! But if only nothing can be proven to exist, then does nothing have any meaning, including our inevitable human suffering?

Noble Suffering?

The Tremendous Emotional and psychological investment (in assigning meaning) that our collective group consciousness carries forward is like a ponderous weight. One which increases from generation to generation.

Nearly "countless" numbers of battling men and women have (often bravely) died; fathers, brothers, sons, uncles and nephews, in nearly countless skirmishes and wars!

Mothers and children from before recorded history have struggled to survive. Acts of tremendous heroism have transpired, and horrible tragedies have taken place! Thus the history of mankind is "bittersweet" with laughter and tears.

Therefore a great and powerful existential vacuum exists, demanding that the vicissitudes of our life must have some sort of durable meaning. The attempt to provide this is woven into our family and cultural histories, and perhaps is nowhere present in greater amounts than in our noble "religious" attempts to explain it!

Our recorded histories are thus quite often a confusing emotional tangle comprised of half-truths and wishful thinking. Which are actually humankinds' perhaps "noble", but nonetheless ineffectual attempts to lend meaning, and provide healing to the otherwise meaningless (and sometimes singularly tragic) events of our perceived past.

Would that it were only ignorance which holds men back from seeing the truth of their own consciousness, and its role in creating and maintaining this egregious sort of apparently "historical" mischief, within the manifestation of the universe (in our group consciousness)!

It is this feeling of duty to the hearsay of the past, combined with the anxiety of a suspected meaninglessness, and empowered by our emotions, which are like "nails in the coffin" of histories illusion. Nonetheless, people often willingly give away their innate freedom, believing that it is their responsibility to do so, and sometimes even attack those who display a hint of disagreement with the ridiculous notion of the "noble suffering" of the human race!

Rather, would it not be better to simply focus upon the tragic events themselves without denying that the events transpired, or trying to rewrite them. And learn how to mitigate their repeatedly tragic appearance in our experience, by focusing upon a deeper interpretation of them in the ever-flowing river of our individual and collective group consciousnesses?

Because while we may have little, or even no control over the automatically unfolding events themselves, we most certainly do have control over our interpretation of them!

Endless Beginnings

In The Beginning something really rather mysterious transpired that we may never fully understand! But in any case, from the force filled arrangements of elemental materials within an ever expanding space-time, the universe eventually became capable of utilizing energy over duration, in various dimensional ways.

Thereby eventually accomplishing the amazing ability to make such an observation and inquiry of its Self!

Indeed, the inordinate power that our mind wields over its contents is also evidently impossible to fully explain. First our mind interprets what the body presents to it and then weaves it into the concepts that comprise our persons and universe, and finally identifies with both.

But we will never be fast enough to catch what is being assembled by our subject-object consciousness from within a spacetimeless Universal-Consciousness-Force, so why bother trying!

It may be better to simply (very clearly) understand our truly intimate (and clearly "virtual") space-time situation within Universal-Consciousness. And so mindfully to dwell therein, until the knots of our evident "bondage" are gently loosened and finally washed entirely away within the ever disappearing sentient "wave" of our own essential witnessing Selves.

A Disappearing Wave

We Are What we don't see, either coming, or going. We are simply a collection of ever-disappearing sentient waves. A phenomenon of consciousness that unaccountably appears to be a binary person and Universe, within an otherwise insentient biological computer.

Boundless Buddhas

The Unbounded Supreme unaccountably gives rise to Universal Consciousness. Which is then carried boundlessly forth upon gentle waves of endless possibilities. Thus creating the illusion of changing form(s), by their sequential insertion within an ideational space-time; to an apparently independently experiencing "subject".

But as it was in the beginning, so it will be in the end, because only the Supreme finally abides. Understanding this, we have only to let go of identifying with our body-mind. Apparently there is nothing else to do!

Some opine that even just hearing (and clearly understanding) this makes illumination almost a foregone conclusion. This is the unmitigated power of the essential truth that appears to "free" our ever-unbounded Essence.

Evidently once the process of Awakening begins there is no stopping it, and moreover, apparently another "living Buddha" sits patiently waiting within the spacetimelessness that quietly awaits us all in the end!

Daylight Stars

EMERGING FROM THE DARK SILHOUETTE of a tall California Live Oak tree, a small mousy-brown *Phryganidia californica* "Oak Moth" flutters its tiny gossamer wings against the dark blue of the late afternoon sky.

And with an abrupt flash, the intrepid little moth emerges into the golden evening sunlight, where it is suddenly transformed into a little living jewel! Its tiny tan wings now glow like sparkling diamonds as they catch the sunlight and hold its brightness in their translucent, tissue-thin membranes.

Soon more Oak Moths emerge, then even more! And before long, the dark blue evening sky is alive with thousands of brilliantly scintillating, fluttering little moths. They look like a convocation of tiny living stars twinkling above, in the slowly darkening sky.

The enchanting little Phryganidias soon shine so brightly in the fading reddish light of the setting sun, that they can easily be seen flitting around the shadowy tops of even the tallest distant redwood trees! But very soon, the scintillating little daylight stars begin to "wink out", one by one, as the retreating sun slowly sinks behind the blue horizon of the nearby Pacific Ocean.

The glowing little "stars" are miraculously metamorphosed by Mother Nature's "smoke and mirror" legerdemain. They are turned back again into mere mousy brown moths, imperceptibly winging their lonely way across a slowly darkening sky.

And so, on our last day together in the forest primeval, I am compelled to ask, "Which is the real, the brightly shining star, or the little mousy brown moth?" But in the end we decide what is real, and which is just an illusion!

"The clearest way into the universe is through a forest."

~ John Muir ~

The End

Epilogue

Breaking Camp is generally much easier than setting it up. Tent poles slip easily from their sockets and with little fuss our gray domed nylon tent is quickly back on the ground. With our camping gear now carefully stored away in the trunk of "Argo", my trusty nineteen-sixty-five midnight blue Mustang מרכבה *merkaba* "chariot", I carefully negotiate the narrow paved road that meanders out through the redwood rainforest.

But there is always a part of me that never wants to leave the haunting beauty of the rugged (yet climatically benign) Santa Cruz Mountains behind! So I drive out slowly, like a snail inching away from a tasty flower, following back along the way that we initially came in, savoring the parting beauty of the redwood wilderness as we go.

With a parting wave to the rangers at the front gate I turn Argo onto the squiggly black ribbon of highway that stretches over the Santa Cruz Mountains to Silicon Valley, and we are soon happily homeward bound!

But just a few turns up the long winding road we suddenly encounter three shy White Tail Deer, *Odocoileus virginianus*, nonchalantly jaywalking in an obtuse (simple-minded) angle across the hard black asphalt. Highway etiquette means nothing to them, so we are compelled to come to an abrupt halt to let them pass by undisturbed.

They are two bucks and a doe, a "forked horn" and a "five-pointer". The "rack" of horns distinguishes the male deer from the females and the more small pointed horns that are present on his rack, the older the buck generally is.

Magnificent creatures of grace, power and speed, the White Tail Deer suddenly disappear from our sight in a few extraordinary bounds! All that remains to mark their passing are a few shuddering twigs and some fluttering leaves, alongside the shadowy mountain road.

Smiling at our good fortune, I gently press my foot down on Argo's accelerator and all too soon we leave behind the eloquent serenity of the Santa Cruz Mountain's "Forest Primeval"!

END NOTES

1. Betty Bethards, *The Dream Book* (Mass: Element Books Inc, 1995), Preface: Dictionary "Guidance, Guide", p. 113.

2. Wei Wu Wei, *Fingers Pointing Towards the Moon* (Colorado: Sentient Publications LLC, 2003), see: "The Screen of Time", p.p. 18-20.

3. Robert Powell, *The Experience of Nothingness* (California: Blue Dove Press, 1996), "Who Knows That I Am?" see: Maharaja's response, pgs. 123 and 124.

4. The Art Information Resource, *Vincent Van Gough Gallery,* (HYPERLINK "http://www/" http://www.vangoghgallery.com/misc/bio.html. 2010), "...the artist was completely absorbed in the effort to explain either his struggle against madness or his comprehension of the spiritual essence of man and nature".

5. A.A.P., *Some Things You Should Know about Preventing Teen Suicide* (Illinois: American Academy of Pediatrics, 2010), "Suicide is the third leading cause of death for young people aged 15 to 24." (In one recent survey of high school students, 60 % said they had thought about killing themselves!).

6. Swami Nikhilananda, *The Gospel of Sri Ramakrishna* (New York: Ramakrishna-Vivekananda Center, 1977), see: "The parable of the fish and the net", p. 164.

7. Ekhart Tolle, *The Power of Now* (California: New World Library, 1999), "A Quantum Leap in the Evolution of Consciousness", p. 67.

8. Tom Griffith, *Plato The Republic* (New York: Cambridge University Press, 2001), Book 6, 484a.

9. Albert Einstein, *Out Of My Later Years* (New York: Philosophical Library, Inc. 1950), see "The Menace of Mass Destruction", p.p. 189-191.

10. Michael Talbot, *Beyond the Quantum* (New York: Macmillan Publishing Company, 1986), p.p. 50-52.

11. G.D. Coughlan and J.E. Dodd, *The Ideas of Particle Physics an Introduction for Scientists* (New York: Cambridge University Press, 1991), see: "Summary and Outlook", p. 208.

12. Richard Morris, *Cosmic Questions* (New York: John Wiley and Sons, 1993), "Traversable Wormholes", p.p. 156, and 157.

13. H. Garcia-Compeán *et al, Quantum Cosmology of Tachyons in String Theory* (Journal of Physics: Conference Series, doi 10.1088/1742-6596/24/1/019), Vol. 24, p. 161.

14. Re. "The Children": *The Holy Bible – King James Version* (Mathew 18:1, 2-4; Luke 10:21); *The Holy Koran* (Hadith: Sahih Muslim: Sahih al-Bukhari: Durriya); *The Holy Torah* (Midrash Rabba: Song of Songs 1:4), "Because of them I will give the Torah".

15. Wei Wu Wei, *Fingers Pointing Towards the Moon* (Colorado: Sentient Publications LLC, 2003), "Physics and Metaphysics", R.E. "Dimensions": Section(s) 25, 27, 29, and p. 87.

16. Robert Powell, *The Ultimate Medicine* (Delhi: Motalil Banarsidass Publishers Pvt. Ltd. 2004), R.E. "Reincarnation": p. 35; pgs. 199 and 200.

17. Ekhart Tolle, *The Power of Now* (California: New World Library, 1999), see: p. 46, last paragraph.

18. B. Misra and E. Sudarshan, *The Zeno's Paradox in Quantum Theory* (Journal of Mathematical Physics, 1977), Volume 18, Issue 4, p.p. 756-763.

19. SIAI, *What is the Singularity?* (San Francisco: Singularity Institute for Artificial Intelligence, 2010), Overview, HYPERLINK "http://singinst.org/overview/" http://singinst.org/overview/what is the singularity.

20. John D. Barrow, *The Origin of the Universe* (New York: Basic Books, Harper Collins Publishers, 1994), pgs. 106 and 107.

21. Shigo, *Modern Arboriculture* (Durham, NH: Shigo and Trees, Associates, 1991) "Roots Hairs and the Rhizosphere", p. 250; "Micorrhizae", p. 251; "Concept of Survival" p. 379; and "On the Intelligence of Natural Systems", p.p. 381-383.

22. M.R. Franks, *The Universe and Multiple Reality*, (New York: iUniverse, Inc. 2003), pgs. 6 and 7.

23. Paul Davies, *About Time, Einstein's Unfinished Revolution* (New York: Simon and Shuster, 1995), see "Einstein's Time", pgs. 32 and 33.

24. Fred Hoyle, *The Intelligent Universe*, (New York: Holt, Rinehart and Winston, 1984), p. 161.

25. Wei Wu Wei, *Posthumous Pieces*, (Colorado: Sentient Publications LLC, 2003), see: "Taking Time by the Forelock", pgs. 6 and 7.

26. Wei Wu Wei, *Open Secret* (Colorado: Sentient Publications LLC, 2004), p.p. 27-29.

27. W.E.H. Tanner, *Traditional Aboriginal Society* (Melbourne: Macmillan, 1987), see: "The Dreaming", p. 225.

28. Wei Wu Wei, *Posthumous Pieces* (Colorado: Sentient Publications LLC, 2004), see: "Dreaming Mind", pgs. 212 and 213.

29. Z'ev ben Shimon Halevi, *A Kabbalistic Universe* (York Beach, ME: Samuel Weiser, Inc. 1992), p.p. 1-45.

30. Gunther W. Plaut, *The Torah* (New York: Jewish Publication Society, 1974), Genesis 2, Par. Vay. 28:16-17… "And Jacob awakened out of his sleep and he said; 'surely, the Supreme is in this place and I knew it not!' And he was amazed, and said; 'How full of awe this place is, it is none other than the house of G-d, and this is the gateway of heaven."

31. Virginia Satir, *The New Peoplemaking* (Mountain View, CA: Science and Behavior Books, Inc. 1988), see: "The Family in the Larger Society", p. 360.

32. Shmuley Boteach, *Judaism for Everyone* (New York: Basic Books, Perseus Books Group, 2002), p. 242.

33. Elizabeth Kübler-Ross, *On Death and Dying* (Harvard: Ingersoll), see: "Lectures on Human Immortality".

34. Sri Ramanasramam © 2000, *Talks with Ramana Maharshi* (Carlsbad, CA: Inner Directions Publishing, 2001), p. 118.

35. Robert Powell, *The Experience of Nothingness* (San Diego, CA: Blue Dove Press, 1996), see: "Who Knows That I Am?" Maharaja's response, pgs. 125 and 126.

36. Gunther W. Plaut, *the Torah* (New York: Jewish Publication Society, 1974), Exodus 3: 13-15.

37. Edward Horowitz, *How the Hebrew Language Grew* (Brooklyn: KTAV Publishing House Inc. 1960), p. 5.

38. Caldwell et al., *Phantom Energy and Cosmic Doomsday* ("Physical Review Letters" - 91: 071301).

39. Mario Livio, *The Accelerating Universe* (New York: John Wiley and Sons Inc. 2000), see: "Expansion", pgs. 80 and 81.

40. Maurice Frydman, *I Am That* (North Carolina: Durham Press, 1999), see: "The case of the blind boy", pgs. 138 and 139.

41. Paul Reps, *Zen Flesh, Zen Bones* (Rutland VT: Charles E. Tuttle Co. 1980), see: "An Oak Tree in the Garden", p. 150.

42. Martin Rees, *Just Six Numbers- the deep forces that shape the universe* (New York: Basic Books – Perseus Book Group, 2000), p.p. 1-4.

43. George Thompson, *The Bhagavad Gita – a new translation* (New York: North Point Press a division of Farrar, Straus and Giroux, 2008), Ch 2, p. 9, stanzas 6-12.

44. Helena Cobban, *The Moral Architecture of World Peace – Nobel Laureates Discuss Our Global Future* (Charlottesville: University Press of Virginia, 2000), see: "Systems of Violence at the Personal Level", "Muller's story", pgs. 116 and 117.

45. Dan Neely and Gary Watson, *The Landscape Below Ground II* (Champaign, IL: International Society of Arboriculture, 1998), Thomas O. Perry, "Keynote Address".

46. Wei Wu Wei, *Open Secret* (Boulder, CO: Sentient Publications, 2004), "Geometrically Regarded", p.p. 11-13.

47. Robert Powell, *The Nectar of Immortality* (San Diego, CA: Blue Dove Press, 2001), p. 59.

48. Wayne Liquorman, *Never Mind* (Redondo Beach, CA: Advaita Press, 2004), p. 116.

49. Sri Aurobindo, *The Life Divine* (Twin Lake, WI: Lotus Press, 2000), p. 566.

50. J.H. Conway and N.J. A. Sloane, *Sphere Packing, Lattices and Groups* (New York: Springer-Verlag, 1988), Ch. 30.

51. Adin Steinsaltz, *The Talmud* (New York: Random House, 1999), "All is in the hands of Heaven, except the fear of Heaven", Meg. 25a: Ber. 33b.

52. N. Matheny, J. Clarke, *Evaluation of Hazard Trees in Urban Areas* (Savoy, Il: International Society of Arboriculture, 1994), pgs. 1 and 2.

53. Maureen Caudill, *In Our Own Image – Building an Artificial Person* (New York: Oxford University Press, 1992), pgs. 223 and 224.

54. Geoff Simons, *Robots – the Quest for Living Machines* (London: Cassell, Villers House, 1992), p. 208.

55. Wei Wu Wei, *Why Lazarus Laughed* (Colorado: Sentient Publications LLC, 2003), see: "Absolutely Us", p. 99.

56. Gunther W. Plaut, *The Torah* (New York: Jewish Publication Society, 1974), Ecclesiastes 1:2.

57. Z'ev ben Shimon Halevi, *The Way of Kabbalah* (York Beach, ME: Samuel Weiser, Inc. 1991), p. 161 (Note: "Tiferet" is analogous to Brahman, and to Universal-Consciousness).

58. Stanley I. Greenspan, *The Growth of the Mind* (Menlo Park, CA: A Merloyd Lawrence book, Addison-Wesley Publishing Co. 1997), pgs. 95 and 96.

59. Rita Carter, *Exploring Consciousness* (Berkley: University of California Press, 2002), p. 135.

60. J. Allan Hobson, *Consciousness* (New York: Scientific American Library, W.H. Freeman and Co., 1998), p.p. 153-155.

61. Antonio Damasio, *The Feeling of What Happens – Body and Emotion in the Making of Consciousness* (New York: Harcort Brace and Co., 1999), pgs. 18 and 19.

62. Jeffrey M. Schwartz, Sharon Begley, *The Mind, and the Brain – Neuroplasticity and the Power of Mental Force* (New York: Harper Collins Publishers Inc., 2002), p.p. 284-286.

63. Wei Wu Wei, *Ask the Awakened,* (Colorado: Sentient Publications LLC, 2002), p. 111.

64. Victor J. Stenger, *Timeless Reality – Symmetry, Simplicity and Multiple Universes* (Amherst, NY: Prometheus Books, 2000), p. 93.

65. William Karush, *The Crescent Dictionary of Mathematics* (Palo Alto, CA: Dale Seymour Publications,© 1962.), p.p's. 9, 36, 37, and 291.

66. James H. Austin, *Zen and the Brain* (Cambridge: The MIT Press, 1999), p.p. 561-567.

67. Igor Aleksander, Piers Burnett, *Reinventing Man – The Robot Becomes Reality* (New York: Holt, Rinehart and Winston, 1983), see: "Automata Finding the Mind in the Machine", p. 151.

68. Z'ev ben Shimon Halevi, *Adam and the Kabbalistic Tree* (York Beach, ME: Samuel Weiser, Inc. 1990), p.p. 96-100.

69. Joseph Campbell, *Pathways to Bliss – Mythology and Personal Transformation* (Novato, CA: New World Library, 2004), see: "Myth Through Time – The Surface and Substance of Myth", p. 21.

70. Wei Wu Wei, *Ask the Awakened* (Colorado: Sentient Publications LLC, 2002), see: "Principal and Function", pgs. 170 and 171.

71. M. Sheng-Yen, Dan Stevenson, *Hoofprint of the Ox* (New York, NY: Oxford University Press, 2001), see: "Ch'an and Emptiness", p.p. 20-25.

72. Francisco J. Ayala, *Darwin's Gift to Science and Religion* (Washington, DC: Joseph Henry Press, 2007), see: "Evolution or Religious Beliefs? The Arrogance of Exclusivity", p.p. 171-176.

73. Abraham H. Maslow, *Toward a Psychology of Being* (New York, NY: John Wiley and Sons, 1998), p.p. 31-33.

74. M.R. Franks, *The Universe and Multiple Reality*, (New York: iUniverse, Inc. 2003), pgs. 11 and 12.

75. Edited by Mathew Greenblatt, *The Wisdom Teachings of Nisargadatta Maharaj* (Carlsbad, CA: Inner Directions Publishing, 2003), see: "The Primordial Illusion", p. 37.

76. Ramesh S. Balsekar, *A duet of One – The Ashtavakra Gita Dialogue* (Redondo Beach, CA: Advaita Press, 1989), Verse 16, p. 31, and Verse 29, p. 41.

77. Wei Wu Wei, *The Tenth Man – The Great Joke, Which Made Lazarus Laugh* (Boulder, CO: Sentient Publications LLC, 2003), see: "Inside the Within", pgs. 184 and 185.

78. Nyogen Senzaki, Paul Reps, *Zen Flesh Zen Bones* (Rutland VT: Charles E. Tuttle Co. 1980), see: "The Gateless Gate" by Mu-mon, "Gutei's Finger" Verse 3, p. 119.

79. The Dali Lama, *The Middle Way – Faith Grounded in Reason* (Somerville, MA: Wisdom Publications, 2009), see: "Compassion", p.p. 1-5.

80. William H. Calvin, *How Brains Think* (New York: Basic Books, Perseus Books Group, The Science Masters Series, 1996), p. 32.

81. Richard Fortey, *Life – A Natural History of the First Four Billion Years of Life on Earth* (New York: Random House, Alfred A. Knopf Publishers, 1977), p. 27.

82. Robert Aitken, *Original Dwelling Place – Zen Buddhist Essays* (Washington, DC: Counterpoint, 1996), see: "Ultimate Reality and the Experience of Reality", p. 90: "Koans and Their Study", p. 103.

83. Robert Powell, *The Nectar of Immortality* (San Diego, CA: Blue Dove Press, 1996), p.p. 94-97.

84. Sri Aurobindo, *The Life Divine* (Twin Lakes, WI: Lotus Press, 2000), see: "The Materialist Denial", p. 19.

85. Steven Rose, *The Future of the Brain – The Promise and Perils of Tomorrow's Neuroscience* (New York: Oxford University Press, Inc. 2005).

86. Abraham Harold Maslow, *The Farthest Reaches of Human Nature* (New York: Arkana, 1993), see: "Notes on Innocent Cognition", p.p. 241-248.

87. Frank M. Stanger, *Sawmills in the Redwoods: Logging on the San Francisco Peninsula, 1849 - 1967* – (CA: San Mateo County Historical Assn., 1967), see: general history.

88. Roy Porter, *Madness – A Brief History* (Oxford: Oxford University Press, 2002), see: "Christian Madness", p.p. 17-25.

89. John Hunter Thomas, *Flora of the Santa Cruz Mountains of California – A Manual of the Vascular Plants* (Stanford, CA: Stanford University Press, 1961), p.p. 16-22.

90. Francesca Fremantle, *Luminous Emptiness – Understanding the Tibetan Book of the Dead* (Boston: Shambala Publications, Inc. 2001), pgs. 199 and 200.

91. Ed Rosenthal, *Why Marijuana Should Be Legal* (New York: Thunder's Mouth Press, 1996), p.p. 1-5.

92. Daniel M. Perrine, *The Chemistry of Mind Altering Drugs* (Washington, DC: American Chemical Society Books, 1996), pgs. 358 and 359.

93. Brink Lindsey, *The Age of Abundance – How Prosperity Transformed America's Politics and Culture* (New York: Harper Collins Publishers, 2007), pgs. 341 and 342.

94. James Mooney, *The Swimmer Manuscript Cherokee Sacred Formulas and Medical Prescriptions* – Smithsonian Institution, Bureau of American Ethnology, Bulletin 99 (Oklahoma City, OK: Reprinted by Noski Press, 2005).

95. Marilyn Tower Oliver, *Drugs – Should They Be Legalized?* (Springfield, NJ: Enslow Publishers, Inc. 1996), see: "Drugs: Should they be Legalized", p.p. 5-10; and "The War on Drugs", p.p. 60-70.

96. Wei Wu Wei, *The Tenth Man – The Great Joke, Which Made Lazarus Laugh* (Boulder, CO: Sentient Publications LLC, 2003), see: "The All Embracing Measure", p. 23.

97. Brian Greene, *The Fabric of the Cosmos – Space, Time, and the Texture of Reality* (New York: Random House, Alfred A. Knopf Publishers, 2004), see: "Roads to Reality", p.p. 17-19.

98. Master Shen-yen, Dan Stevenson, *Hoofprint of the Ox* (New York: Oxford University Press, 2001), pgs. 23 and 24.

99. George Thompson, *The Bhagavad Gita – A New Translation* (New York: North Point Press, 2008, Verse 2:16, p. 10.

100. Alan Watts, *The Watercourse Way* (New York: Pantheon Books, Random House, 1975), see: "Tao", p. 55.

101. Alan Watts, *The Watercourse Way* (New York: Pantheon Books, Random House, 1975), see: "Shi-shi wu ai", p. 35.

102. Alex Shigo, *Tree Anatomy* (New Hampshire: Sherwin/Dodge, Littleton, 1994), see: "Oxyen18 isotope in fogwater", p. 13.

ACKNOWLEDGEMENTS

SINCERE THANKS MUST go to my editor and life-long *nonpareil* best friend, Professor Robert Brown, emeritus, of U. C. Santa Cruz. And to my inspirational teachers of many years ago, the indefatigable Rabbi's Berk and Bloom, תודה רבה אחים שלי!.

Thanks is also amply merited by my lovely wife Gloria Jackson, dedicated teacher, for suggesting "A Tree Trimmer's Guide to the Universe" *übershrift* "heading"; *"Who can find a good woman? She is precious beyond all things. Her husband's heart trusts her completely. She is his best reward."*– Proverbs 31:10.

I am also grateful for her longsuffering patience and faith in me during מיין לאנג גוט וועסטלינג *meyn lang got gerangl* "my long G-d wrestling" with the more "metaphysical" aspects of this book; *"Whither thou goest I will go, and where thou lodgest I will lodge, thy people shall be my people, and thy G-d my G-d... "* – Ruth 1:16.

Thanks as well go to my long-time Stepp's Tree Service client, and good neighbor, retired Stanford Particle Physicist, David "J", Dr.Sc. For his open minded friendship, and also for the example of his unforgettable devotion to his partner and avocation, invaluable informal physics lessons, and infectious light of inspiration!

And a special thank you goes out to Joe Thibodeaux Esq., Santa Clara County ADA (now retired), and to Robert Montez, Palo Alto Police Department, Narcotics Division Detective (retired), for looking over the law enforcement aspects of Forest Primeval.

Thanks also to Sabra "Diane Hayes", for her Modern Hebrew language editorial assistance, and to the (now, truly ephemeral) Taoist Sifu "Wei Wu Wei" for his Ancient Chinese Language translation.

And a hearty heartfelt thank you to Master Arborist Ray Morneau, J.D., for his expert review of Forest Primeval's tree biology, natural science, arboricultural, and "Tree-Law" portions.

Thanks also to another long-time Stepp's Tree Service client, astute bibliophile and kind founder of the "Friends of the Palo Alto Animal Shelter", Scottie Zimmerman. For her friendship, her steadfast kindness to our animal fiends, and for her helpful advice in making Forest Primeval's formerly rather complex book Index more "user friendly".

My special gratitude also goes out to Dr. Deidre Stegman MD, for her review of Forest Primeval's physiology and medical portions, and for her expert medical advice and excellent care, over many memorable years.

She has been fighting for me like a true warrior, and running point "like a marathoner" along the Front Lines of the battle for my welfare and survival. (It's a comfort to know that you've always "got my back", thank you!).

Also, my undying thanks goes out to the compassionate and gifted (Medical) Doctors Eugene Kim, and Derrick Wong. And especially to the *Palo Alto Medical Foundation*, for showing the wisdom to hire (and retain!) doctors with such superb skill, and boundless compassion.

And to the brilliant and kind Dr. Robert Sinha, and his incomparable Staff, of El Camino Hospital (who also seem to love paper airplanes just about as much as I do!), I literally *owe my life* to you, for helping me defeat my most formidable foe, during the difficult early stages of writing this book.

Words are just simply inadequate to convey the depth of my gratitude to you all, for making the unbearable possible, and the seemingly almost "impossible", even more than probable! תודה מעומק ליבי "Thank you all, from the bottom of my heart!"

I

The owners and the staff of the "Artists haven" Palo Alto Coffee Shop also deserve recognition for their kind encouragement, friendly Persian *Farsi* over-the-counter lessons, occasional warmhearted Spanish *la chanza*, and excellent قهوه *ghahve-ye*, merci!

Over the years, I've been impressed with your honesty, hard work and compassion, especially to your elderly, handicapped, and otherwise health "challenged" customers.

As a profitable small Service Business owner, and human being, for a good many years, I can see that you not only know the losses and hardship that we must all eventually endure, but even more importantly, you let your neighbors know that you care! For some, who have outlived everyone that they once loved, or may be facing certain death all alone, this can make all the difference in the world!

"Y'all" seem almost like family now, after all these years. And, I suppose that you are (maybe even more than we both know!). Because, at least as I see it, you're an important part of both my community and *our* great big "American Country Family" now, and I'm mighty proud to say so, and to have you as my neighbor too!

And, if you'll please pardon my using this Acknowledgement section as a "Bully Pulpit", I'd like to offer a few words of friendly advice to some of my "less-well-informed" fellow Americans. To sum up the recalcitrant problem of racial prejudice (and profiling), in a less gentle American "Country" sort-of way, *"Don't believe everything that you hear!"*

To be a good citizen, you need to stand up, start thinking for yourself, and begin vetting your supposed experts "facts" for yourself. If you're looking for a good place to start, talk to a veteran whose actually spent some time in the Mideast, if you really want to know what the story is.

Because, if you've got a problem with loyal American Muslims like these, well then, I guess you've got a problem with me too brother! Because Americans are supposed to protect the backs of their fellow Americans, instead of stabbing a knife into it!

At least that's what our own courageous young military teaches us, at the risk of their own precious lives! And when you don't, just what kind of an "American" do you think that you are? When their freedom is what our sons and daughters have fought and even died for!" Enough said.

And although it is generally not acknowledge, if not almost unknown in the West, science owes a tremendous debt to an عراقي *al-'Irāqi* "Iraqi" physicist (Ibn al-Haytham, 965 CE) who "discovered" the scientific method, and to a few other enlightened Middle Eastern Muslims who wisely preserved the collective knowledge of humanity throughout the long and barbaric "Dark Ages" of the West.

The modern name "Iraq" most likely came from a 3,300 year old, "time and circumstance" ruined, Sumerian city once called *Uruk*, that *reliable* historians say actually invented the written language! In other words, whenever we discuss 'most any Middle Eastern "Semitic" people, we are generally talking about a very ancient and venerable culture, that I believe Americans need to know much better; for our own good, if nothing else!

Indeed, Coffee Shops enjoy a rich intellectual history. And where coffee itself is concerned, were it not for a casual conversation in an unpresupposing British coffee shop, Sir Isaac Newton may never have written his *Principia*, which laid the mathematical foundation for all of modern physics. Evidently, the contribution of coffee to the creative process, if not to our health in general, is not to be underestimated!

II

I also must thank the International Society of Arboriculture, particularly the Western Chapter, for all that they have done to promote the arboricultural profession, the science of safe and effective tree care, and the important preservation of trees against the ominous "rising tide" of Global Warming.

My thanks as well to the brave men and women of the American Armed Forces who daily put themselves in harm's way to protect our precious and hard earned civil freedoms. The American Armed Forces have been there for all of my family's veterans, to the very end. And even said a sad "final goodbye", to every one of them, with the defiant report of an

Honor Guard Salute, to let everyone know that you are removing another fallen American hero from the final battlefield of life.

I was always there with you, quietly watching, while you lovingly draped them in “Old Glory”. I tried my best to stand tall and proud. But I could never hold back my tears as I watched you reverently fold Her back up, into a tight triangle of blue sky and bright stars, with deep respect, for one last time.

And I watched as you put Her, with loving gratitude, where She really belongs. Back into the hands of someone that they were willing to die for, if need be.

But I could also always see the sad understanding, even in their young eyes, that the one left holding the flag, instead of their loved ones hand, was probably the last thing on their minds. You will always have my deepest gratitude, for honoring their service to our country.

Notwithstanding the fact that I haven’t always agreed with our own government’s occasionally erroneous policies, and sometimes disastrous decisions, it was always an honor to serve with you, albeit during a rather complicated and troubled time.

Mark Garvin, former editor of *Tree Care Industry Magazine*, also deserves my sincere thanks for encouraging my writing and for publishing my first magazine articles, many years ago. (We shared some fun stories about my brilliant, but “quirky”, aunt Billie (Wilma) Lee Sturm, who was the editor and publisher of *Pacific Logger and Lumberman West* in the 1950’s, and my first inspiration as a writer.).

My thanks also goes to Dr. Larry Costello (now retired) and to his assistant Katherine Jones of the University of California Agricultural Extension Service of Half Moon Bay. For their invaluable expertise and assistance with an award winning, and life-saving, Caltrans funded “Resistograph ®” decay detection study.

And, for Larry’s important creation of the California Tree Failure Report Program and Database. Which continues to yield an ever growing collection of invaluable field data about tree safety, and has since even gone "international"!

My thanks as well to Dr. Klaus Mattheck, of Karlsruhe Research Center in Germany. The Resistographs ® genius inventor, affable Arborist educator, and pioneer researcher into the biomechanics of trees and tree failures around the world.

I am also thankful to the US Forestry Service of Yosemite National Park, and to Bruce Hagen (now retired) of CDF (CalFire). For their outstanding service to our "Golden State's" trees, citizens, and to the dedicated people who maintain them. Bruce gave his "Forestry" input very early on. Which, probably unbeknownst to him, played an important role in securing the funds for this important pioneering decay detection study!

And to, then Caltrans Highway Maintenance Superintendent, Carl Biancini, for going the extra distance within the always tight-budgeted Department, "to make it so". Also to now retired Superintendent Ed Kaiser, who stuck by my side while it was being fought for, and was a true-blue friend, to the bitter end!

In this time of growing awareness of the sacrifices that are being made by our nations loyal civil servants, I also want to recognize the "unsung heroes" of the California Department of Transportation Tree Maintenance Crews, that literally "go out on a limb" and put their lives on the line every day through traffic, storms, fires, earthquakes, highway accidents, "hell and high water" to keep our beautiful California Highways safe for our travelling public.

(Like my good friend, the late Caltrans Tree Maintenance Supervisor Don Lichtlighter, another hero who gave his life in service to the people of the great State of California.). Before eventually going into private practice, it was my privilege to serve with Caltrans for many memorable years.

III

Because trees cannot cover the cost of their own care, every tree trimmer also understands that they owe a debt of gratitude to the people that are willing to do it for them. And in my private practice, among clients and neighbors, I have been blessed with the very best! But from those who have finally passed on (as we all eventually do) I have discovered that (in addition to my "worldly" pay) you have also given me an unexpected treasure, the "timeless" happy memory of knowing *you*!

But I especially thank my brother and sister "caretakers of the trees". We are only a miniscule portion of the current seven billion people on planet earth, and it is a wonder that we are able to manage its trees as effectively as we do, with the labor and talent of such a very few!

Nonetheless, evidently our most important job, which we have perhaps even unbeknownst to ourselves been preparing for as we have labored to create a safe and effective Arborist profession that is based upon science rather than mere anecdotal knowledge, is to assist in the preservation and proliferation of our planet's precious trees.

Our job has never been more important than it is now, as we grab our tools and step up to the microscope, as well as the "saw and spade", to take our place beside the scientists who are working to educate the public and to convince the governments and gigantic corporate businesses of the world to invest the necessary money and resources to begin to reverse the ever growing effects of Global Warming.

Evidently, without my even knowing it, this book has been covertly "writing itself" in the secret "deeper places" of my mind and heart for a very long time now! Consequently, I am grateful to so many people, who have inspired, taught and assisted me through our shared experiences in life, that it is sadly impossible to mention them all here.

To the many who are not here acknowledged, teachers, colleagues, clients, "brothers in arms", family and friends, let me just say that you not only have my undying gratitude, but that you will also never be forgotten (especially those of you who are no longer in this beautiful, yet פעקכתה *fekochta,* world with me!). At least for as long as I continue to sentiently inhabit my functioning body and mind.

And last but not least, I wish to thank my "Guides", whether living, or transitioned back into the fundamental mix of things. Not only because you know "Who" you are, but also because you care enough to remind us all of just "Who" we really are too. Because, we are all really "Avatars in disguise", just as much as "You"!

Note: The opinions expressed herein are the authors alone and are not necessarily shared by the author's professional, religious, or personal affiliations, the books publisher, distributor(s), or any person, government, institution or authority, who may have been mentioned and/or assisted in its preparation.

Glossary Language Key:

Ara; = Aramaic

Arb; = Arabic

Che; = Cherokee

Chi; = Chinese

Frn; = French

Ger; = German

Grk; = Greek

Heb; = Hebrew

Hin; = Hindi

Hwi; = Hawaiian

Jap; = Japanese

Lat; = Latin

Rus; = Russian

San; = Sanskrit

Glossary

In This special “hand crafted” Glossary, the etymology of certain uncommon and “key” words that are occasionally used in Forest Primeval is explained, because a words’ history often helps to clarify its essential meaning.

Additionally, a Language Key is provided on the previous page, to assist you in understanding the abbreviations that are being used at the beginning of the Glossary definitions.

It should also be noted that many Greek and Latin terms stem from Proto-Indo-European roots, which may provide an important link to their ancient metaphysical and philosophical meanings.

Accordingly, certain words may have been “repurposed” to reflect more accurately their root meanings. But more importantly, it also explains how they are being used in Forest Primeval.

Also, it is probably best to keep in mind that even the most compelling of words are always just a rapidly evaporating stream of transient thoughts. And moreover, that the true origin of every word that we think into existence, lies fundamentally within the machinations of our own consciousness.

Hence, even more important than the stream of words that constantly fills our heads, is the eternal Witness that grants sentience to the (ego) knower of our words, thus illuminating them with the “reflected” sentient light of the knower’s cognition. (Words requiring further clarification are explained in the Appendix, on page 517).

Aba: *Heb;* “father”.

Abraham: *Heb; Ab* “father” + *am* “people” = patriarch of the Jewish people.

Adonai Echad: *Heb;* “G-d is One.”

Advaita: *San; A* "without" + *dvaita* = division. A branch of Hindu "Vedanta" philosophy that espouses non-duality.

Adventitious bud: *Lat; adventicius* "accidental". A latent bud.

Affective: *Lat; affectus,* the "emotional" aspect of cognition.

Akash: *San;* "Space", more specifically an aware "Superspace".

Alchemy: *Arb; al-kh miya* "chemistry". The symbolic transmutation of "baser metal into gold" that represents the transference of identification away from our body-mind, and back to our truer impersonal Self.

Algorithm: *Arb; al-Khwārizmī,* a set of instructions to accomplish a complex task or problem.

Allah: *Arb;* "G-d".

Amrita: *San;* "Ambrosia", the sustenance of "eternal life".

Ananda: *San;* inner "Bliss", or abiding "Peace".

Anandamide: *San; Ananda* "bliss" + *(Lat) Amide* "ammonia" = a psychoactive compound found in cannabis, involved with memory.

Antediluvian: *Lat; ante* "before" + *diluvium* "deluge" = "before the flood". Also, very old or antiquated.

Anthropomorphic: *Grk; anthrōpomorphos* "of human form". Anthropomorphism, for example, is the attribution of human qualities to the Divine.

Aphelion: *Grk; aphelion* "away from" + *helion* "sun" = the farthest point of an orbiting object from the sun, while *perihelion* is the closest point of an orbiting object to the sun.

Apperception: *Frn; apperceptiōn* "conscious with full awareness".

Arboreal dendrite: *Grk; déndron* "tree". The branched projections of a neuron that conduct electrochemical stimulation in our brain.

Arborist: *Lat;* from *arbor* "tree", a certified tree expert, or researcher.

Asceticism: *Grk; askētikos* "austere", the practice of rigorous austerity.

Avatar: *San; avatarana* "descent" of the Divine, in a rare continuity of Self-Awareness from womb to grave.

Awareness: *Ger; wachen* "to be awake", also *gewahr* from *ge* + *wær* "to be aware". A disidentification with our experiential body-mind, and reintegration of our identity with our essential Awareness (when capitalized it refers to the Supreme, source of the Superverse).

Bar Mitzvah: *Heb;* "Son of the Commandment(s)". The ceremonial assumption, at thirteen years of age, of adult Jewish religious responsibilities, for girls it is called a "Baht" Mitzvah.

Benoni: *Heb;* "In between".

Bhagavad-Gita: *San;* A portion of the Hindu epic *Mahabharata*, literally the "Song of G-d".

Bifurcation: *Lat; bis* "twice" + *furca* "fork" = splitting or dividing into two parts.

Bodhi: *San;* "Consciousness", specifically a Universal-Consciousness.

Bodhisattva: *San; Bodhi* "Consciousness" + *Sattva* "True" = a person indelibly awakened to his/her true Nature.

Brahman: *San;* Divine "Universal-Consciousness".

Birkenau: *Ger; Birken* "birch", the Auschwitz II death camp built near birch trees in 1941 Poland by Nazi controlled Germany.

Cambium: *Lat;* "change", a layer of undifferentiated tissue in plants that provides undifferentiated cells that can become bark, roots, shoots, buds, flowers etc.

Cerebellum: *Lat;* "little brain", the lower back portion of our brain that is involved in motor control, cognitive functions and emotions.

Cerebral cortex: *Lat; cortic* "bark", a thin mantle of gray matter that covers the surface of our brain.

Ch'an: *Chi; Chán* from Sanskrit *dhyana,* "Zen" in Japanese.

Chi: *Chi;* "Life-Force", pronounced *Ki* in Japanese.

Chitti: *San;* "Universal-Consciousness Force".

Cognition: *Lat; cognition* "mental processes", involved in achieving knowledge and comprehension.

Concept(s): *Lat; conceptum,* "ideations" that appear within our consciousness that comprise our experiences of person and universe.
Conifer: *Lat; conus* "cone" + *ferre* "bearing" = "held within a cone", the *gymnosperm* "external seed" genus, generally an "evergreen" tree.
Consciousness: *Lat; conscious* "aware". The essential nature of the universe, responsible for its appearance and that of life and mind within it (i.e. with a capital "C").
Contiguous: *Lat; contingere* "to touch", as in adjoining or bordering, sentience for example is contiguous with the appearance of mind, mind with that of life, and life with the appearance of contiguous energy, space and matter.
Continuum: *Lat; contine* "continuous", such as a field, range, area or sphere.
Creationism: *Frn; creacion* "created thing". A widely held belief that the universe and what it contains is created by fundamentally divine, rather than strictly natural, processes.
Daedalus: *Grk; daidalos* "clever worker", see pg. 524.
Daven: *Heb;* "Pray".
Deliquescent: *Lat;* "Spreading", a growth form descriptive of certain trees.
Dharmakaya: *San;* "Buddha Mind", the universe's substrate of Universal-Consciousness.
Digineli: *Che;* "Brothers", *Digineli* in formal "oratorical construct" form. And (ᎠᎾᏓᏅᏟ) *Anadanvtli* "brothers", in general usage form.
E. regnans: *Lat; regnare* "rule" or "be king", regal name notwithstanding, its common usage name (prosaically) remains "Swamp Gum" eucalyptus!
E. viminalis: *Lat; viminalia* "Willow-like", having long flexible branches, its common name is nonetheless "White Gum" eucalyptus.
Emet: *Heb;* "Truth".
Elohim: *Heb; El* "G-d" + *im* (a plurality suffix) = Universal-Consciousness.
Emet Elyōn: *Heb;* "Ultimate Truth".

Endorphins: *Fr; endorphine*, from "endogenous" + "morphine" = the body's own "opium", consisting of neuropeptides produced in our pituitary gland and central nervous system.
Epistemology: *Grk; episteme* "knowledge" + *logos* "reason" = the theory, nature and limitations of knowledge, in other words, a philosophical examination of the principles of knowledge and order in our universe.
E. globulus: *Lat; globus* "globe". Also, refers to the large spherical seed pods of the behemoth "Blue Gum" eucalypti.
Ex nihilio: *Lat; ex* "out of" + *nihil* "nothing" = "from nothingness".
Exoteric: *Grk; exo* "outer" + *trechein* "to run" = "from the outside". An explanation of the universe from "outside of" the mind.
Fastigate: *Lat; fastigiatus* "peaked". Descriptive of trees with a narrow triangular shaped "crown".
Gan Eden: *Heb;* "Garden of Eden". A metaphor for our natural state of independent Awareness, before the "fall from grace" comes upon us, after our birth, when we begin to identify with our body and mind.
G-d (god): *Ger; gudan* from root *ghau (ə)* "to call", or "invoke".
Gehinom: *Heb; Gey* "valley" + *Hinom* = the "Valley of Hinom" or "Pit of Hell", located adjacent to the walled city of old Jerusalem in Israel, a literal juxtaposition of good and evil!
Genome: *Ger; pangen'* "gene" + *om,* from *soma* "body" = genome; *pangen* is from the Greek word *pantos* "all".
Gesher: *Heb;* "Bridge".
Gnosticism: *Grk; gnostikos* "knowing, being able to discern".
Grace: *Lat; Gratia* "kindness, good will".
Gulag: *Rus; "Gulag"* is an acronym for the Soviet bureaucratic institution *Glavnoe Upravlenie ispravitel'no-trudovykh Lagerei,* "Main Administration of Corrective Labor Camps".
Guna(s): *San;* "Attribute". According to East Indian Vedantic philosophy, the three fundamental aspects, or "forces" of the universe.
Hellenistic: *Grk; hellenistikos*, relating to the "Hellenistic period".

Heuristic: *Grk; heuretikos* "inventive, to discover". Also, "common usage" problem solving, learning, and discovery.
Hippocampus: *Grk; hippos* "horse" + *kampos* = "sea monster". In Greek mythology, a sea horse with two forefeet and a body ending in the tail of a fish, also part of the limbic system (see pg. 549).
Hologram: *Grk; holos* "whole" + *gramme* "letter".
Hridaya: *San;* "heart", the universal center of our being.
Ideational: *Lat; ideatio* "conception of ideas by the mind", consisting of or referring to ideas or thoughts of objects not immediately present to our senses.
Ima: *Heb;* "mother".
Infinity: *Grk; apeiria* "infinity". From *apeiros* "endlessness", the boundless nature of endless time, space, or quantity.
Insentient: *Lat; in* "lacking" + *sentire* "perceive, feel, experience" = not sentient, without life, consciousness, or perception.
Jijimuge: *Jap;* "interdiffusion"; as found in the Kegon Sutra.
Jñana: *San;* "Wisdom". Direct knowing of the *Atman* "True Self".
Jñana Marga: *San; Jñana* "wisdom" + *Marga* "path" = "Wisdom Path". The awakening of *Prajna* "intuition" and the apperception of "Self-Awareness".
Kabbalah: *Heb;* from *Lekabel* "to receive directly".
Kahuna: *Hwi;* "Shaman" or "Holy Man".
Kallipolis: *Grk;* Plato's fabled utopian city, ruled by "philosopher kings".
Keiki: *Hwi;* "child".
Kundalini: *Skt;* The adjective *kuṇḍalin* means "circular". And *kundalini* refers to the mixture of a supposed *Pranic* "Life Force" that we take in when we breathe, with a *Sat Guna* "Identity principle" in the food that we consume, and its storage in a *chakra* "wheel-like" center at the base of our spine, as a purported "Serpent Power" which unwinds and ascends to an ostensible "Crown Chakra" with our eventual experience of *Samadhi* "Enlightenment".
Kwatz!: *Jap;* Similar to a *Kiai!* The powerful "meeting of chi" in Martial Arts. In Zen, it is the sudden meeting of the Life Force and primeval

Awareness in spacetimeless intuitive spontaneity, expressed by this (Life) Force-filled exclamation.

Lamed Vav: *Heb;* לו "thirty six", according to esoteric Jewish tradition, is the minimum number of fully Awakened "Avatars" that are necessary to balance the destructive impact of humanities ignorance on the world.

Lila: *Heb;* "night".

Magicus naturalis: *Lat;* "Natural Magic".

Maharaj: *San;* "Great King" or "Great Sage".

Maya: *San;* "Illusion", the veiling and projecting force of the Divine.

Metacognition: *Grk; Meta* "beyond" + cognition = "cognition of cognition", or "knowing about knowing", a "witness perspective" that incorporates all aspects of cognition.

Metanoesis: *Grk; Metanoea* "changing one's mind", a shifting of the locus of identity away from the false self and back to our true "Witness-Self".

Micorrhizæ: *Grk; Micr* "tiny" + *rhiza* "roots" = tiny roots, are *proprietary* "particular to each species" and *symbiotic* "beneficially interconnected" fungi that inhabit and connect a plants tiny "feeder" roots to the soil, providing nutrients and moisture to the host plants roots, and carbohydrates to the guest fungi from the host plant.

Mirakuru: *Jap;* "miracle".

Midrash: *Heb; mi* = "from" + *derash* = "drawing forth".

Morpheus: *Grk;* "God of Sleep".

Mouna: *San;* The "Grand Mystery" of Nirvana, and the Kabbalistic "Veil of the Absolute" (see *"Vimalakirti",* pg. 515).

Nano: *Grk;* From *Nanos* "dwarf". Also, 10^9 or one billionth of the adjoined unit, this is to say that one nanosecond is one billionth of a second!

Nefesh: *Heb;* "Soul".

Newt: *Lat; Ambystomatidae macrodactylum* "Long-toed Salamander", a native of Santa Cruz Mountain streams in Northern California.

Nihilism: *Lat; Nihil* "nothing". The view that Divinity does not exist outside of human conceptualization, that nothing including the contemplator exists, and everything is thus devoid not only of existence but of meaning as well!

Nikkud: *Heb;* "Dot, period, point". The Kabalistic "Seed" of the Tree of Life, also, the primeval singularity before the "Big Bang".

Nirvana: *San; nir* "out" + *va* "to blow" = the "disappearance" of our subject-object consciousness and the revelation of our primal pure Awareness, which is essentially "void" of consciousness.

Nisarga: *San;* The "natural state" of Nirvana, for example *Nisargadatta* "dwelling in Nirvana" is the Yogic epithet of an influential Jñani Sage.

Noumenality: *Grk; noein* "to apprehend". The opposite of phenomenality, the person and universe are phenomena that arise in individual subject-object consciousness from *Noumenon* (i.e. transcendent pure "Awareness" or "Presence", see "Emmanuel Kant", pg. 530).

Objectification: *Lat; Objectum* "thing put before". The act of attributing name and form to the objects of consciousness (i.e. objects comprised of consciousness).

Outer Cambium: *Lat; Cambire* "to exchange". The actively dividing water and carbohydrate conducting tissues, located just under the bark of woody plants.

Palisade cells: *Frn; Palissa* "stake". Cells in the cortex of a leaf or bark of woody plants that contain the carbohydrate producing substance "chlorophyll" that gives plants their characteristic green color.

Parabrahman: *San; Para* "beyond" + *Brahman* "god" = G-d the Supreme.

Parajñana: *San; Para* "beyond" + *Jñana* "knowledge" = the deepest, "ineffable" Source of all consciousness and wisdom.

Phenomenality: *Grk;* from *nooúmenon* "thought of". The manifestation of the universe within our individual consciousness.

Phloem: *Grk; phloos* "husk, bark". The food conducting tissue of trees and vascular plants, on the inner side of their bark.

Photosynthesis: *Ger; Photo* "light" + *synthese* "synthesis" = the ability of plants to turn sunlight into food thereby making life possible on our planet earth.

Plenum: *Lat; Plenum* "fullness", as opposed to *vacuum* "emptiness". In this work, plenum refers to an Einstein-Bose-like concentrate of "fullness"of infinite potential in the "Superspacetimelessness" from which the Superverse emerges, and it also refers to the Yogi's *Akash*.

Plutocracy: *Grk; ploutos* "wealth" + *arkhia* "rule" = "rule by the wealthy".

Polar: *Grk; Polus* "axis" of rotation, for example, a wheel has two "poles", the outer rim and the inner rim, both of which rotate around the axis of the wheels hub, the wheel is one unit with two sides, without which it could not act as a "wheel". Also, having two opposing sides like a magnet or a dipole electrical device. The fundamental polarity of human consciousness is subject and object.

Pons: *Lat;* "bridge", the swelled portion of the brain stem at the top of our spinal cord above the *foreamen magnum* opening at the bottom of our skull.

Positivist: *Lat; positivus* "settled by arbitrary agreement", as opposed to *naturalis* "natural". Belief in the objective reality of the universe and the person as its subject/object.

Pragmatism: *Grk; Pragma* "deed, act". A philosophical view that claims a proposition is true only if it works satisfactorily, and that the meaning or validity of an idea depends upon the practical consequences of accepting it, rather than the strength or logic of the idea in itself (i.e. "Idealism").

Prajna: *San;* "Consciousness", more specifically, the active aspect of Universal-Consciousness (i.e. in *Forest Primeval* it is called "Consciousness-Force", see "Likewise", Consciousness-Force, on pg. 448).

Prakriti: *San;* "Nature", more specifically, "Mother Nature".

Prana: *San;* "Life-Force" (see "Prana" pg. 568).

Psychoactive: *Grk; psyche* "mind" + *Lat; activus* "state of activity" = a chemical compound that affects our perception.

Psychology: *Grk; psyché* "mind" + *logia* "study of" = "study of the mind". Relating to the mental processes that give rise to our conglomerate and interactive concepts of person and other.

Psycho-sensorial: *Grk; Psyché* "mind" + *Lat; sentire* "perceive" = the confluence of mind and senses in our consciousness.
Psychosomatic: *Grk; Psyche* "mind" + *soma* "body" = our "mind-body" (syn. body-mind).
Purgatory: *Lat; purgare* "purify" or "place of cleansing". The insertion of a fundamentalist exoteric-objectivist concept into esoteric-objectivist religion.
Purusha: *San; prusha* "Cosmic Man", the Kabbalistic "Adam Kadmon", our essential "Universal-Consciousness".
Purushottatma: *San;* essential "Awareness", (i.e. יהוה *Yahweh,* الله *Allah, Deüs* "the Supreme", etc.).
Qualia: *Lat; qualificare* "to attribute a quality". In philosophy the inner experiential content of our consciousness.
Quandary: *Lat; quando* "when". A dilemma, a difficult or seemingly insolvable problem.
Quantum: *Lat; quantum* "how much", introduced into physics by Max Planck in 1900.
Quark: Physicist Murray Gell-Mann, in 1964, borrowed a nonsense word from James Joyce's "Finnegan's Wake" to describe an elementary quantum particle that is a fundamental constituent of matter.
Reality: *Lat; reālis* "real or actual", from *rēs* "matter or thing". The idea that there exists a quality of true existence that differentiates actual things from imaginary, or idealistic things.
Relativism: *Lat; relativus* "having reference or relation to". Relativism holds that whatever exists does so only in relation to whatever else exists.
Ribosome: *Eng;* "Ribonucleic acid" + *Grk; soma* "body" = Ribosome.
Rimōn: *Heb;* "Pomegranate", also (rather "soberingly") in Modern Hebrew, "Hand Grenade".
Roshi: *Jap;* A Zen Buddhist Master teacher.
Rostral: *Lat; rostralis* "beak". In creatures with a distinct head, the *anterior* "before" end, which is the "beak" or mouth end of the body.
Sa-Hahm: *San;* "I Am". The Divine's first thought of individual existence.

Samsara: *San; sam* "to complete" + *sr* "movement" = "wandering through", the insufficient concept of our actual "physical" birth, death and rebirth. More likely the Buddha was referring to being trapped in a repetitive cycle of phenomenal manifestation within dualistic consciousness, which is said to be the result of believing that we are a body, or an embodied individual soul.

One way, for example, in which this could happen, is by (our) endless "lateral" interdimensional travel, from one parallel universe into another, within our endless and ever-present "Superverse". Indeed, this could as well be what is impersonally going on in Consciousness, beneath our rather superficial, if not illusory, "personal" space-time experiences of life!

Satori: *Jap;* viz. "Awakening".

Sensei: *Jap;* "Teacher".

Sentience: *Lat; sentiens* "to feel", knowing that we exist and that we know, which gives us the perspectives of individuality and volition, all living things are sentient to the degree that their evolution allows.

Serotonin: *Lat; serum* "watery fluid" + *Grk; tonos* "tension" = "Serotonin", which is derived in our intestines from the amino acid tryptophan, is also involved in involuntary muscle contraction and positive mood states; hence when our mind is upset, often our gut is too!

Shaqwi: An American Indian word for "rattlesnake".

Shema: *Heb;* Literally "hear", or "listen". Also the name of a prayer that embodies the fundamental Jewish concept of *monotheism* "one G-d".

Shoah: *Heb;* "Holocaust".

Sifu: *Chi;* Master Spiritual Teacher, Kung Fu Master, or simply "teacher".

Singularity: *Lat; singularis* "single". A monumental event having no precedent.

Skandha(s): *San;* "aggregates". Theravada Buddhism divides these experiential aggregates that together comprise the *dharmas* "realities" of person and universe into five categories (see pg. 577).

Solipsism: *Lat; solus* "alone" + *ipse* "self" = "solipsism". The proposition that we can only be sure of our own minds existence, which cannot be proven either, leading to the absurd conclusion of our inadvertent self-annihilation!
Sotto voce: *Lat;* "Quiet voice".
Space: *Lat; spatium* "place". An apparently enveloping and internal, but invisible to us, local and (simultaneously) non-local four dimensional "place" without which nothing could be perceived (see pg. 578).
Stomata: *Grk;* στόμα *stoma* "mouth", openings on the bottom sides of leaves that exchange water vapor and gas with the atmosphere and close by utilizing unique hydraulics in hot dry conditions.
Subliminal: *Lat; sub* "below" + *limen* "threshold" = "below the level of conscious awareness". Subliminal ideation is the initiation of thoughts below the level of our mental awareness, which then emerge into our consciousness.
Sunnyasin: *San;* A wandering ascetic in search of enlightenment.
Supreme: *Lat; Supremus* "highest". The ultimate Source of the universe.
Talmud: *Arm; talmid* "student", and *talmud* "teachings received by a student". A voluminous Rabbinic compilation of scholarly Torah related discussions and debates.
Tantra: *San;* "to weave", (see pg. 585).
Tao: *Chi;* "Way". A metanoesis of perspective, occasioned by the shifting of our point of identity back to the essential Source of its sentience, in harmony with our true nature.
Tao Te Ching: *Chi; Tao* "Way" + *te* "virtue" + *ching* "classic" = a classical Chinese treatise on the Tao, written by Lao Tsu in the sixth century BCE.
Tekiah: *Heb;* "blast". A loud single long tone that is blown on a *shofar* "rams horn" during certain Jewish religious ceremonies, such as the High Holy Days of *Yom Kippur* "Day of Atonement" and *Rosh Hashanah* "New Year".
Telencephelon: *Grk; telos* "end" + *egekephalos* "brain" = *telencephelon.* The cerebrum together with the diencephalon or "posterior part" of our forebrain.

Temporal lobe: *Lat;* from *tempus* "time". Lobes located inside our skull, on both sides of our brain just above our ears, involved in hearing, speech, comprehension, verbal memory, naming, language function, object perception and recognition, short and long term memory, episodic functions and translating neural input and cognitive functions into a space-time orientation.

Thalamic-region: *Grk; Thalamos* "inner chamber". The thalamus is a centrally located structure in our brain which relays information to its cortex.

Torah: *Heb;* "Law". The first five books of the Jewish Bible.

Totality: *Lat; totus "entire".* The entire universe, including everything outside of its apparent manifestation within our individual consciousness.

Über: *Ger;* "Over". A superlative, as in *Über Mensch* "Super Man", or *Über Geist* "Over Mind".

Upanishads: *San; Upa* "near" + *ni* "down" = sitting near a *Sat Guru* "Master teacher" to receive the final, or ultimate, teachings about the nature of man and the universe. The core Hindu teachings of the *Vedanta* "final", or "ultimate", scriptures.

Utopia: *Grk; ou* "not" + *topos* "place" = "nowhere". Coined in 1516 by Sir Thomas Moore as the name of a mythical island with a perfect legal, political, and social system in his book, with the same name, that explores the subject of an ideal society.

Vedanta: *San;* "Final". The "final" portion of the Vedic Hindu scriptures.

Vimalakirti: *San;* A lay Buddhist in the time of Gautama Buddha, during the fifth and sixth centuries, credited with expounding a *Sutra* "Scripture" that bears his name "V. Nirdeśa Sutra", which is a seminal Mahayana Buddhist exposition of non-dualism, culminating in the teaching of *sunyata* "emptiness", and the wordless teaching of *mouna* "sentient silence" (For a brief comparison of early "Theravada" and later Mahayana Buddhism see "Skandhas" on page 577).

Vritti: *San;* The smallest possible point of pure Consciousness-Energy, a sentient "consciousness wave", that is the smallest possible constituent of a thought, which is a mental modification similar in function to a computer "bit", except that it is also fundamentally sentient (as occasioned by its own necessary Principle of Identity).

Wavicle: *Ger; waben* "undulate". Neither strictly a particle nor a wave, a wavicle is a combination of the *Sat* "Identity Principle" in the singularity of its "particle" aspect, and *Chit* "Universal-Consciousness Force" in the implicate "dualism" of its "wave" aspect (see pg. 596).

Wu: *Chi; Wu* "aware nothingness" is what remains when we re-identify with the true Witness (Awareness) of our minds, by realizing that our essential sentient Awareness (as the Supreme) always resides in the space-timeless "gap" that is always present between each frame of our apparent subject-object experience.

Yahweh: *Heb;* the "Supreme" (syn's "Awareness" and "Witness", see pg. 596).

Yesod: *Heb;* "Foundation". The ninth *spherõt* "sphere" of manifestation in the Kabbalistic *Etz Chayim* "Tree of Life" paradigm of our universe.

Zen: *Ch'an* in Chinese. A Japanese branch of East Indian Buddhism deeply rooted in Chinese Taoism, (see pg. 597).

"You ask me why I dwell in the green mountain, I smile and make no reply for my heart is free of care. As the peach blossom floats down-stream and is gone into the unknown, I live in a world apart that is not among men."

Master Po, 701-762 CE

Appendix

In Contradistinction to the routine appendix's "dictionary definition" of concordant words, this Appendix, which was written while we were still camping in the redwoods (you've always been here with me), is also, at times, a conversation about the deeper meaning of the terms that are being defined. Hence, you may also, on occasion, find some references to helpful explanatory *dialectics* (discussions) that are located elsewhere within this book.

As you read, imagine that we're still sitting here together discussing the terms, next to a warm campfire, under a dark Santa Cruz Mountain sky that's chockfull of bright "childhood" stars!

Aphelion: *Aphelion* is the farthest point from the sun of an orbiting object, while *perihelion* is the nearest.

Apperception: As used by Emmanuel Kant in his "Critique of Pure Reason", apperception may be either *a priori* "pure", or *a postiori* "empirical". In this work it refers to the "pure" awareness, or non-dualistic consciousness, which is present prior to our bifurcation of elaborated thought concepts into an assumed subject and ideational objects.

Ayn Rand: Alisa Zinov'yevna Rosenbaum, 1905 – 1982, was born in Russia and moved to America in 1926. She is the elitist author of the controversial books "Atlas Shrugged" and "Fountainhead", and is credited with developing the philosophy of Objectivism.

Bal Shem Tōv: The Bal Shem Tōv, 1698 – 1760, is revered for his "democratization" of the Jewish religion, and for the mystically based "Chasidic" movement that claims him as its primary founder. But as with most awakened Sages, it is difficult to say how he would personally feel about his latter day followers interpretation of his intimate occult teachings. However, the unusual "name" which has been given to him speaks volumes in itself!

This is to say, that in Hebrew his rather inscrutable name means "Master of the 'Good' – as in "superlative" – Name." And it seems not incorrect to suspect that the "Shem Tōv" (Good Name) that is being referred to is יהוה *Yahweh* the "Supreme"! In which case the "mastery" to which it refers, is likely no different in essence than that of the Buddha, Jesus, Mohammed, or any of the other fully awakened masters that we have been discussing in Forest Primeval.

BASHŌ: Matsuo Bashō was a Zen poet during the Edo period in Japan; 1644 – 1694. This better-known poem (which I believe was either written or inspired by Bashō and much later introduced to the world by Allan Watts) was evidently my subliminal inspiration for the moon observation on page 235 of Forest Primeval: "Luminous fruit hanging from pine bow? No, the Moon!" Bashō is not only saying that things are not always as they may at first seem, but that the form and name of things are actually assigned by, and appear within, our minds!

BIOLOGICAL COMPUTERS: In 1965 Silicon Valley researcher Gordon Moore correctly predicted that the number of resistors on silicon chips would double biennially. His prediction has withstood the test of time, but we have finally reached the point of diminishing returns. This is not only because silicon based computer chips will soon reach the limit of resistors that we are able to etch onto their surfaces, but also because it will continue to cost ever more money to advance the technology that allows us to do so!

However, recent advancements in the use of GaN (Gallium nitride), which is ten times more efficient than silicon, is both lowering the cost and extending the inevitable end point in the dilation of diminutive chip design. Nonetheless, with a minor mathematical adjustment for Gallium nitride's increased efficiency, Mohr's Law still applies.

Thus even GaN enhanced chips will soon reach their limits too! Further, even if carbon "nano towers" of atomic sized arrays are used, the Heisenberg Uncertainty Principle creates likely insurmountable engineering problems. Even *Leptons* such as photons and electrons are comprised of subatomic wavicles, which may also serve to carry the information from one atom to another, but are nonetheless, evidently by their very nature, too unpredictable! Their inherent uncertainty in location and speed, or space and time, make them unlikely candidates for carrying accurate binary information on the quantum level.

Therefore biological computers which use DNA sequences as carriers for binary codes are being considered. Their capacity to enable the solving of complex problems by massive "parallel processing" with trillions of reactions taking place simultaneously, far outstrips even the tremendous speed of modern silicon based computers.

Flexibility in design and the ability to interact with the DNA in human cells may also promise a bright future for their use in medicine. But in the search for true Artificial Intelligence, "A.I." capable of learning, self-repair, sentience, and continual "self-heuristic" improvement in complex pattern recognition and multifaceted problem solving, these hybrids will probably not go the full distance either.

However, the already ongoing process of engineering living organisms, by growing brain cells on artificial surfaces, might actually succeed! By carefully controlling the development of hybrid computer technology it may even be possible to create an organism with the capacity to evolve enough to finally reach the fabled "Strong AI Singularity" plateau.

But this is also, quite literally, creating "a brain in a bottle"! Which brings up some rather serious moral and ethical questions. Indeed, many difficulties lurk along the way, and many more should we succeed.

Problematically, where Strong AI's anticipated problems and inevitably unforeseen complications are concerned, it is not so much a matter of "if" we can accomplish this, but *when!* Our best hope in minimizing them is in preparing for them with competent development guidelines, such as Eric Drexler suggests that we do with our rapidly developing nano technology, and then wisely implementing them as AI is developed (and eventually begins to refine and pursue its own development!). But if past results are any indicator of future success, "playing god" may harbor serious unforeseen consequences!

Alternately, "qubit" (no relation to the biblical *c*ubit) computing, with super-cold quantum computers, could also hold the key to the unlimited externalization of our brain's intelligent biological growth, and the inevitable concomitant ability to gain an ever greater control over our psychosomatic selves and our universe. In any case, we now stand near the brink of bone fide AI, and only time will tell the story of its fuller emergence on planet earth!

BIOLOGICAL PROGRESSION: "Biological Progression" is the tendency of a tree species to recapitulate its ascendancy in a biome. For example, following a forest fire grasses are replaced by shrubs, shrubs by "sun loving" pine trees and pines by a climax forest of shade tolerant evergreen and hardwood trees.

BODHISATTVA'S VOW: This is the Bodhisattva's vow; *"I vow to liberate all beings without number"* But, certainly this is an impossible task! So what could the Mahayana Buddhists possibly mean when they take it? Likely this, because the entire universe is essentially comprised of the (quantum) potential filled "emptiness" of impersonal Universal-Consciousness, there is nothing substantial to be liberated from!

And because there exists no substantial beings to be liberated, there is actually no one to "liberate" as well! Therefore, in liberating themselves from the ignorant perspective that the universe is materially substantial, all beings are liberated, and the Bodhisattva's vow is fulfilled.

However, countless insubstantial beings certainly do remain ignorant, despite the Bodhisattva's enlightenment. To say differently, begs reason! This is because all unaware "apparent" subjects, remain the ideational objects of other ignorant "suppositional" subjects. Therefore, those who still believe that they are "entities", sadly, remain subject to suffering. For the sake of these, compassion in the form of guidance to end their ignorance is given, as needed, for so long as the enlightened Bodhisattva remains embodied. For the *Sadhaka* "Spiritual Warrior/Seeker" this essentially means, "No person left behind

However, even from this perspective, each "Living Buddha" furthers the evolution of the universe, which is the activity of Universal-Consciousness and not the Bodhisattva. (I.e. We are all "dead" – or sleeping – Buddhas until we wake up to the fact of who it is "in Essence" that we really are!). "Free will" is thus a myth born of an ignorant conditional perspective, which is subsumed by the non-contingent provenance of Universal-Consciousness. In other words, as Sri Krishna once stated, *"Whatever is, was, and shall forever, Be";* even if we mistakenly believe that we are a volitional agent, in a spontaneously unfolding Superverse!

CHAKRAPANI: *Chakrapani* "flywheel" is a Sanskrit term that describes the considerable power that our binary human mind has in convincing us that its cognitive representations of our persons and universe are real. Like the heavy, rapidly spinning flywheel of a tree trimmer's brush chipper, from the advent of the primordial Big Bang, the considerable force of our rapidly expanding universe is actually behind our every thought!

And like an inertia empowered flywheel, the further that we move away from our witnessing center and into our thought-constructed universe, the more powerful its convincing force is upon us! The measured stillness of regular meditation is thus an effective method of damping the inertia of our cognitive "flywheels" down.

Chief Seattle: "Chief Seattle" was a prophetically aware Duwamish Indian chief (1780-1866), pronounced *Si'ahl* in the *Lushootseed* language, who was born on the Black River in Northwest Washington State. He was an eloquent spokesman for his people and for the continued welfare of the American wilderness, and is credited with an impressive array of insightful observations, for example: "Man did not weave the web of life; he is merely a strand in it. Whatever he does to the web, he does to himself."

Cerebral Cortex: The "cerebral cortex" is comprised of a hierarchical arrangement of neurons and synapses that are connected to the rest of our brain. It is responsible for many of our higher brain functions, such as sensation, voluntary muscle movement, thought, reasoning, and memory.

Cognition: Cognition is a series of psychophysiological processes that create an endless cavalcade of concepts, which appear serially in ideational space and time. All that is known of our universe is contained in these thought concepts.

There is a simultaneity of identity and cognition, which allows the ongoing concepts of "person" and "universe" to become "real" in the flow of our consciousness. Our reality is thus comprised of concepts.

But while the actuality of Universal-Consciousness on the fundamental pre-subatomic string and precognitive levels remains elusive, it is apparently not impossible to re-identify with it, at the level of our precognitive, purely Supreme identity.

There evidently exists a "forceful continuity" of impersonal Universal-Consciousness between these two (i.e. Supreme, and individual-consciousness), at the most fundamental levels of undifferentiated Cosmic Energy.

This is evidenced by the interaction of our conscious intention with our universe's foundational subatomic wavicles, and their interaction with one another over nonlocal (unlimited) space-time.

Such as, but not limited to, "simultaneity" (i.e. quantum entanglement) wherein paired leptons simultaneously change the direction of their "spin" regardless of any apparent space-time distances.

In other words, everything, including our consciousness and the force of our consciousness, are merely portions of an infinite, spacetimeless and universal quantum "sea" (i.e. the Multiverse).

Cows: Cows quite naturally live a quiet life that is neither boring, worrisome, nor painful. When left undisturbed, it is a life that could even be described as happy. As Nietzsche euphemistically opined; "A human being may well ask an animal, 'Why do you not speak to me of your happiness but only stand and gaze at me?' The animal would like to answer, and say; 'The reason is I always forget what I was going to say' – but then he forgot this answer too, and stayed silent; so that the human being was left wondering..."

CRISPR: "Clustered Regularly Interspaced Palindromic Repeats" is a rather quaint term for a unique naturally occurring bacteria that was first discovered in 1987. Since that time it has been turned into a microscopic sized "biological robot" that has been made quite efficient at creating and using a naturally occurring enzyme to cut and splice human genes. Because it utilizes RNA as a *sensor*, has a computer to *decide*, and an enzyme that acts like an *actuator*, it has the three basic components that qualify it as a "robot".

Lead "Human Genome Project" researcher George Church, of the Harvard Medical School, predicts that CRISPR will eventually allow us to edit the entire human genome. Just a few of the astounding possibilities are; encoding the data of the entire Internet into the human brain; creating a direct wireless interface with the Internet and other human brains; curing and preventing diseases; repairing damaged or failing human bodies; extending the human life span; expanding the range of human intelligence;

and engineering our human body to withstand the rigors of interstellar space travel, (just to name a few!).

Dr. Church is also of the opinion that all of these developments are, in fact, the products of the natural evolution of life on our planet, and hence not simply the artifacts of our own devising; "If our planet gets struck someday by an asteroid and it wipes out every form of life on earth, from a Darwinian perspective it would be a terrible failure!"

But if we are able to one day explore amongst the stars and take life out into the Cosmos, it will be nothing less than a great Darwinian success! As it turns out, while the gifts of science and technology are certainly somewhat open to abuse, they may not be "unnatural" at all. And the greatest gift that we can give to our children is the same one that Mother Nature grants to us all, the ability to not only endure, but to also manifest a better future for all!

"If we can begin to change every atom, the possibilities are limitless!"

George Church

DAEDALUS: In Greek legend, Daedalus' son Icarus ignored his father's warning, flew too high, and fell to his death in the sea, because the sun melted the wax that held his wing feathers in place.

DIENCEPHALON: The *diencephalon* is composed primarily of two structures, the *thalamus* and the *hypothalamus*, and serves as a relay system between sensory input neurons and other parts of our brain. It operates as an interactive site for our central nervous and *endocrine* gland systems, and also works together with our *limbic* system, which is involved in our brain's emotional and memory functions.

Dopamine: Dopamine is our brain's very own naturally produced "cocaine"! It is involved in our motivation, desire and pleasure, our ability to execute smooth controlled movements, the flow of information to the frontal lobes of our brain from other areas of our brain, memory, attention, problem solving and other neurocognitive functions.

Dopamine reuptake inhibitors (DRI "antidepressants"), for example, block the action of the dopamine active transporter "DAT" that pumps dopamine out of a neuron's synapse and back into its vesicle for future use.

Diogenes: Diogenes (412-323 C.E.) was a "Stoic" Hellenistic philosopher who is credited with being the father of the philosophy of "Cynicism". He purportedly carried a lantern around in the light of day, as he searched in vain for an "honest man".

Doppler Effect: The "Doppler Effect" is the difference in the wave frequency of light when it is measured in relationship to the position of its observer. For example, whenever light travels great distances across our universe it begins to shift into the red, or less energetic lower end of its spectrum. This is because our universe is continuing to accelerate as it expands outward from its nascent position of original oneness in the Primeval Singularity that was present prior to the "Big Bang".

Indeed, we cannot even see the majority of this expansion! Because its constituents have been moved so far away at FTL inflationary speeds, that our entire universe will likely end, before its photons could even begin to traverse the vast distance from our universes "leading edge" to where we reside, in a rather smallish Solar System, on a scimitar-like arm, in a far-away corner of our Universe's rapidly spinning "Milky Way" Galaxy!

Dystopia vs Utopia, And Universal Opposition: "Dystopia vs Utopia" is an iconic literary example of polar opposition. Everything in the universe is either in stasis (balance), or opposition (opposite position). Three dimensional "motion" in our universe, actually an energetic change of location (and often of position as well), could even be said to be caused by the natural attraction of emerging dissimilarities and opposites as they actively seek a renewed balance in space, as measured over its extension in four dimensional time.

Thus, when more energy is introduced into a moving energy-object its location begins to change more rapidly as it tries to reestablish stasis within the ever-extending "void" of space and time.

However, our familiar properties of motion emerged from an atypical "fifth state", that of the "Elsewhere" from which the Primeval Singularity coalesces. The perfect stasis (Supersymmetry) of the universe's Primeval Singularity somehow "broke" into a brief spacetimeless chaos, where the laws of physics did not strictly apply.

But within immeasurably short moments, the cycle of motion began as superhot pure energy began to expand concurrently with the space that contained it, and then began to cool and coalesce into multiple quantum expressions of force and form, but from constantly changing new positions.

Thus polar opposition and space-time work together through an apparent change in position and location that is brought about by introducing or subtracting energy. But if so, could this teach us anything about the more subtle manner in which our ever extending multiverse, which is evidently comprised of constantly emerging parallel universes, functions with the addition of the energy of our observing consciousness into the quantum substrate of the penultimate Superverse that it comprises?

In a "Static Parallel Lattice Superverse" paradigm, when the initiating inflationary force of the Big Bang overpowers the capacity of our expanding universe to contain its (perhaps infinite) force by continually changing, its new position is likely in an adjacent universe.

Differing by one quantum particle, the new adjacent universe is separated by the shortest distance that is possible to measure in our universe, which is purportedly one *Planck* or 1.616199×10^{-35} meters (a "meter" is 3.28 ft.).

Moreover it is also likely briefly connected in Planck time (10^{-43} seconds), which is evidently the shortest possible measurement of time in our universe, by a quantum level "vortex" of energy, which forms out of a briefly intervening chaotic state.

But this is also a phenomenal interpretation of impersonal Universal-Consciousness' occult actions that is brought about by our apparently personal individual consciousness' unseen interaction with the universe at large, on a quantum level. Which, unnoticed by us, brings about a cascade of parallel universes, manifested by positive, negative, neutral, "exotic" and "dark" forces (i.e. "poorly understood" and "unobservable" forces, respectively).

And within our individual consciousnesses, which are manifested in our body-minds by and through these forces, there appears an apparently unchanging and independent "subject", and an evidently endless stream of observable "objects"!

This allows our spatio-temporal experience of, and by, the principal spacetimeless Awareness of the Supreme, from which the nascent force of impersonal Universal-Consciousness emerges and coalesces into the space-time and subatomic strings that make up our universe (and us!).

But the true position of our independent Supreme "Witness", in the primeval spacetimelessness of the pre-singularity Elsewhere, is of course prior even to the appearance of the psychosomatic experiential "lens" of our body-mind.

And it is through our body-minds that Universal-Consciousness experiences an ongoing stream of apparent diversity. While our individual consciousness (almost simultaneously) becomes imbued with sentience as it is being witnessed by the ever present independent Supreme.

This is the purported "Gateless Gateway" of Zen that opens inwardly into the "fifth dimension" of our deeper, ever witnessing Self, which sometimes leads to a reintegration of perspective that is often called

"Awakening" (syn. Buddhist *Nirvana*, Kabbalistic *Tiferet*, or Yogic trance-less "eyes-open" *Nirvikalpa Samadhi*).

But nothing has changed except our point of perspective, which modifies Universal-Consciousness' identification with the body-mind instrument of the Supreme's apparently dualistic experience, which we commonly, and incorrectly, believe is our very own "personal" body-mind!

EINSTEIN'S ORIGINAL SPEED OF LIGHT THOUGHT EXPERIMENT: After graduating from college Albert Einstein could not find a job in his field as a physicist and so with the help of his father he took a job as a clerk in a patent attorney's firm in Bern Switzerland. One day, while commuting home from work on a bus, he happened to look back at the city's large clock tower.

As the bus sped away from the tower he imagined that the bus continued to accelerate. As the bus began to approach the speed of light, he noticed that the movement of the hands of the clock on the tower began to slow down. When the bus finally reached the speed of light, the hands of the clock stopped moving entirely! Einstein realized that although time stood still in his relationship to the clocks position in space and time, that on the bus that he was travelling on it moved in its usual way.

All at once he realized that space and time must exist as a single unified substance or "field" that was being effected or "warped" by the presence and momentum of the bus! Then he suddenly understood that Isaac Newton was incorrect in assuming that gravity was an attractive force, like magnetism.

Instead, the presence of a large mass, like a planet, is actually curving or "warping" the space-time around it so that it is actually space-time that is "pushing down" upon our heads, which creates the sensation of an "attractive" force of gravity!

Einstein also proposed another "thought experiment" to explain his Special Theory of Relativity: If the light from a light source located inside a speeding train emits a beam of light simultaneously "towards the front and the back" of the speeding train, and it is carefully measured from outside of the train, we will discover that it arrives at the back of the train first! And the faster that the train goes, the longer it will take the light to arrive at the front of the train, relative to an outside observer.

But to a passenger on the train, the light will continue to travel at the same speed in both directions, no matter how fast that the train is going! Hence, the speed of light is apparently absolute and unchanging, while we remain the prisoners of our own ever present "moment" in time (which is also the real evidently *physical* "Power of Now"). Consequently, our previously misunderstood present "moment" now takes on a rather different meaning!

Moreover, a clock located inside the speeding train will move ever slower as it approaches the speed of light, relative to an outside clock. Indeed, if both the clocks and the train are located in empty space, we actually have no way of judging which clock is correct.

Hence, there are no absolutes when it comes to time! In other words, the space within the train is relative, or proportional, to the motion of the time that it contains, just as the space outside of the train is relative to the clock located outside of the train. Thus, there are no absolutes when it comes to space as well!

EINSTEIN'S POSTHUMOUS "MUON PROOF": The "muon" is a recently discovered extremely small subatomic particle that was unknown to Einstein. But it nonetheless has acquired the distinction of proving the veracity of his "Special Theory of Relativity". When cosmic rays strike our earth's atmosphere from within "deep space" they sometimes hit a muon.

We can now measure this event and have discovered that the impacted muon then travels at 660 meters in 2 millionths of a second, which is around 99% of the speed of light. But to us the muon appears to travel 32 kilometers in the same amount of time! Thus, time "expands" for the muon, which proves that the duration of time is relative to its experiencer, evidently in much the same way that its existence is!

Therefore, like Einstein, we are left to wonder, just how much of our place within, as well our experience of, any ostensible "reality", is determined by the relative extremities of the motion of its energies, as well as their relative (atomically expressed) space-time positions within it (i.e. the Universe).

EMMANUEL KANT: Emmanuel Kant was an 18th century German philosopher who correctly reasoned that our mind shapes and defines the limits of whatever we experience. Therefore subject-object perception, causality, and space-time are methods of consciousness whereby our mind perceives, and they do not exist in this way outside of our perception. In other words, our universe is essentially "noumenal".

Classically, "*noumenon*" refers to a phenomenal object of cognition, in other words, to objects or events that are independent of our senses and known only to our minds. The obvious conclusion is that; "Noumena cannot be an object, which emits the phenomenon".

However, Kant evidently missed the fact that this "split", of noumenon and phenomenon, is likely an intellectual refinement of our mind's perceptive dualistic limitations. Yet a "perceiver" is clearly necessary for any sort of perception to occur!

The spacetimelessly independent Perceiver (with a large *P*) of our space-time involved subject/object perceiver (i.e. ego) is thus most certainly not only necessary, but is necessarily "noumenal" as well! Thus even our perceptually evident (but actually impossible!) "subject/object" perceiver is itself merely a perception!

But it is useless to simply point out the differences between the perceiver and the perceived, as some degree of separation is obviously necessary for any act of perception to occur. Furthermore, whatever is perceived is eventually done so from within a "self-aware" (i.e. sentient) noumenal perspective.

However, this situation also raises far more questions than it answers, not the least of which is the infamous "Gap Quandary" (see dialectic on pg. 537). But since both our universe and its experiencer (i.e. "us") are present, yet both are beyond the ability of our mind to perceive directly, the seminal question remains; "What is the nature of the Noumenon?" And the only acceptable answer is evidently the silence of our own observing Sentience!

ENDORPHINS: Endorphins are *"endogenous opioid polypeptide neurotransmitters"* that are produced by our pituitary and hypothalamus glands, primarily during exercise and sex. Endorphins are our brains own "feel good" drug. Curiously opium fits snugly into the "feel-good" endorphin receptors that are located throughout our body and brain, which certainly explains why refined opiates (like heroin) are so terribly addictive!

ENTROPY: "Entropy" is the tendency of systems and objects eventually to fail, and as such is simply a more technically useful scientific expression of the ancient East Indian Yogic paradigm of universal "Gunas". Thus *Tamas* "negative entropy" is simply another name for the distribution of entropy from a system in order to preserve it (see "Gunas"; *"Tamas"*, view dialectics on pg. 540).

ESOTERIC-OBJECTIVIST: It is impossible for an incorporeal soul and a corporeal body to interact, as Emmanuel Kant correctly pointed out. Therefore "Esoteric-Objectivism" promotes either an unlikely pluralistic "spiritually/material", or an equally unlikely "materially/spiritual" universe, either of which could only act in accord through an unlikely set of randomly related coincidences!

EVE: Quite simply put, the veridical "Eve" of all terrestrial life is Mother Nature. While the progenitor of all hominids was evidently a strong but diminutive black woman who lived in Africa around 150, 000 years ago! Hence, humanities once (ignorantly) touted racial differences are in genetic fact only "skin deep".

When we are all finally able to admit to the obvious truth of this fact and lovingly embrace her as our very own "Mother", perhaps one day we will even be able to stop killing and persecuting each other over our trivial and imagined differences. Then by working together, as an interdependent and caring "human family", we can one day complete Martin Luther King's prophetic dream, and our slowly healing world will truly live as One!

אַלע מענשן זײַנען ברידער

By Y. L. Peretz (1852-1915)

אַלע מענשן זײַנען ברידער:
ברוינע, געלע, שוורצע, ווײַסע...
מישט די פֿאַרבן אויס צוזאַמען!
אַלע מענשוײַנען ברידער
... פֿון איין מאמען.

Ale mentschen zaynen brider:
Broyne, gele, shvartze, vayse…
Misht di farben oys tsuzamen!
Ale mentishen zaynen brider
…fun eyn mamen!

All men are brothers:
Brown, yellow, black, [and] white…
Mix the colors all together!
All men are brothers
…from one mother!

In loving memory of Arthur Jackowitz, and Faye Jacobowitz. And all of our precious loved ones who could not escape the pitiless flames of the Nazi Holocaust.

EXISTENTIAL DILEMMA: Classical "Existentialism" holds that "objective truths", such as mathematics, are simply too detached (or "observational") to effectively describe or explain our human experience. Thus our "dilemma" is described as the quiet human struggle to make sense of, and cope with, the apparent meaninglessness of life.

However, as used in this work, our so-called "existential dilemma" begins with the appearance of our illusory subject/object egos (i.e. our ego is also an object of our experience), by which the objects of our consciousness are conceptually experienced as if they constitute an internally experienced, but objectively real and completely independent, wholly "external" universe.

Thus our "existential dilemma" is initiated by an inaccurate perspective which, like a mirage of water in the hot desert sun, seems real until its cause is finally revealed.

EXOTERIC-OBJECTIVISM: "Exoteric-Objectivism" promotes a "positivistic" perspective, in that it posits the objective reality of the universe and our person as its subject/object ego. In other words, its person-subject (us) must also be an object with the ability to fulfill a subjective role in an objectively real universe. Nonetheless, because it is clearly impossible for a subject to simultaneously be its own object, there are demonstrable discrepancies in this hypothesis!

EXOTERIC RELIGIONS: "Exoteric religions" are primitive attempts to explain the universe in predominantly objective, but nonetheless supernatural terms. An exoteric god is surmised to be a substantial entity that creates an objectively real universe, universal polarity is ascribed a moral sphere, and man must exercise his "free will" to choose its "good" aspects. Evil and misfortune are explained by the existence of a malevolent demigod who interferes in the lives of men, by prompting or tricking them into doing "wrong". The "righteous", who choose the good aspects, are then presumably rewarded by the good god, and the wrongdoers, who do not, are punished by the evil demigod.

Forces, Laws, and Causality: Emerging from a primeval state of undifferentiated pure energy, by way of probability, the universe obeys fundamental and mathematically predictable laws that order the unfolding of its constituent energy, space and time.

It can be argued that, from the beginning of space-time, an underlying unification of law and force determined the fundamental direction that the universe took in probable causality in the first few "Planck" moments of the rapidly expanding Primeval Singularity.

Thus, as it began to cool and separate into primary particle waves, space and time, the universe followed a definite path leading to our present moment.

This process will presumably continue until the temporal end of our spatially extended universe is finally reached. Therefore what appears to us as our wholly independent "free will" is only the inevitable extension of the universe's primeval "most probable" course, which is most likely accomplished by an automatically unfolding causality. Nonetheless, this does not mean that anything goes!

The lines between our common experiences of "inner" and "outer", for instance, are indefinite because the unfolding of interrelated occurances in apparent causality includes not only inevitable lapses into chaos and reordering, but also their unfolding as experiential "events" in the unavoidable emergence of our phenomenally interactive consciousness' as well!

Indeed, outside of consciousness it is difficult, if not impossible, for anything to "exist" in the common sense of the word! Its existence is therefore only probable. This often manifests psychologically in our human experience as a seeming battle between independent forces of "good" and "evil" (and possibly even subliminally within our emotional responses as well). This is clearly an erroneous projection of our space-time limited psyches. Yet what we interpret as "good" will likely inevitably triumph, as higher levels of consciousness and consciously directed order continue to unfold - as they "Will". In effect, everything is indeed already "written", and it is this ever extending probability of order that science is attempting to understand.

FORGETTING: "Forgetting" serves the important function of selectively limiting the constant cascade of phenomenal information that we almost instantly receive from our senses. It also aids in the essential process of storing our apparently "present" experiences in memory. But it is also necessary for our experience of episodic space-time to occur.

The cognitive process of forgetting likely allows this by generating the gaps of unconsciousness that separate one present moment from another, as our experiences are quietly shuttled away into our ever-growing storehouse of memories.

But forgetting is also a thought process and as such it too, like our dream and "deep sleep" dreamless states, appears before the ever-witnessing Supreme Awareness that is our own essential Nature. Thus, awakening to our Awareness of forgetting has the power to open the intuitive *(Bodhi)* doorway to our essential "Satori" Awareness.

FOURFOLD SUFFERING: Legend has it that, once upon a time, a young East Indian prince named Siddhartha Gautama left the luxurious compound where he was being purposefully sheltered by his father, the ruling king, from the pains of the world. (Before Siddhartha's birth, the king's wise men had warned him that the future prince would renounce the world in search of enlightenment if he ever encountered the suffering of humankind.)

Outside the compound, Prince Siddhartha saw a groaning woman give birth to a crying infant. Further on he met a woefully deformed leper and on his way to the Ganges River he saw an old man suffering his last moments as he gasped for air on his deathbed.

When he at last came to the river's edge, he saw people mourning and praying together around a funeral pyre for their departed loved ones. And as predicted by his father's soothsayers, that very night Siddhartha secretly scaled the palace walls and eventually made his way into the wilderness in search of a solution to the suffering of humankind.

Much later, after becoming enlightened, Gautama Buddha taught that enlightenment is "An end to suffering". A fundamental tenent of his teaching is the so-called "Fourfold Suffering" of birth, sickness, old age and death, which are inescapable for as long as we identify with our bodies.

But since everything is clearly known through our mind, it seems likely that the suffering of experience as an illusory "entity" is the crux of his meaning.

GAIA: "Gaia" is a hypothesis that was presented by environmentalist James Lovelock and microbiologist Lynn Margulis in the 1970's proposing that living organisms evolve as a community that interacts with their environment to create and sustain a life friendly planet.

GNOSTICISM: Gnosticism is the teaching that the cosmos is created by an imperfect god, a demiurge with some of a supreme gods *pneuma* "breath" or "soul". The superior god is referred to as having *pleroma* "fullness".

Sadly, in the prejudicial past, a few misguided proponents of a nonsensical sort of baseless "spiritual" Gnosticism actually promoted the outrageous myth that the purportedly "inferior god" is the god of Abraham, as opposed to the supposedly "higher god" of Christianity!

But in this science and metaphysically themed (more contemporary) work, the dynamic pneuma portion (Brahman) is referred to as the Supreme, יהוה *Yahweh*, الله *Allah*, *Deüs "G-d"*, or simply as "Awareness."

While the Supremes' demiurgic aspect (Brahma), is referred to as אלוהים *Elohim* "Universal-Consciousness".

GAP QUANDARY: The apparent gap between consciousness and its bio-chemioelectric substrate has led many brilliant philosophers into a logical quandary, the so-called "mind-body" or "gap quandary" problem.

GOOGOLPLEX: A "googol" is represented by a 1 that is immediately followed by *one hundred* zeroes, and a "googolplex" is a "googol of googols"! But Google® is a visionary Silicon Valley company, just down the street from where I live, in climatically congenial Mountain View, California.

However, a googolplex is also a number that is so very large that it could never actually be written! But, even a googolplex was soon discovered by Cosmologists, and other scientists who commonly deal with very big numbers, to not be nearly large enough to measure things on the Multi and Superverse levels. So an even larger number was required!

Obligingly, if not inadvertently, genius mathematicians with the aid of modern computers started to provide them. First the *Graham Number* appeared, with a bit of true irony within the arcane machinations of the investing world! But it was quickly followed by the much larger *Skewe's Number*, and even larger *Moser's Number(s)*.

Nonetheless, even the smaller "Graham Number" remains highly speculative! Because, even if we use Planck length symbols to represent each of its numbers, it is still too large to write within the physical confines of our visible universe.

To solve this problem mathematicians are currently working with *Transfinite Numbers*. Which are numbers that are larger than all finite numbers, but nonetheless are not necessarily "supremely infinite"!

(In other words, Transfinite Numbers are so large that they are no longer exactly "finite". But they are also not necessarily "infinite" either). Indeed, an algorithm has recently been devised that uses such large representative numbers, that a solution to the entire Graham Number sequence has already been found!

Moreover, by taking advantage of quantum level features, like "simultaneity", and the self-directive learning capacity of a rapidly evolving A.I., with faster "super cold" quantum level computing (at FTL speeds) apparently, "the sky is the limit"!

Therefore the best that we can do is to predict that such large numbers could *possibly* exist. Indeed, because a googolplex, or even a Graham, Skewe, or Moser's Number, like everything else in our experiential (i.e. phenomenal) universe, is really just a concept comprised of thoughts, quandaries will inadvertently appear when we attempt to apply them to mathematically represent an almost infinitely large Superverse. (Which is comprised of an almost infinite number of "parallel" Multiverses!).

Moreover, since even a *googolplex* cannot be proven to exist, both it and any subsequently larger number, must remain infinitely, just so! This being the veridical case, such grand number sequences will most likely forever remain, except for their many revelatory "footprints" in our cosmos, unprovable theories.

But the same can also be said for any nearly "open ended" equation, since the necessary "reality" that is said to contain it is also just another unprovable hypothesis posing as a provable theory.

Hence, when persistently pursued, both will invariably lead to the equivalent of mathematical "infinities" (i.e. errors). Nonetheless, this also does not mean that such speculative numbers do not have, and will continue to have, profound heuristic applications when we are dealing with "near infinite" numerical conglomerations.

This is to say, that simply because the effects of consciousness are subtle does not mean that they do not exist, in the same manner that the natural process of radioactive decay once appeared to violate the universal law of the "conservation of energy", until the almost invisible neutrino was eventually discovered.

But just because its presence is occult (i.e. "hidden"), does not mean that some sort of supernatural phenomenon is going on. Indeed, Young's curious "Interference Experiment", Michelson-Morley's rather mysterious "Double-Slit Experiment", and Schrödinger's temporally suspended half-alive/half-dead "cat in a box" thought experiment, certainly seem to indicate otherwise!

In other words, viewing something, when measured on the quantum level, measurably effects not only its present state, but even the temporally "non-local" spatial outcome of its quantum interactions, such as the peculiar property of so-called "quantum entanglement". And unlike the purely intellectual pseudo-theories of "reality" and "googolplexes", this interaction can be experimentally proven!

We evidently need more than a mere statistical method, like the nonetheless incredibly accurate Heisenberg Uncertainty Principle, to more accurately predict certain of the (only evidently anomalous) independent and interactive behaviors of subatomic particles on the quantum level.

And moreover, we will not be able to do so, until we can find a method to accurately include the interactions of our own consciousness in the equation of *Experiential Manifestation* on the quantum level!

GRACE: Many think of "grace" as being granted the Divine gift of spiritual strength to endure suffering. This is, in a fashion, doubtlessly true, but it also has a far grander, if not more original meaning! Because Grace, in its more ancient Hebrew origins, is synonymous with *teshuva* the "return" to *Pardes* "Paradise" (literally the "Garden", a word adopted long ago from Persian Farsi). This is "returning" to the awareness of already being in the "Garden of Eden".

The important thing to understand is that there is nothing that we can do to return the "reflected" sentient awareness, which temporally appears within our individual consciousness, back to the Supreme "G-d Source" from which it never really left!

Therefore, our incomplete perspective, of being an illusory individual must die; "None may look upon the face of G-d and live" – שמות *Shemot* "Exodus" 3:20. This "return" to Awareness is often characterized as an act of Divine "Grace".

GUNAS: There are three gunas, *Rajas* the "Force of Expansion", *Tamas* the "Force of Contraction", and *Sattva* the "Force of Equilibrium", which collectively generate the universal principal of "Identity". In physics and cosmology they are correspondingly the positive, negative and neutral forces.

Likewise, in psychological terms they are action, ignorance and harmony, respectively. In Yogic parlance they are, *Ida Nadi* the "right hand" subtle Life-Force channel (LFC) to the right of the spinal column, *Pingala Nadi* the "left hand" LFC to the left of the spinal column, and *Shushmna Nadi* the "central" LFC that follows along the spinal cord to our brain.

And in the same arrangement of expansive, contractive, and neutral forces, in Kabbalistic terms they are the ostensible "Three Pillars of the Absolute" that support the Jewish metaphysical *Etz Chayim* "Tree of Life" paradigm.

Purportedly, as latent consciousness emerges from within insentient matter the gunas become increasingly subtle, but nonetheless remain ubiquitous fundamental principles throughout our entire manifest universe.

But while their actions are well documented by today's physics, and even the presence of an inherent "sub-quantum" Consciousness as their quantum initiator has likely already been found, its presence is nonetheless in the early stages of a more reliable experimental verification.

CHANA SENESH: Near the end of World War II Chana Senesh was parachuted by our British allies into Nazi held Yugoslavia, to fight for the liberation of her fellow Jews from the unimaginable horrors of Hitler's nearby Auschwitz concentration camp. But sadly she was captured, tortured, and subsequently put to death by the Nazis. Nonetheless she lives on in perpetuity as an inspirational icon of selfless courage and grace under fire.

"Grace under fire" is the rare balance of unshakable courage and divine compassion, on a sliding scale of violence and munificence, as the situation demands. But compassion is always the greater of the two, for it sprouts forth from our innermost core.

It is also the root source of the preternatural calm and indefatigable courage that is often associated with "grace under fire". (Irrespective of the sometimes disturbing violence of certain "uncertain circumstances"!) Because its veridical "Source" is *agape* "impersonal Divine Love". Which is also the veridical nature of our innermost being.

Rather than the usual callous anger and bravado which often arise from survival fear when we are threatened, this fierce and unselfish love is the true source of "grace under fire". Which my fellow Jews so often (prompted by their generally compassionate and justice loving natures) engage when innocent lives are threatened.

And moreover, mortally engaged Jews also tend to almost "instinctively" embrace grace as the puissant experience of ים-אנו-אל *Emmanuel,* which literally means "G-d dwells with[in] us". The Kabbalists of old called this greater truth, which is the independent witnessing of experiential events by our innermost being, the presence of the *Schenah* "percipient Divine".

This is to say that "Emmanuel" is the sudden realization of oneness with our deeper, truer Natures, which imbues its experiencer with both intrepid courage, and abiding שלום *Shalōm*. This is the veridical source of the dauntless strength which is firmly anchored in our innermost being. It is the mysterious שלום "Peace which passeth all understanding" that invariably terrifies and confuses the heartless tormentors of the occult Divine.

As such, it is also clearly the "property" of all persons of good will, be they Jews, Christians, Muslims, Buddhists, Hindus, ethical agnostics, scientific atheists, (or adherents of any other reasoned and "life-affirming" belief system)!

The following is a poem written by dear brave Chana Senesh which not only uplifts and inspires, but also reminds us of the tremendous depth of courage, compassion and understanding that is the limitless power of selfless, divine Love:

אשרי הגפרור

אשרי הגפרור הנשרף והצית להבות,
אשרי הלהבה שבערה בסתרי לבבות.
אשרי הלבבות שידעו לחדול בכבוד...
אשרי הגפרור שנשרף והצית להבות.

חנה שנש

"Blessed is the Match"

"Blessed is the match consumed in kindling flame.
Blessed is the flame that burns in the secret fastness of the pain.
Blessed is the heart with strength to stop its beating for honor's name.
Blessed is the match consumed in kindling flame."

Chana Senesh

July 17th, 1921 – November 7th, 1944

HEURISTIC: In computer science, a heuristic algorithm, or simply "heuristic", is a method of finding an acceptable (but not necessarily optimal) algorithmic set of commands to accomplish a task within a given set of constraints, such as space and time.

For example, when we pack up a backpack in preparation for a wilderness adventure, we do not attempt to discover the perfect "fit" for each and every item. Large items, which take up most of the room, are put in first, then the smaller things are packed into the remaining spaces.

The result is "good enough", but not necessarily optimal. Which is exactly the manner in which nature generally functions, and is evidently also why bumblebees can fly, despite the fact that they appear to be so poorly designed for the task!

HIGGS FIELD: The "Higgs Field", and the particle by the same name that carries its charge, are believed to be responsible for breaking the pure energy "Supersymmetry" (perfect Oneness) of the Primeval Singularity, and thus initiating the Big Bang.

They transfer mass to the sub-atomic particles (quarks) that make up matter, allowing the force of gravity to act upon them, and are thus responsible for the physical appearance in space and time of our universe. They were only recently discovered at the Large Hadron Collider in Switzerland and represent one of the greatest discoveries of modern physics.

But whether they were indeed responsible for breaking the initial "supersymmetry" of the universe is yet to be substantiated. Yet in a strictly physical fashion, although it is also clearly physically impossible, it does seem quite likely!

Yet if it did actually transpire, as it apparently did in some fashion, we are still obviously missing some critical ingredient in our assessment of the event. No-doubt, what we are probably missing, is the trustworthy detection of a subtle universal energy of some sort.

But, we are also most likely missing the fact that it is the directive force of an impersonal "Consciousness". By which it is not only compelled to exist by the fundamental principle of its own identity, but is also known (through its eventual expression as "us") to have transpired, and therefore to "exist"!

HIPPOCAMPUS: The *hippocampus* looks like a "sea horse". It lies under the medial temporal lobes on each side of our brain, and is involved in the complex processes of forming, sorting, and storing our memories.

HOLOGRAM: "Holography" is a technique that reconstructs the recorded light that is scattered from an object, so that it appears as if the object is in the same position that it was in when the reflected light was recorded, relative to the recording device.

Suspended in space, the projected light image changes as the orientation and position of the viewing system changes, in the same way as if the actual object were present, making the recorded image or "hologram" appear to be three-dimensional.

Indeed, our familiar "objects of consciousness" are essentially the equivalent of mental holograms that are assembled from different neurological points of perspective!

HWANG PO: Huang-po Hsi-yün (Jap; Ōbaku Kiun) was an Awakened Ch'an master, who was born in Fujian China during the Tang dynasty (618-907 CE).

IMAGINATION: The classical neurological interpretation of our imagination is that it is an aspect of our higher cognitive function, which allows images to be combined to produce novel representations. This permits the formulation of imaginary scenarios that allow the prediction of outcomes, which are only discretely "true" in relationship to our decision making capacity, but which nonetheless favor our survival!

However, in this work the term is applied in a wider sense to every thought that comprises our inner universe. This is true whether it arises from an automatic function that enters into the field of our attention, or to an inner creation of our mind, independent of any somatic input.

But in the end, whatever thoughts we have are a combination of our mind, brain and body functions, which of course have to be viewed by an apparently independent self or "subject".

Many neurologists try to explain this interaction of subject and object as a phenomenon that somehow arises solely from the actions of a body and brain, within an "emergent" mind. In this work, we explore alternative scenarios.

I AM THAT I AM: "I Am That I Am" is a simple palindrome in Hebrew *"Ehyeh Asher Ehyeh"* and in each direction it is identical in its meaning. This is to say that using the "polite demonstrative" of *Ehyeh* (as an adjective of *Asher*) demonstrates a great deal of respect for its Divine Subject!

Which is in itself an intriguing if not scholarly use of *Asher* as a malleable noun! Thus, "Semitically speaking", these words are meant to convey both the power and the awe of *Yahweh* (i.e. "Parabrahman", G-d", or "Allah"…) the Supreme.

INFINITE REGRESSION: In mathematics "infinite regression" is considered to be a mistake. In philosophy and debate, it is a method of argument, *reducto ad absurdum* (reduction into absurdity), which is used to reduce your opponents propositions into nonsense. But when rigorous mathematical computations yield "infinities", forging ahead is a waste of time, because somewhere in the beginning, or along the way, a mistake has apparently been made that leads us into pure nonsense!

A Zen or Advaita Master says that by diligently observing our mind we may eventually understand that we are not merely our body, our mind, or its contents. But when they say that one day we may also realize that we are actually observing the observer of our mind, this does not necessarily suggest an "infinite regression", which is occasioned by a vain attempt to reach a primary observer. Instead it reveals a faulty premise borne of our illusory subject-object perspective, which leads to regressive nonsense!

Our mind requires a brain, and it is the mind that observes our body. Yet the observer of our mind first requires a sense of being, which it acquires "reflectively" by observing the thoughts that fill our mind. But the observed is not the observer, since it must appear before (i.e. "in front of") the observer in order to be observed. Therefore, the primary observer (which is our authentic Self) exists even prior to our (ego's) psychosomatic sense of existing, and thus before the consciousness that is necessarily antecedent to (follows after) our sense of being.

Were this not so, the illusion of infinite regression that appears when we stand between two mirrors would not be an illusion, and universal nonsense would prevail! This is to say that the orderly episodic "serial" progression, which allows for both implied division and comparison, and reflective addition and summation, ceases when only our pure fundamental presence as the independent Supreme witness of our minds remains.

In other words, when both of the mirrors that an observer is standing between are removed the observer remains, but he can no longer see himself! And because our principal Awareness is in essence spacetimeless and infinite, we are in actuality the emergent and end point of "zero" and all other numbers, which are subsumed in our essential Monism. This rightly appears to our dualistic subject-object perspective as infinity or infinite regression, but because it is an artifact of our perception, it is also necessarily neither!

INSTRUMENTALISM: In the philosophy of science, "Instrumentalism" posits that a concept or theory should be evaluated by how effectively it explains and predicts phenomena, rather than how accurately it describes an objective reality.

JOHN LOCKE: (1632-1704) John Locke was a politically insightful British Empiricist (i.e. "all knowledge derives from experience") who theorized that the "Natural State" is a moral state and that the "Political State" should reflect the natural state. In his "Two Treatises on Government" Locke proposes that human beings are endowed by nature with certain inalienable rights, "...to life, liberty and the pursuit of happiness".

JÑANA MARGA: Jñana Marga is enlightenment through study, self-inquiry, guidance from an awakened Sage, and meditation. As opposed to *Bhakti*

Marga, the "Path of Devotion" to an enlightened Guru, whereby selflessness and *Prem* "impersonal Love" are utilized to reintegrate our misplaced identity with the *Atman,* which is our "true Self".

JÑANESHWAR: Jñaneshwar was a yogi born in 13th-century India who evidenced whole-body tissue regeneration during an extended period of "Samadhi" (a trance-state of greatly slowed homeostasis), demonstrating that such things are evidently possible!

Due to advances in science over the past 200 years, the human lifespan has doubled (from around 40 to 80 years). The laws of physics do allow for our conditional "immortality" but *entropy* (the ubiquitous disintegration of organization) must first be overcome. This can be done by introducing energy into our failing systems through genetic engineering.

By removing the R.A.S.2 and S.C.H.9 senescing genes from laboratory mice, for instance, their lifespan is roughly "doubled". We share 90% of the same genes with mice, and by removing just these two genes, we may soon be able to double our average lifespan to over 200 years!

Moreover, by matching the genes of millions of young and old people, through repeated comparison and subtraction, we can reveal the genes where aging is concentrated. So far we have already discovered over 60, but who wants to live for a thousand years with an enfeebled body and a progressively weakening mind! So the search is on for flesh decomposing bacteria that can degrade the bits and pieces of resistant DNA and materials that accumulate in our cells lysosomes and contribute to our aging.

Simply by "taking out the trash", we could reverse the aging process in 60-70 year olds and prolong our youth for up to 30 years! And the chances are better than 50/50 that in the next 25 years we will be able to accomplish this by splicing bacterial genes that produce enzymes capable of degrading all of the "bio-junk" that accumulates in our bodies (i.e. inside and between our living cells), into the bacterial cells that already inhabit our bodies.

Synthetic biology offers another solution, by finding functions in living organisms that are useful for fighting aging and increasing our lifespans, and splicing them into bacterial genes, and perhaps even into our own cells with CRISPR (see, pg. 523). Eventually, when the gene sequences are understood, we will be able to reprogram the bacteria (and our cells) to do our bidding in the same fashion that we program a computer.

Then we will be able to take control of our own genes and program them for enhanced youth, longevity and increased physical and mental capabilities. As incredible as it sounds, we presently stand on the threshold of our own conditional "immortality". Much further down the road, we will come to understand the entire human genome and will then be able to take control of our own evolution.

JUST HOW BAD?: With each passing day we are losing around .1% of our planets 8.7 million species of life, and even if extinction is a naturally occurring event in nature, with the exception of several past planet-wide extinction events, the rapid current decline is not only unprecedented, but directly related to human actions!

With all of the hyperbole that environmental concerns are generating in the media today many people justifiably wonder, "If we do nothing at all, just how bad will 'Global Warming' eventually become on our planet Earth"?

The good news is that we will most likely never experience a runaway greenhouse effect like the planet Venus experienced, which eventually turned it into an inhospitable, searing wasteland. This is because all of the fossil fuels that are present below the ground on our planet were at one time aloft in our planet's atmosphere.

But the bad news is that this was during the Jurassic Era when dinosaurs still freely roamed our planet, and endless forests of massive redwood trees still covered its primeval temperate regions under an endless forest canopy of luxuriant green. Which was constantly helping to absorb the massive amounts carbon dioxide that was still aloft in our planet's atmosphere!

However, unfortunately, what this also means, is that our planet's polar ice caps and glaciers will all melt, which will cause all of the oceans on our planet to rise. In fact, they will eventually rise up to about the level of our Statue of Liberty's elbow!

This means that the large cities of the world, which are all located near coastlines, river mouths, and low lying bodies of water, will all have to move inland an average of about one hundred miles.

The increasing world-wide temperature will have a tremendous negative impact on agriculture, making it much more difficult to feed our ever-growing worldwide population. Warmer oceans will give rise to global Superstorms that will likely make our current Hurricanes look like "a tempest in a teapot"!

A plethora of other problems, including the probable extinction of a great many more species of flora and fauna around the world will most likely also continue to occur. Hence, while the situation is ominous, especially if we do nothing to reduce our dependence on fossil fuels, it is not hopeless. But who wants to live on an increasingly inhospitable planet, especially when we have the ways and means at our disposal to keep our ever-warming globe from warming too far!

LIFE-FORCE: The "Force of Life" was present in potential even before the appearance of the Primeval Singularity, and after the expansion of our universe it brought forth the life latent within insentient matter and continues to advance its expression in emerging consciousness.

LIMBIC SYSTEM: The "Limbic System" is our *paleopallium* "old mammalian" brain, which consists of primitive neural structures on top of our brain stem that are involved in our experiences of emotions and pleasure.

Lin Quan Yuan: "Lin Quan Yuan" is a Taoist temple that still lies (sadly, at this time) in ruins, in Southern China.

Memetics: Memetics is an interesting theory of mental content and transfer by way of informational *memes* (a word based on "genes"), that was modeled on Darwinian evolutionary theory by Richard Dawkins, author of *The Selfish Gene*, published in 1976.

Metacognition: Psychologist John Flavell coined the term in 1976; "Metacognition refers to one's knowledge concerning one's own cognitive processes or anything related to them, e.g., the learning-relevant properties of information or data. For example, I am engaging in metacognition if I notice that I am having more trouble learning A than B; if it strikes me that I should double check C before accepting it as fact".

In A.I. computer science, metacognition is considered to be the threshold of Artificial Intelligence success, while so-called "Strong AI" moves the bar beyond the limits of human intelligence. However, human creative abilities such as unbounded inquiry, intuition, imagination and art pose serious obstacles that bring into doubt the completeness of the classical Metacognitive Theory in explaining human awareness.

In other words, a highly advanced computer may be able to contemplate the universe in ways far exceeding our human mind, but still be unable to enjoy the interior content of a sentient awareness which exceeds the processes by which it is contemplating thoughts, and from which the enjoyment of unexpected, new, and unpredictable outcomes can arise.

Therefore in this work, "meta-cognition" refers to the unique capacities and independent nature of human Awareness, rather than to the mere biomechanical aspects of human intelligence, which inarguably underlies all of our aspects of cognition.

Microtubules: For over a decade, we have known that the neurons in our brains contain microtubules. They provide an internal structure to our brain cells much like our bones do to our bodies. But advances in instrumentation reveal that something more is happening on a deeper level. In 1996 Stuart Hameroff MD and physicist Roger Penrose proposed that these microtubules also process information on the quantum level. This allows our brains to function like a quantum computer, with the ability to process information at FTL (Faster Than Light) speeds.

No doubt, this is what allows our minds to "spacetimelessly" disassemble and reassemble our experiential universes into what we automatically cognize as the constant unidirectional flow of time, before we can even sense its occurrence. Moreover, because the entire universe is interconnected on the quantum level, this means that our brains are too!

Which seems like a plausible enough explaination for the remarkably accurate cosmic insight of (sometimes tragically misunderstood!) so-called "Seers" (i.e. visionaries) like poor immolated Frater Giordano Bruno.

Our sentient consciousness likely emerges from these connections, and since space and time are not only malleable but even somewhat reversible on the quantum level, our intuition and its ability to process non-local and meta-temporal "psychic phenomenon" are no doubt contained therein as well.

Thus our consciousnesses are connected through the ubiquitous simultaneity of quantum entanglement and we can probably communicate and even mentally "evolve" as a group by subliminally sharing our individual life experiences. But we must wait to substantiate this more directly, until the not-too-distant invention of the neutrino microscope which will likely allow us to view our universe on a sub-electron scale.

MIDRASH: It is a curious fact of life that much is sometimes "hidden in plain sight" within deceptively simple things. In order to tease out what may be hidden away within the very beginning of the Jewish Bible (the "Torah"), we can apply an ancient technique called *Midrash* "to draw forth from within".

With the five "books" (really divisions of a single scroll) that comprise the main body of the Jewish Bible, for instance, a refined Hebrew translation of the Torah scroll reveals that:

תורה Torah = "Law": But the non-Hebrew word "bible" is from the Greek word *biblia,* which means, "books".

בראשת Bereshit = "In the beginning": While *genesis* is a Greek word, which means, "creation".

שמות Shemōt = "Names": Yet "Exodus" comes from the Greek word, *exōdus,* which means, "going out".

ויקרא Vayikra = "And He will call": However "Leviticus" is from the Greek word *Leutikon,* which means, "Levites".

במדבר Bamidbar = "In the wilderness": But "Numbers" is from the Greek word *arithmoi*, which is ironically, based upon a census taken in this portion of the Torah scroll!

דבדים Devarim = "Things": Yet "Deuteronomy" originates from the Greek *leuteronomium,* meaning "second law", since a cursory reading of the scroll seems to be a mere repetition of the previous Numbers portion of the Torah scroll.

Now having more accurately defined our terms, we can be sure from מי *mi* "what or whom", we are דרש *derash* "drawing forth" an סוד *sōd* "occult" (i.e. hidden) "midrashic" meaning. From the mystical perspective of the Kabbalah, the Torah deals with the "esoteric laws" by which the universe and humanity are manifested.

While from a more intellectual "Rabbinic" perspective, the Talmud Torah deals with the *Halacha* or "everyday laws" that many orthodox Jews still live by. Hence, from an esoteric perspective:

Bereshith is concerned with the manner in which consciousness completes the manifestation of the universe within the dualistic experience of our subject-object consciousness. It provides an array of secret "keys" to unlock our identification with the biological machinery that makes all of our experiences possible, yet causes us endless suffering, until our primal state of pure, sentient Awareness is thoroughly understood!

Shemōt is concerned with shifting our sentient Awareness away from its mistaken identification with our subject/object (ego), which has caused us to "fall" into enslavement, within an illusory (experiential) universe of name and form.

Vayikra is involved with revealing the manner in which we are called by Divine "Grace", through no technique or method of our own, to reawaken from the illusion that we are an impossible "subject/object" that is capable of independent volition, in an objectively real external universe.

Bamidbar describes our long and painful wandering through the "inner" wilderness of our illusory individual consciousness, which always lies on the spacetimeless cusp of our sentient impersonal Awareness, as our identification with our illusory ego and contents of our memory and imagination at last begins to subside.

Devarim points to the challenging final metanoesis of our point of perspective, as we are guided into the "Promised Land" of our true and natural state of undivided pure Awareness.

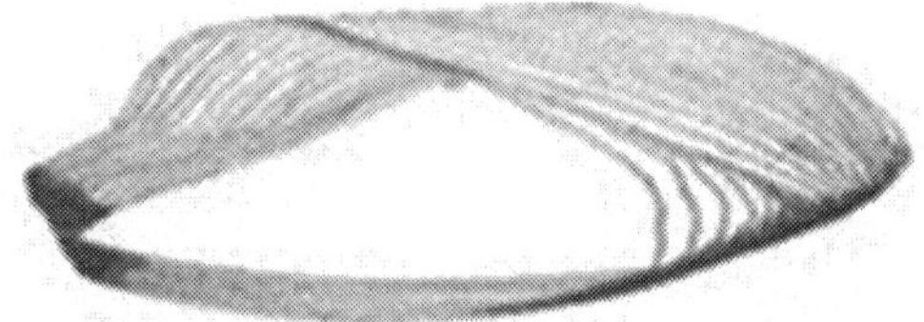

A Möbius 'String' Paradigm of Space-Time

Our Experiential Universe is composed of Möbius-Strip-like, quantum sized *Vritti* "Möbius Strings", if the strings are thought of as being comprised of the same basic "Stuff" as the original Primeval Singularity, that is, of impersonal Universal-Consciousness Force. As the primeval singularity expanded into the Big Bang, it had to go somewhere.

To allow this, space-time simultaneously appeared, in an FTL (faster than light) period of hyperinflation. Everything is composed of this same original "Stuff", so even our consciousness displays its essential characteristics; if it did not, it (and we) could not exist!

This universal principle, in the continuity of presence or "identity", can be represented by placing conceptual *time* on one surface of the binary (i.e. two-dimensional) Möbius Strip/Strings (which here represents the sentient reflection in our apparently individual consciousness of the elaborated substrate of Universal-Consciousness), and conceptual *space* upon the other.

In our conscious experience the symbolic three dimensional objects that are comprised of algorithmic-like neuronal "codes" appear as our cognition assembles them into conceptual existence. Which it does by serially inserting four dimensional gaps of ideational space-time around, within, and between them.

But this does not mean that they are not still quantumly interconnected throughout our universe! And, of course, space-time and whatever energy, or energy expressed as "matter", that they may contain, did not exist (except in potential) before our universe began to inflate.

Moreover, since the extremely (if not infinitely) compressed nascent Primeval Singularity also evidently contained a veritable, but perfectly "balanced", infinity of possible things, including the space-time that is necessary for anything to appear, moving it at all would likely take an infinite amount of force, if not time!

Hence, even the first few Planck sub space-time moments could take almost forever to happen (see, "Singularity Generation and Supersymmetry Breaking", on pg. 302). Which, given the endlessness of the resulting Superverse, is apparently about how long its appearance from a perfect not "not-nothingness" state of spacetimeless infinite potential requires!

In other words, our universe remains forever One, even in the apparent multiplicity of its ever extending parallel dimensions. Thus, it is always up to us to discover how our consciousness figures into its eventual appearance, as and within our hominid minds. And how our dualistic minds relate to the universe at large. As we go about the scientific business of gaining a deeper understanding of its physics, physical laws, and the technological innovations to make better use of it.

Muwekma Ohlone: The Muwekma Ohlone are a California, San Francisco Bay Area based, American Indian tribe.

Natural Selection: "Natural Selection" is a term that Charles Darwin used in his book *The Origin of Species*, which introduced a theory of natural evolution, often called "survival of the fittest", to the world.

Neural Nuclei: Neural nuclei are membrane-enclosed structures that are found in the *soma* "body" of a neuron, which contain the nucleolus and chromosomes that govern the production of proteins in our nerve cells. The nucleolus of a cell's nucleus produces *ribosomes* (RNA), which assemble proteins from amino acids.

Genetic information is encoded within long strands of *deoxyribonucleic acid* (DNA) from which our chromosomes are made. Our DNA is composed of two long chains, which are intertwined in a dual helical arrangement.

When activated, the gene subunits of the chromosome initiate the production of *messenger ribonucleic acid* or "mRNA", which is a duplicate of the information contained in the gene. The mRNA then exits the nucleus and attaches to a ribosome where it acts as a blueprint for the production of a protein molecule.

Without this complex *genome* "biological blueprint" life as we know it would not be possible! In this way, Mother Nature remembers and records "Her" successes, and thus life continues and evolves. But Awareness must be present for this to occur, although it is most certainly noumenal!

Thus life has emerged from within insentient matter. And the creative successes of our essential spacetimeless Awareness are memorialized within a plethora (abundance) of ever evolving biological arrangements of matter and energy.

Therefore, Awareness is always present in its extended form as dynamic Universal-Consciousness throughout our entire universe. It constitutes the immortal "Stuff" of which we are made, and by way of its percipience it is the initiator of sentience within our transitory spatio-temporal experience(s) of life.

NEUTRINOS: Neutrinos are so tiny that even an electron has a million times more mass than they do! They are so miniscule that they almost never interact with ordinary matter, and are thus almost undetectable. In fact a neutrino could pass through tens of thousands of light years of dense iron and never strike an atom, or even begin to slow down!

They were first predicted in 1936 by the researcher Wolfgang Pauli to explain the mysterious loss of energy as an electron escapes a decaying radioactive element. Decades later their existence was substantiated when they were revealed in the radiation emissions of a nuclear reactor.

But today they are revealed by flashes of light in the Japanese *Sūpah Kamiokande* neutrino detector. The "Super" Kamiokande is a light detector lined well that is filled with one hundred million gallons of purified water.

It is buried a half mile below the surface of the ground in order to screen out the more massive "cosmic rays" (mostly protons and electrons) that constantly rain down on our planet but are greatly reduced in number and harmful intensity by our planet's magnetic field. Nonetheless, these cosmic rays occasionally produce the eerie nighttime spectacle of the colorful Aurora Borealis, during times of intense solar activity.

Tremendous amounts of neutrinos are formed when a giant "blue star" finally exhausts all of the nuclear fuel that once balanced it against the gravity of its own mass and it suddenly collapses inward onto itself. When this happens its constituent atoms internal protons and neutrons are forced together at near light speed, thus creating an exploding "Super Nova" that is even brighter than an entire galaxy!

Countless numbers of neutrinos were formed in the first "Planck" moments of the Primeval Singularity. But unlike the leptons that could not escape as photons of light until our infant universe had expanded and cooled sufficiently (around 300,000 years after the "Big Bang"), completely undeterred, the infinitesimal neutrinos immediately began to escape at near light speed into the universes rapidly expanding space.

Whenever we gaze out into space with our most powerful micro-wave telescopes, we are actually travelling backwards in time, and we eventually can go no further when we finally reach the photonic "Red-Shift" moment when our universes light photons were first released.

To answer this challenge, our newest deep-space telescope, the James Webb or "Hubble 2" infrared telescope, will soon allow us to get our first glimpse inside our universes (heretofore impenetrable) cosmic "Red Wall"!

But when we (in the not-too-distant future) create a neutrino telescope, we will most likely be able to look past this literal "Veil of the Absolute" to see the very first moments of the Primeval Singularities appearance.

Moreover, on the scale of the very small, when we are finally able to build a neutrino microscope, we will likely be able to also then see into the mysterious quantum world that underlies our visible universe. Thus there remain endless wonders in our universe, just waiting to be seen!

NON-ACTION: "Non-action" is non-participation in the illusion of individual will, and should not be confused with not taking an action, which is also an action!

NOTHINGNESS: Incredibly, our entire universe began from nothing! One way that this could happen is if "nothing" somehow became "something". Indeed, this can even be demonstrated mathematically, since everything can be shown to cancel out ultimately into zero.

To understand this theory let's begin by imagining the condition of "nothingness", which prevailed before the Big Bang took on the appearance of being *something* by suddenly taking on just a single dimension. Subsequently, and quite explosively, it then expanded its all-encompassing monism into endless other dimensions and parallel universes!

But now the positively displaced nothingness, that became our universe, left a "hole" in the infinite nothingness that it came from, like building a sandcastle on a sandy beach. (See the dialectics for, "Before the Beyond", on pg. 300).

Some theoretical astrophysicists believe that the invisible hole in the "quantum sand" that our universe left behind is evidenced by the invisible dark energy and matter that surrounds and balances it, simultaneously holding it together, while driving it apart.

Thus the underlying symmetry of the observed physical laws, that allow our universe to exist, are born and persist throughout the dimensional extension of its essential nothingness, into an apparent "somethingness" that is comprised primarily of energy and space. However, the universe effecting presence of an unseen, and evidently even much larger, amount of so-called "dark matter" and "dark energy" could as well simply be the presence and influence of the rest of the unseen multiverse that we are only a minuscule part of.

A lot more work is needed before we can even begin to claim that we have more than a rudimentary understanding of how our universe actually began, or even how it continues to appear to exist in an infinite field of primeval not "not-nothingness" (i.e. neither some-thing or no-thing). But any theory that does not also include the consciousness by which the universe understands its own existence, is also clearly incomplete!

Objectivism: "Objectivism" proposes that our mind cannot create reality. It attempts to offer a means to discover an independently existing reality, which it suggests must exist first. It opines that our consciousness cannot be conscious in and of itself, and therefore it must be conscious of something external, or prior to itself.

Hence, action and causation are related to a physical entity, rather than to a consciousness that is independent of our body, brain and senses. Consciousness, it claims, merely responds to its environment, and if the physical entity is different, so will its actions be different.

This results ultimately in a rejection of anything transcendent to, or independent of, our physical existence, including the attribution of any intrinsic sanctity, or necessarily equal valuation of human life. Therefore, the lives of some may be considered to be more important or valuable than others, which leaves the determination of their worth to others.

As a result "Elitism", eugenics, selective repression, and even genocide may result if ethically based human rights are not carefully reasoned into existence, or are purposefully ignored to promote the wellbeing of those deemed to be the "most important".

Objects Of Consciousness: Whatever we can know of our person and universe appear to us as thought-objects, which are thus "objects of consciousness".

Objectivist: An "Objectivist" is a person who believes that the objects which he experiences in his mind, must exist in an objectively real "material" universe, which is located outside of his mind. But the resulting problem which the Objectivist has ignored, is that his ostensive "person and universe" are actually an illusory objectification that is accomplished by creating apparent objects of individual consciousness, which are only ideationally separated within his consciousness.

This is accomplished by an artful action of focus that creates the multifaceted illusions of different dimensional perspectives of which the objects of our consciousness are continually comprised. These are the purported "objects" of our consciousness which we persistently experience in three ideational dimensions, from within a four-dimensional "space-time" perspective.

This is because our brain functions much like a holographic projector, creating objects of thought and concept, which are phenomenally illuminated as they are experienced, while the Supreme witnesses them through its noumenal aspect of Universal-Consciousness. Thus the illusion of an independently subjective entity is born. In this respect, all phenomenal objects and subject/objects (i.e. "egos") are merely artful Cosmic illusions!

OM MANE PADME HUM: "Om Mane Padme Hum" is a phrase from the Buddhist sutras (scriptures). *Om* is the "unstruck sound" from which all sound arises as vrittis "mental modifications". *Mane* or *manas* refers to our "thinking faculty" or mind.

Padme "lotus" refers to our centerless center, which is synonymous with the Rosicrucian and Masonic Orders' "All-Seeing-Eye" that not only graces our countries currency, but also symbolizes the true Supreme Witness of our every thought!

And *hum* refers to the gungun "consciousness vibration" (an East Indian Marathi phrase), which the ancient Kabbalists referred to as Hōd "reverberation"; an aspect of which allows our internal experience of phenomenal sound.

Thus the Kabbalist's *Hōd* "reverberation" of *Zimzum* "vibrations" arises from the unstruck sound of Om. Which exists only in the space-timeless, eternally present palpable moment before it is cognized into our common experience of "sound".

Hence, sound can be followed back to its penultimate experiential source within our ever immanent Universal-Consciousness, as the percipient original sound of "Om".

This is the essence of chanting and meditation upon *Para* the "source of sound", the opposite pole of which is referred to in the Torah as G-d's "calling forth of the Creation". Thus, our apparent "reality" is actually comprised of our mind's constituent "objects of consciousness", which are automatically assembled within our mind! Evidently, couched within this colorful archaic Buddhist terminology, there remarkably exists some rather profoundly relevant contemporary science!

Our Address In The Superverse: For the sake of achieving a more enlightened "cosmic perspective", it seems important to first understand our true place in the grand cosmic scheme of things.

Thus we are located on a planet called "Earth", in a planetary Solar System that is embedded far out on a spiral arm of the Milky Way Galaxy, some 27,000 light years away from a rather massive Black Hole that is located in its very center!

Our galaxy is in turn located in a vast so-called "Local Group" of other rapidly spinning galaxies, which are situated within an even larger cluster of galaxy groups that is called the "Virgo Super Cluster".

But even our vast Virgo Super Cluster is only one among a nearly countless number of other super clusters of galaxies that together make up the grand cosmic web of our detectable universe.

Yet even our gigantic universe is actually just one amongst an innumerable number of others. Which are in turn located within a "Multiverse" of universes that are scattered across the immeasurable space and time of countless other parallel dimensions that together comprise our unimaginably vast "Superverse".

Moreover, contrary to our usual experience of things, our ever-inflating Superverse also has no center and no edge! Evidently, even our most reliable mathematics of Euclidian geometry becomes nearly irrelevant in our inevitably vain attempts to accurately measure such an infinitely large and endlessly interdimensional structure!

And while it is true that the Superverse that we are a only a miniscule portion of is much greater than anything that we could ever imagine, the very fact that we are presently contemplating it at all is perhaps the greatest wonder of them all!

OUSPENSKY: Pyotr Demianovich Ouspensky (1848-1947) was a Russian Philosopher who used Euclidian and non-Euclidean geometry to explain our mind, and the dimensions that lie outside of our mind's normal perceptive ability.

A student of the mystic, George Gurdjieff, Ouspensky held that there exist three basic methods of spiritual growth that demand seclusion from the world, the "Way of the Fakir" that deals with the physical body, the "Way of the Monk" that he characterized as an emotional approach, and the "Way of the Yogi", which relates to the mind.

But he also taught a "Fourth Way", where growth takes place in the midst of ordinary life, manifesting the natural harmony of body, emotions and mind. In this work, we also indorse an (enhanced and updated) "Fourth Way".

OUTER CAMBIUM: An "outer cambium" is the vascular cambium that's under the tough outer bark of trees, which divides and produces *xylem* on the inside, and *phloem* on the outside.

The xylem vessels conduct water and nutrients upwards from the trees roots to its food producing leaves, and the phloem vessels move the elaborated food produced in the leaves downward, to feed the tree.

But trees do not heal like people, because if their cambium or xylem is injured they must grow new tissue in new places, to replace their damaged cells.

Correspondingly, trees also do not stretch upward as they grow! The branches stay on the same level as the tree grows taller. Which it does by dividing existing cells and "stacking" the new cells on the old, like building a fireplace with cellular "bricks" and *lignin* "mortar".

On the inside of the tree, this process is responsible for the annual growth rings that look like the circular grooves of an old fashioned vinyl record. Which you can readily see when you cut a tree's trunk (horizontally) in half.

In the spring, the new cells of most non tropical trees are larger than the ones that form later in the summer, which creates the smaller, darker looking "annual rings".

PHOTOSYNTHESIS: On planet Earth, except for green plants that through the process of "photosynthesis" utilize the energy that photons deliver from our sun, extremophiles and a few ancient fungi that use energy from other non-living sources, all life exists by feeding upon other life.

This is euphemistically called the "food chain". Thankfully we are at the top of the food chain! Nonetheless our bodies are food for other life forms. Indeed, our bodies are actually "food bodies", comprised of the elaborated food taken in by our parents and ancestors, and more recently, by ourselves.

Yet many believe that we somehow "choose" to be born. But, when our body and brain are not present, neither is the consciousness that they support, and in its absence any decision making is surely impossible! And moreover, who would knowingly enter into a clearly carnivorous food chain?

It seems much more likely that consciousness emerges from within life, just as life has emerged from within insentient matter, and matter from within the Primeval Singularity that somehow emerged from within, what many theoretical physicists are now calling an "Elsewhere". However, many persons also believe that there must be a separate, incorporeal soul.

But is it not more likely that if such exists it is a physically contingent and therefore temporary phenomenal soul. A "psychic soul" comprised of the energy that the force of Universal-Consciousness has brought into conglomerate being, out of the endless potential of the primeval cosmic Elsewhere.

But photosynthesis also holds a puissant key to assist in the long term sustainability of our human civilization, because it utilizes carbon dioxide and water molecules to make carbohydrates. In fact, plants produce almost six times the total amount of power that is currently being produced annually by humanity!

If we can unlock the "trade secrets" of nature that allow plants to do this so much more efficiently than we can currently do in the laboratory, photosynthesis has the potential to not only supply a tremendous amount of food to the world, but clean electrical power as well.

And if we can scale an efficient artificial photosynthetic process up to a planet-wide industrial level, it could also help to "scrub" the excess carbon dioxide (that is currently overwarming our atmosphere) from the air much more efficiently than we can currently do in the laboratory.

PLANCK TIME: The smallest time measurement that we can succeed in making is $t\text{pl} = 10^{-43}$ seconds, which is .10 preceded by 43 zeroes, which go between the period (•) and the 10 (i.e. t = time, and pl = planck).

And of course, "something" must be minimally present in time to quantify, and its measurement is evidently $r\text{pl} = 10^{-35}$ grams, which is represented by .10 preceded by 35 zeroes, in the same configuration! (Note: r = distance, and there are 28 grams in one ounce).

These are the minimal "Planck" measurements of our four dimensional universe. But while these measurements are mathematically demonstrable in three dimensions, because they are so minuscule (i.e. a Planck length is 10^{-20} times – smaller than – the diameter of a proton) they cannot actually be performed to ascertain their veracity!

The best that we can currently do is to predict quantum sized particles, that are really "wavicles", by tracking their energy signatures within a certain range, with graphic computerized representations, as protons are split apart in high energy collisions. (See "Wavicle", on pg. 596). But we cannot actually see the quantum particles, we simply know that "something" must be there by the telltale "footprints" that they leave in space-time!

In Forest Primeval we suggest that the "vrittis" of Mahayana Buddhism and "skandhas" of ancient Servastivadin Buddhism conceptually represent the same scale of energetic expression, except that they are also concerned with the aspects of consciousness by which we are cognizant of our universe. (I.e. ut hoc loco "as here used"; quantum scale, Mobius-strip-like, space-time bounded "strings", comprised of Universal-Consciousness/Force).

As the Guatama Buddha explained them, seen as apparent objects they are a nonexistent "void of annihilation", and seen as phenomenal subjects they are "void of prajna" (see "Void" dialectic, 3rd paragraph, on pg. 463).

This is to say that they are comprised of a pure subjective Awareness, which neither exists nor does not exist, but stands apart from either, so that our consciousness can spatially "measure" the universe into a temporal existence (within our mind).

Thus "existence" can only happen within our "binary" (subject-object) individual consciousness. But it should also be kept in mind that there exists a solution of continuity on the most fundamental level of our universe.

In other words, everything is in a sense "united" through the foundational Supreme Awareness that is shared between every form of Universal-Consciousness and its aspects as space-time and energy. This is to say that our ostensibly "individual" consciousnesses are also comprised of the same Universal-Consciousness that comprises the entire universe!

PLASTICITY: In neuroscience, "plasticity" is a term that describes our brain's remarkable ability to adapt and change. For instance, it has been discovered that the human brain has the ability to generate new brain cells in the area around the hippocampus.

We don't currently understand exactly how this happens, but the possibilities that exist when we do, since our brains could readily transfer any stored information from "older" neurons to newer neurons, is truly staggering!

POSITIVIST: "Exoteric-Objectivists" believe that the universe and our persons are objectively real, and that our body, mind and ego are our real self and constitute the subject/object (i.e. a "subject" that is also somehow an object) of a veritable external universe.

"Esoteric-Objectivists", on the other hand, believe that they are essentially an incorporeal soul, and that their experience of person and universe are objectively real. But that their objective persons and universe(s) nonetheless enjoy a bona fide subject-object relationship, while their wholly incorporeal "soul" remains their true subject. Both are variations of what is referred to in this work as the "Positivist" perspective.

PRAGMATISM: In the philosophy of science "Instrumentalism" is a type of "Pragmatism", which holds that a concept or theory should be evaluated by how effectively it explains and predicts phenomena.

Pragmatism has been widely accepted as a synonym for "level headedness" (i.e. practicality). While its true philosophical implications, as a means to gage truth and order in our life and actions, are not generally well understood.

For example, one problem with a literal interpretation of pragmatism is its tendency to restrict our imagination and creativity, by dismissing the validity and importance of things that cannot be readily tested, in terms of their practical consequences.

PRAJNA: "Prajna" is our percipient Universal-Consciousness in action, but primeval Universal Consciousness arises, as if uncalled, from within a more essential principle of Supreme Awareness.

But both were spacetimelessly present before the presence of our purportedly strictly physical universe, and of our bodies through which it is known. Moreover, both are also precognitively present before the appearance of the universe within our minds.

Hence, both share in the same fundamental qualities of pure presence and awareness. But our individual consciousness is comprised of knowledge arising from experiential subject-object concepts. However, these are actually interdependent dualistic concepts, in essence comprised of "thoughts". Which are understood to exist only because of the underlying percipient presence of Universal Consciousness, and the absolute, fundamental presence of a principal Awareness that lends its Supreme sentience to our consciousness!

In other words, prajna could as well be said to be the dynamic aspect of Universal-Consciousness that brings the underlying presence of Supreme Awareness into our everyday minds. Which is accomplished by illuminating the operations of our brain's neural network with the inherent sentience of the universe's pre-existential, principal Awareness.

We "emotionally" sense the occult operation of prajna as intuition. Therefore it has a much different, complete, and immediate quality than the temporary and incomplete knowledge that spontaneously manifests as our everyday, subject-object consciousness.

Knowledge, for instance, is primarily dependent upon division and comparison, and the spontaneous subliminal mental functions of our memory, imagination, and anticipation that automatically arise from within the subliminal functions of our bodies and brains.

Therefore metacognition may well be possible in AI (Artificial Intelligence) if a metacognitive processor is invented and appended to an artificially cognitive system that successfully reproduces the various functions of our human brain.

But just as within our human brain's algorithmic (symbolic) representations of "person and universe", the fundamentally ubiquitous subquantum presence of a Supreme Awareness, and quantum presence of a Universal-Consciousness, is most likely responsible for the post quantum emergence within space-time of our Universe.

Moreover, it is also most likely responsible for its own eventual inclusion in the elaborative processes of the Superverse, from within the impenetrable mystery of our pre-Primeval Singularity's "Elsewhere".

One thing is certain, humankind is most certainly not the terminus of Nature's ongoing evolutionary process!

Prana: The universe is coalesced by *Prakriti* "Mother Nature", through *Purusha* the "Principle of Identity" and *Prana* "Life-Force", into living forms. The primal substrate of Supreme Awareness, and the Prajna(ic) Force of Universal-Consciousness are introduced ("reflected") into living bodies as the Atman (our so-called Soul).

But there is never any factual question of an independent soul, since the Atman may seem to be separate, but is evidently not only conditional, but changeable and perishable as well!

Succinctly, the "Identity Principle" is ostensively present in the food that we consume. And the pranic "Life-Force" is said to be present in the *Akash* "Primal Superspace" that we breathe into our bodies.

Thus, although our personal experience of the resulting spatially dependent ideations of our subject/object persons and temporally sentient "psychic souls" may well be quite compelling, they are nonetheless psychosomatically contingent, and are therefore only transitory phenomena.

Moreover, perhaps a more exact interpretation is that the concepts of Prana and Akash are an alternate explanation of phenomenal space-times interconnectedness with the totality of the Superverse.

Which is also scientifically verifiable, when we begin to factor consciousness into our equations and experimentation. Indeed, at least where AI is concerned, we are effectively already there!

Prefrontal Cortex: The "prefrontal cortex" is the *anterior* "forward" part of the two frontal lobes of our brain. Our prefrontal cortex is responsible for our brain's executive functions of decision making by comparing and choosing among variables regarding possible and probable future outcomes.

Psychosomatic: Psychosomatic literally means "body-mind". Both our body and brain must be present for our mind and the consciousness that it contains to arise. Sentient consciousness is "reflected", or arises in our mind, due to the ubiquitous (i.e. "everywhere") presence of Universal-Consciousness and the universe's ever present primeval Supreme Awareness (which was also present as the Elsewhere from which the Primeval Singularity arose) that passively functions as its Witness.

This is analogous to the phenomenon of light arising from the intense chemical reaction (oxidation) of fire. The light may arise, but its "brightness" is only seen when it is present in our minds!

PURGATORY: "Purgatory" is the belief that there exists an objectively real "place" wherein souls that have done about an equal amount of good and evil deeds while embodied reside until their "final judgment".

Final judgment is either arrived at by a god, or is debated between an evil demigod and a good demigod, or "archangel", and the soul is eventually sent (perhaps forever!) to either "heaven" or to "hell".

Some religions entertain the concept that the prayers of the living for those in purgatory can have an ameliorating effect on their final judgment, or can even release them from hell through a supernatural process of Divine clemency.

QUALITATIVE GAP: The infamous "Qualitative Gap" is the difference in quality that exists between objectively "real" material things and their nonobjective subjects. It is the logical basis of the philosopher's recalcitrant "mind-body" dilemma.

QUANTUM UNCERTAINTY: The "uncertainty" principle of quantum mechanics was developed in 1925 by Werner Heisenberg, a German theoretical physicist. It is a theory proving, among other rather remarkable things, that it is actually impossible to determine simultaneously both the position and velocity of any subatomic particle!

Because the "particle" is also a wave, its location is spread across the wave, and when its position in particle form is predicted, the momentum of its wave form "spreads out" making it impossible to accurately determine its exact location.

The theory demonstrates, by "predictive" mathematical probability, that they are apparently inversely proportional. The "uncertainty" problem lies in trying to make the wavicle conform to the dimensional limitations of our dualistic perception.

To solve this, an equation was developed to predict its "standard deviation", based upon the statistical probability of its distribution in space and time, which is the square root of the variance of its possible locations.

REALITY: The problem with the theory of "Reality", is that whatever we can know, is purely conceptual. This means that the reality of our person and universe, or subject and object, is a matter of conjecture. Therefore plausibility may (at present) still be our best guide to ascertain the truer nature of things.

Science, for instance, is based upon an empirical method of theoretical plausibility that is considered to be demonstrable by rigorous mathematical analysis and "unvarying" experimental results, which tend to support the contemporary theory. But in the final analysis "Reality" is a theory that is impossible to prove, except by consensus of agreement.

The actuality of things simply lies beyond our current ability to conclusively substantiate, because it evidently lies outside of the realm of the ideas and thoughts by which anything is known!

Indeed, even the dualistic consciousness, by which we perceive anything, is actually always one, which includes both subject and object in order for our cognition to transpire. Thus, whatever we cognize must appear as a composite of interdependent subject-objects to our truer selves, its independent universal Witness (i.e. the Supreme).

"Consciousness is never experienced in the plural,
only in the singular."

Irwin Schrödinger

Reincarnation: The concept of "reincarnation" (rebirth), which depends upon an extraphysical continuity of our person(s), is implausible after the body and brain, which support the concepts and thoughts that constitute our ideational persons, die and disperse back into the elements from which they arose into an experiential temporality.

However, this is not to say that a near infinite number of progressively differing simultaneous lives may not be present in multiple parallel universes. Which also does not preclude the possibility of our fundamental percipient Awareness' moving "laterally" amongst them, perhaps even unknowingly?

Moreover, it also seems likely enough that when we die, we might even summarily, from within spacetimelessness, "wake up" as *Everything* all at once! (Now being solely comprised of, personally unconscious, impersonal, pure Awareness).

Nonetheless, if we then become trapped again, by identifying with another body-mind, it clearly has nothing whatsoever to do with any other transitory, fundamentally "impersonal", universal processes of so-called "previous" births!

Consequently, if there is a "continuation", it can only be a continuity of impersonal universal Forces and principles of Identity, and not that of an individual "person" or personality. However, since there exists a potentially infinite number of progressively different parallel universes, it is also possible that some distant iteration of ourselves may even exist as an *immortal being* in one of them!

Which is not really so very strange, since this simply means that in this particular universe the Divine has finally taken control of the quantum material of its own universal expression! But what is truly strange, is that it also indicates that whatever exists may continue to "potentially" exist in perpetuity, even though we can no longer sense its presence!

Relativism: Relativism "begs the question" by assuming facts that are not present in the evidence, because if there is nothing objectively "present" to relate to, whatever exists may or may not really exist, since it cannot be experientially (i.e. "subjectively") measured!

Since relativism misses the fundamental fact that whatever "exists" does so only within our dualistic consciousness, it also misses the verity that the apparently relative universe of our experience, is actually tantamount (equivalent) to an illusory projection!

Renee Descarte: Renee Descarte was a French mathematician, scientist, and Early-Modern philosopher (1596-1650) who rejected uncompromising religious authority in the quest for knowledge and meaning.

Revisionism: Revisionism is the onerous rewriting of history in an attempt to change people's perception of past events. For more information please visit the "Museum of Tolerance" at HYPERLINK "http://meuseumoftolerance.com" *museumoftolerance.com,* the "United States Holocaust Museum" at HYPERLINK "http://*ushmm.com" ushmm.com,* or the State of Israel's "Yad VaShem Holocaust Museum" at HYPERLINK "http://www.*yadvashem.or" yadvashem.or*.

Samadhi: Samadhi is a Yogic (Sanskrit) term meaning "Enlightenment". It is said to be of two sorts, *Savikalpa* which is a "breathless trance state", and *Nirvikalpa* which is purportedly, "an enlightenment that forever abides, with eyes open or closed".

However, the syncope and resulting cerebral hypoxia of the former casts serious doubt on both its safety and perhaps even its authenticity as a "bodiless" state of awakening!

"But in that state [of Samadhi] his body does not last many days. He remains unconscious to the outer world. If milk is poured into his mouth, it runs out. Dwelling on this plain of consciousness, he gives up his body in twenty-one days." – *The Gospel of Sri Ramakrishna,* by Swami Nikhilananda.

SCHRÖDINGER'S CAT: A thought experiment was proposed by Erwin Schrödinger in 1935, wherein a cat is placed in a box with a canister of poison gas. The gas is then released by a quantum event, the number of particles detected by a Geiger counter in a given time period. If the number is odd the gas is released, if positive, the cat lives.

A "quantum event" now determines the poor cat's fate! If the elapsed time happens to coincide with a negative count, according to quantum physics until the box is opened and the cat is seen, the cat now remains in two superimposed states, one "half alive", and the other "half dead"!

But when the box is finally opened and the cat is seen by the experimenter, objective reality collapses back in on the events. Moreover, that "reality" also simultaneously collapses back in time to reunite the moment of the fateful count, the gases interim release, and the cat's present death!

Indeed, it is impossible to even determine the proper temporal progression of events, because the cat exists in a state of "limbo" until the count begins.

This thought experiment lead to notions of the universe splitting into two copies, "...one containing a live cat, and one a dead cat, each containing the experimenter as well" (Everett). To which, of course, was added the concept that, "...each universe is changing in content" (D. Deutsch).

Then, by hypothetical extension; "In general, if a quantum system is in superimposition of, say η quantum states, then, on measurement, the universe will split into "η" copies. In most cases η is infinite.

Hence we must accept that there are an infinity of 'parallel worlds' co-existing alongside the one we see at any instant. Moreover, there are an infinity of individuals, more or less identical with each of us, inhabiting these worlds" (Everett).

Apparently expanding upon this hypothesis, M.R. Franks has also postulated the existence of a googolplex of "parallel static universes". (A "googolplex" is also said to be the rather large number of quantum particles that comprise our universe). A "googol" is 1 with a hundred zeroes after it. While a *googolplex* is not simply one with 100 x 100 zeroes behind it, it is 1 with a googol of zeroes behind it!

This is a very large, but still finite number. But if you traveled to the farthest star visible from the orbiting Hubble Telescope and wrote average sized zeroes all the way, you still wouldn't have space enough to write them all! Nonetheless, it is not nearly the largest number that, with the aid of modern computers, we have been able to mathematically "create" (although *predict* may be a better descriptor!). (See "Googolplex", "Graham Number", 3rd paragraph on pg. 537).

Moreover each parallel universe exists, much like orbiting electrons do, separated by a microscopic "membrane" of spacetimelessness. In this manner creating our common experience of continual, causally related change.

The fantastic lengths that science will go to, to preserve the paradigm of a dualistic, subject-object universe, is truly astounding! It seems much more likely that these creatively different, but mathematically defensible views, amount to no more than different perspectives on a universe in which an incomplete dualism is itself the principle illusion.

But it is nonetheless a thoroughly pervasive illusion, which seems to divide infinitely into space-time bounded segments and dimensions, what is essentially always an interconnected One.

Yet, if this primal Oneness is allowed, the present paradoxical plethora of bizarre, but scientifically defensible perspectives may again become generally valid rules, relative to the particular point of perspective that is taken.

"M" Theory, for instance, is a recent "Theory of Everything" that attempts to arrive at just such an essential Oneness. But it is being hampered by our currently limited scientific paradigm and instrumentation.

However, as we continue to expand our premises to accurately include the consciousness by which our universe is known to exist, and to advance our technology in order to substantiate our premises empirically, no doubt our understanding of the universe will also continue to deepen! (See "Singularity", the "Hawking Singularity Theorem", 2nd paragraph on pg. 576; and "String Theory", the "M Theory", on pg. 581).

SEROTONIN: Serotonin is found primarily in the *enterochromaffin* cells of our gut where it is used to regulate intestinal movement. About 20% of our body's serotonin is produced in *seratoginic neurons* in our central nervous system where it is used to regulate our muscle contractions, appetite, mood, sleep, and even some of our cognitive functions such as memory and learning.

SHAO-LIN: "Shao-Lin" is a Ch'an Mahayana Buddhist Temple in Song Shan, Henan Provence in Deng-fengh China, founded in the fifth century; also a monk of the Shao-Lin order, or the martial arts developed by them.

SHIVA: Shiva is the *Raja Yogi* "King of Yogi's" and Hindu god of "cyclical destruction". But in the deeper spacetimeless actuality of our inherently Conscious Universe "Shiva" is quite simply our own deeper, truer "Self" our *Sat Guru* "True Teacher". The Kabbalah refers to this concept as "Elijah" the נבי *Navi* "Prophet".

Thus, Shiva symbolizes our final disengagement from the thought comprised illusions and delusions that constitute our (generally unexamined) mortal "persons" and universe(s). (I.e. This is impersonally accomplished through a re-identification with our illusory individual "person's" veridical percipient Witness.) Which is actually our very own authentic and immortal (yet ever *impersonal*), foundational Self!

SINGULARITY: In the philosophy of technology, the emergence of an unpredictable, qualitatively different future is called a "Technological Singularity". For example, if a computer can be built that exceeds the capacity of our limited human brain, it could theoretically build a machine superior to itself. Advancing quickly into a super-sentient artificial intelligence, or so-called "A.I. Singularity".

In physics, the "Penrose Singularity Theorem" explains the existence of space and time-like singularities, in non-rotating uncharged, and rotating charged Black Holes.

The "Hawking Singularity Theorem" predicts the primeval existence of a singularity in which the energy was greater than the pressure of nascent gravity. Which led to a brief period of inflation, lasting from around 10^{-36} to 10^{-32} seconds. This is to say (strictly as an example), that around 10^{-36} seconds after the Big Bang, that our universe expanded like a balloon that was being instantly filled with a massive amount of air, which caused it to expand much faster than the speed of light, by a factor of 10^{78} in instantaneous volume!

The primeval singularity's perfect symmetry was likely broken when the emergence of a gravity imbuing "Higgs Field" and "Higgs Boson" led to the separation of the strong nuclear force from the early heat-merged electromagnetic and weak nuclear interaction of forces.

(The existence of the Higgs Field and Higgs Boson have recently been confirmed at the Large Hadron Collider in Switzerland. But just exactly how the perfect symmetry of the Primeval Singularity was broken is yet to be discovered and confirmed.)

This "Electroweak Epoch" lasted about 10^{-12} seconds during which our universe was comprised of dense, hot plasma, and large numbers of exotic particles were created.

When the universe cooled a bit, during the "Quark Epoch", it was filled with a hot *quark-gluon* plasma. The weak interaction became a short range force, and gravity, electromagnetism, and the strong nuclear interaction took their present forms.

After around 10^{-6} seconds the "Hadron Epoch" began when the temperature fell enough to allow the quarks to bind together into *hadron/anti-hadron* pairs. When our rapidly expanding universe cooled further, these hadron positive-matter/antimatter pairs began to annihilate each other.

In about 1 second only a small residue of "positive matter" hadrons remained. And they have been categorized into two families' *baryons* and *mesons* (protons and neutrons are a type of baryon). This is when the "Lepton Epoch" purportedly began.

Leptons are a family of subatomic particles that include the *electron*, *muon* and *tauon*. All of which are subject to electromagnetic force, gravitational force and the "electroweak interaction", but not to the "strong nuclear interaction".

Now, three classes of elementary particles were present, leptons, quarks, and the *gauge bosons* that are thought to carry the fundamental forces of nature.

Our present universe is believed to be largely comprised of these particle families, and also to be powered by their interactions with "fields" (of force) that extend ubiquitously across an ever expanding space-time continuum (see the "Standard Model", pg. 579).

SKANDHAS: In classical Servastivadin (Theravada) Buddhism the "Five Skandhas" are, *Rupa* "Form"; *Vedana* "Sensation"; *Sanna* "Perception"; *Samskara* "Mental Formations"; and *Vinnana* "Consciousness".

These five are the basis of a matrix of dharmas that early Buddhists surmised were the fundamental constituents of reality. Later Buddhists, the Mahayanists, refuted this concept in favor of a *Prajna-paramita* "Immanent-Transcendent" intuitive model.

But it is not necessary to get caught up in the divisive classical rhetoric, if we understand that each of these *dharmas* "mental elements" is comprised of vrittis (see "Wavicles", "vrittis", pg. 596).

Because when vrittis are taken to represent the smallest possible point of pure *Paramita* "transcendent" Universal-Consciousness (UC) energy, they reveal that their veridical existence is simultaneously within and without us, as the UC/quantum string substrate of our entire universe!

At this level all virtual and veritable realities interface, like the opposing sides of a cosmic "coin" that is made up of the same UC "stuff", which was tossed into the ever inflating space-time of our universe by the force of the Big Bang (see "Planck Time" pg. 527, and "String Theory", pg. 581).

Space: "Space" is our mental representation of the *Akash* "superspace/void", which is "filled" with the presence of Universal-Consciousness. It is invisible to us because it is the unshifting (subliminal) subjective point of our (Supreme Witnesses) perspective.

Yet it is only because of this unchanging point (of our Supreme Witnesses perspective) within the apparent "Akash" of Universal-Consciousness that we can even become sentient!

And only when we become sentient, can we perceive the constant four dimensional temporal flow of our individual consciousnesses. Which contain the endless concepts that comprise our apparent three dimensional experiences of person(s) and universe(s).

But trying to observe our superspace's fifth dimensional locale is like trying to gaze into our own eyes without the aid of a mirror.

Thus, all that we can perceive of "space" is our mind's representation of its four dimensional reflection in our consciousness. As it appears to separate our three dimensional subjects and objects of consciousness within space and time.

Which can only thereby appear to move separately through time relative to their position and speed. They are thus enabled to appear within our cognitive representation of reality, as both a "subject/object" and an "object/object". Which are now subject to Einstein's General Theory of Relativity, as it is heuristically represented (as a "biological algorithm") within our consciousness!

Space-Time: Space and time are cognitively inseparable because they are actually just different mental representations of our minds episodic method of measuring (our truly immeasurable "monadic" Superverse)! This is to say, that every apparent "thing" in our universe is only apparently separate (and separable) above the ever interconnected sub-quantum scale.

This is because, on this more fundamental level, our universe's only apparently separate "strings", of seemingly dualistic particle-waveform energy, and evidently fungible intervening space and time, remerge into one universal substrate. And it is this fundamental substance that today's theoretical astrophysicists are attempting to explain, and particle physicists are attempting to reveal.

Our mind "measures" the universe into conscious existence by adding the spatial dimensions that define the limits and dimensions of our perceived objects. It then makes them perceivable by inserting an apparently enduring space, as an only evidently malleable time between them. Because it takes episodes of apparently changing "time" to move through an even non-apparent (but nonetheless experientially durable) space!

And because our illusory experience of changing "space" is also the actual unchanging observational point of our perception, it appears to be invisible, yet even its evident "invisibility" is also a necessary object of our perception!

Therefore the invisible "object" of our apparent space (i.e. "space" is also an object of our perception) also differentiates each moment or frame of our perception in a temporal manner, as our faculties of memory, imagination and anticipation function.

Thus the mental impression of an ever-flowing stream of past, present and future events is subliminally imparted, by our binary individual consciousnesses, to an intrinsically inseparable über Conscious universe!

STANDARD-MODEL: The "Standard Model" of subatomic particle theory posits twelve elementary particles, which are the building blocks of our universe, and four fundamental forces that act upon, within and between them.

Ordinary matter is made up of atoms that are comprised of three basic components; a nucleus composed of protons and neutrons being circled by whirling clouds of potential electrons. These three basic atomic components are themselves comprised of even smaller, subatomic particles.

Six of these particles are called "quarks", which have acquired the curious names of "up, down, charm, strange, bottom and top". Protons, for example, are comprised of two "up" quarks and one "down" quark.

The remaining six elementary particles are leptons, the electron, much heavier muon, its cousin the tauon, and three neutrinos.

As far as we currently know the four fundamental forces of the universe are electromagnetism, gravity, and the strong and weak nuclear forces. These are produced and carried by fundamental particles of force; the "photon" which is both a particle of light and the arbitrator of electromagnetism (i.e. iron filings are attracted by a magnet because both exchange photons); the "graviton" which is believed to be the mediator of gravity, the "strong atomic force" which is borne by eight (rather euphemistically called) "gluon" particles; and the "electroweak force" which is conveyed by three particles termed W+, W-, and Z.

The composition and behavior of all these forces and particles is elegantly explained, with the notable exception of gravity. The formulation of a reliable quantum theory of gravity remains one of the most challenging problems in theoretical physics. Nonetheless, we may be very close to formulating one, with the recent confirmation of the Higgs Boson that apparently transfers the mass to matter for gravity to act upon.

Which is in itself problematical, since gravity is then evidently a rather mysterious sort of quantifiable yet anomalous force that is somehow capable of being transferred to wavicles of energy. That have through the dual principles of identity and *nous* "absence", and forces of attraction and repulsion, have arranged themselves into apparently solid "matter". Thus allowing them to somehow "warp" their invisible (but nonetheless necessary) ostensively enveloping space-time in such a way that it appears to be an attractive force!

STOCKHOLM SYNDROME: "Stockholm Syndrome" is a term that was coined by psychiatrist Nils Bejerot after a Stockholm bank robbery in 1978 where, ironically, after being held for six days, the hostages became emotionally attached to their captors, and astonishingly even defended them after their ordeal was finally over!

According to the FBI approximately 27% of long held victims develop Stockholm Syndrome. Psychiatrist Frank Ochberg further defined the condition, and eventually developed methods to provide remedial aid in hostage situations.

STRANGE LOOP: A "strange loop" is present in both computers and human consciousness if, when we move either forward or backwards in a hierarchical system, such as a computers "software" algorithms or our brains own memory algorithms, we mysteriously arrive right back at the very place where we began! Hence, in a fashion (like our subjective "person") our entire universe is actually a *strange loop*. Because, when we travel far enough in any direction, we eventually end up right back where we started from!

STRING THEORY: "M Theory", a proposed Master Theory of everything, currently seems the most probable answer as to what a purportedly "physical" universe may be composed of. It incorporates all of the more credible theories of physics (Newtonian, Einsteinian and String) as mathematically correct, but nonetheless differing perspectives of cosmic scale.

It also suggests that all of the particles described in the Standard Model (SM) are simply different manifestations of a more basic proto-energy substrate (i.e. fundamental "Strings" that are composed of the perturbations of infinitely expanding probabilities that initiated the Big Bang).

A particle in the SM, for instance, is simply a point with no internal structure, therefore all that it is capable of doing is moving. (Which certainly contributes to the evidently almost "massless" state of photons, in this commonly accepted, yet often insufficiently examined paradigm of "material" manifestation!). Hence its movement and energy are evidently synonymous with its form, and if it stopped moving it would no longer exist!

But if we were able to look inside, from very close up, we would see that the particle is actually not a point, but a tiny loop of cosmic "string". An energy string can vibrate or oscillate at different speeds, or frequencies, like a violin or piano string does.

At one level of vibration, from a distance it appears to be a photon, at another frequency a quark or a gluon, and so on. The splitting and recombination of strings corresponds to what, on a larger scale, appears to be particle emission and absorption.

There are actually five different String Theories, four of which incorporate ten dimensions, and one that has twenty-six! But when we include a superlative "eleventh" dimension, from which to fully view the other ten, the dimensions take on the characteristics of "membranes".

It is within these (now "scalar") dimensional membranes that energetic "strings" coalesce from potential existence into a percipient sort-of existence within space and time. (Curiously, unlike vector location values, scalar "field" values, like the tops of a tall mountain range, not only remain invariant, but do so at every point throughout our entire universe!).

They do so from within the proto-energy substrate, of which the membranes are essentially composed, powered by the motions or vibrations that began with the singularity breaking perturbations that initiated the so-called "Big Bang".

But they evidently only complete their transformation into "existence" when they are experienced within our consciousness. Apparently endless, slightly different iterations of "Parallel Universes" are also created in this Cosmic process of experiential cogeneration.

These dimensional membranes are linked together by the rate of vibration in such a fashion that the strings can be thought of as connecting the membranes together as they emerge from within each dimension. Their interdimensional connections are like the bridge that connects a violin's strings together before they are anchored into the body of its wooden frame.

And while we don't know exactly how this happens it is most likely accomplished in much the same manner that a drum's surface vibrates in harmonic sympathy with an orchestra that is playing music over a radio's vibrating speaker that is located across the room.

But in this work we ask a further question, "Of what is the proto-energy substrate that underlies the membranes, strings and space-time, comprised?" The short answer is that it must be comprised of (for want of a better term) a "Consciousness-Force". Which somehow made its appearance from within the spacetimeless Elsewhere that evidently preceded both the "substance" and "nothingness" of our manifest universe.

It is therefore the subtlest yet most powerful thing in the universe, and most certainly precedes consciousness in the conventional sense, and even whatever our subject-object consciousness may subsequently contain within itself! Words are simply inadequate to convey it. And because the term "god" carries too much baggage, for want of a better descriptor, in this writing we may as well just call it "Universal-Consciousness".

"M" Theory and the Standard Multi-Universe Theory are our most plausible current theories about our universe's existence, and while they disagree about the manner of its generation, they both agree on the existence of multiple parallel universes. Both suggest that they are separated in different dimensions.

"M" Theory posits that every trillion years or so a few of the endless universes that constitute our multiverse, which are distributed in parallel dimensions, are slowly drawn together by so-called "Dark Energy". And when they collide there is another Singularity, and a new universe is born, in a continuing cycle!

But please keep in mind that "parallel", in this usage of the term, does not simply imply a dualistic "side-by-side" distribution of universes! This is because when an event occurs in our experience it not only causes a universe to expand into existence within our evident consciousness, but also in endless (and slightly different) iterations, in every possible "direction." But of these, we can only experience those that occur within three dimensions, from an evidently invisible fourth of apparent space-time, from within an unimaginable "fifth" Dimension!

Indeed, the recent confirmation of "matter waves" also confirms the Cosmic Inflationary Theory. Coupled with the recent confirmation of the Higg's Field and the Higg's Boson, the theory of an endless multiverse (i.e. "Superverse") that is generated from an initial "Primeval Singularity" is more than likely correct. Evidently, once inflation occurs (with nothing to stop it), it simply goes on forever!

Nonetheless it apparently does slow down on occasion due to the differing rates of expansion brought on by the forces of dark matter, gravity, and the strong and electro-weak nuclear forces. This happens just long enough to form a new "Singularity", initiate another "Big Bang", and create another "Pocket Universe" in a new dimension. And then it happens all over again, in an endless succession!

Thus there probably exist countless multiverses, in every possible configuration, in an inestimable and evidently endless Superverse! In Forest Primeval we further suggest that each parallel universe is likely a *chiral* + $^{-}$1ɳ. A "mirror image", like that of neutrons (matter) and positrons (anti-matter), that only differs by one malleable quantum particle (i.e. + "-1ɳ", the above *chiral*) from the previous dimension.

And moreover, that all are interconnected at their extremities of conversion by the subtle but pervasive force of quantum gravity "entanglements"!

But of course, this presumes that other substantive limiting forces do not exist, which only points out the substantial limitations that still exist in our current level of understanding.

SUPREME: A scientific explanation of the "Supreme" is that it is essentially the so-called "Elsewhere" source of the universe's initial perfect Primeval Singularity. And the Primeval Singularity itself is, for want of a better term, most likely some sort of pre quantum-string, spacetimeless and universal, non-conditioned (i.e. pre "chain of causality") Consciousness.

Moreover, the perfect Primeval Singularity most certainly already contained its own potential diversification into the multi-universal infinitude of our presently unfolding Superverse!

Therefore, evidently our universe's foundational identity principle (i.e. "Consciousness") is also fundamentally present in its every inflated, transformed, and subsequently (re)connected aspect.

Which, in its subsequent energetic extension within a constantly inflating "space-time", then followed a mathematically predictable "most likely" causally conditioned course. (Including the presence and contents of our purportedly "individual" consciousnesses).

Hence, our universe's foundational identity principle is also present on the apparent "in", and the evident "out" side of our substantially "illusory" individual consciousnesses.

Moreover, this fundamental sub-quantum Energy, of impersonal Consciousness "in motion", began to operate the quantum machinery of our universe. When the cosmic clock started "ticking". Following our universe's initial inflationary period. Which lasted around 380 thousand years after the Big Bang, when *photons* (particles of light) were finally released.

TANTRA: "Tantra" is a diverse ancient East Indian method of belief and practice, which holds that our universe is the manifestation of a divine Consciousness-Force.

This force, *Shakti* (Syn. *Prakriti* or "Mother Nature") is revered as the symbolic representation of the universe's creative feminine force. Its male aspect, *Shiva,* can also be (not inaccurately) regarded as the essential transcendent source of identity (i.e. *Sattva Guna*) for all manifestation. Which is present as the reflected "soul" or *Atman* within human beings.

In practice *Tantra* "to weave" offers teachings and methods to reintegrate our illusory human identity with its Divine Source (i.e. the Supreme).

TARDIGRADE: *Tardigrades* "Water Bears" are charming little eight-legged microscopic creatures that generally inhabit the dewdrops that cling to the mosses and lichens that often grow on the north side of tree trunks.

Not only are they almost everywhere, but there are also around a billion of them for every one of us! Moreover, they are also so unbelievably tough that they can survive without food or water for years, and can even endure the cold vacuum and cosmic radiation of outer space!

Mother Nature always has a survival strategy for continuing life, even if the worst should transpire, as it has on several occasions in our planet's geological history. No doubt they continue the genetic progression of the survivors of these planet wide extinction events. (See, "The Seeds of Life", immediately below).

If we can ever figure out the specific gene sequences that allow the clever little Tardigrades to accomplish these incredible feats of survival, one day we may be able to transfer them to the genes of our own future astronauts, which will then allow them to do the same!

THE SEEDS OF LIFE: Although life is most certainly fragile, it is also incredibly durable! (As the long-term survival of the microscopic "extremophiles" that were recently placed on the outside of the International Space Station proves!) And evidently the universe has a unique method of seeding our entire cosmos with it!

When large sized asteroids and comets crash into planets and planetary bodies that harbor life, the life on them is generally extinguished in the resulting “firestorm”. But some of its extremophile microbes also survive inside the rocky debris that is thrown out into space, some of which eventually lands on other “life-friendly” planets, moons or planet-like bodies.

Many astrophysicists, cosmologists and theoretical physicists now believe that in this way, over nearly inestimable millennia, life has spread itself throughout our Universe.

Moreover life has probably spread itself throughout the Multiverse as well, if not throughout our entire (mostly unseen but ever-expanding) Superverse, which includes every possible parallel universe as well!

But we cannot see our still-inflating Superverse in its entirety. Because it is so very large that even its photons, which are travelling towards us at our universe’s maximum speed, could take almost “forever” to traverse the vast distances that are necessary to finally reach the lens of even our most powerful (light actuated) telescope!

TIME: “Time” is a function of our receptive apparatus. It appears to be the fourth dimension of space, but it is also perceived by us as the motion of time operating within space. Time is narrow and constricted, because it is our mind’s instrument of measure. But the universe is broad and boundless, and continues to grow, yet our mind wants to contain it.

To do so “space” is created by our mind (in order to have something to measure). “Inner” space gives dimension to the objects of our consciousness, and “outer” space provides the apparent separation that is evidently required between them. Both of which are present in our rather peculiar inside/outside universe that is comprised of the cosmic force of a universally Conscious “Energy”.

Time and space are thus inseparable conceptual aspects of our individual consciousnesses. Without them the universe and person of our common experience can be neither conceived or cognized! For example, it is the space between the marks on a ruler that make it useful.

In like manner our concepts of past, present and future are “marks” of measurement placed by our minds upon the incessant binary (subject-object) flow of our individual consciousness. But our consciousness is only present for as long as the universe’s occult (subtle) Life-Force animates our bodies.

In its undifferentiated state, our dualistic consciousness is only present as our subjective consciousness, which nonetheless appears to our awareness of it as an object of our experience. This can be thought of as the conscious space between our ever shifting moments of experience, before they are measured by our minds into an ideational past, present and future.

Hence consciousness, insofar as we commonly experience it, is really the “reflection” of Awareness into our minds. Thus it also includes the inherent sentience that is necessarily intrinsic to any sort of a self-aware Awareness, in order for it to be “aware”!

But at its perceptive source, before our phenomenal selves and objects are measured into existence, Awareness is always present, but cannot be conscious of the fact! Nonetheless it is there, prior to either our individual consciousnesses, or the Universal-Consciousness that manifests our body, mind and universe.

The Source of all consciousness is thus also “spacetimeless”. Our pure Awareness is the percipient source of Universal-Consciousness, dynamically examining its Oneness in experiential diversity, through the phenomena of our individual consciousnesses.

Our essential (Supreme) Awareness is thus both broad and boundless, existing beyond, before, and in between, the endless measurements and concerns of our individual consciousness as it spontaneously arises in our minds.

“We” are nowhere present. But suffering inevitably results when our body-mind instrument of experience thinks that it is actually in control of the ceaseless flow of experiences that are being automatically generated by our ever expanding, but nonetheless simultaneously “inside/out” and “outside/in”, Universal-Consciousness comprised universe.

When this fabrication is finally given up, the broad and boundless peace of Universal-Consciousness surfaces. Like a babbling brook, our mind continues to chatter on, but it is no longer identified with.

Thus "Enlightenment" is not something to be accomplished, created, or discovered. It is our natural condition, just waiting to be acknowledged, in the quiet moments between our thoughts and experiences.

Metaphorically speaking, the true nature of our universe "shines" in these eternal, changeless, and quiet moments, like a sentient sun shining out from between our temporarily intervening thought clouds.

The continual ideational penumbra (shadows) that are being cast, fashion our impressions of the world and our resulting existential pains. But our suffering (which is born of this experience) disappears when these divisive "clouds" no longer obscure the "Inner Light", by which our universe is known.

(Pea memoriae, the perspicacious mystic, Betty Bethards. Who wisely taught that we humans are actually just, "G-d in a little earth suit". And therefore, "We should always follow our own Inner Light"!).

TIME TRAVEL: Stephen Hawing conducted an interesting test to determine if time travel might become a reality in the future. He hosted a benefit to welcome future travelers to our time, but none showed up. Of course this test is inconclusive, on the grounds that future time travelers might wish to maintain their anonymity to protect themselves from becoming extinct in their own future!

While non-interference seems the only way to do this, there are perhaps more ominous implications. If time travel backwards in time is possible, and we are not visited, it could also mean that humankind (in our universe) might not survive long enough to develop it, which is most definitely "food" for some rather disturbing thoughts!

But of course if time (as we experience it) is indeed an aspect of consciousness, travel within it must also take place within consciousness, in some fashion. Moreover, this could as well be a movement into the past of a "static" parallel dimension.

Which would then not only take place within our own "present" experience, but would also leave our own time-line forever undisturbed by any visitations from any entities that may come to exist within our own future time-line.

In a sense, we travel in time whenever we experience a "past" memory, the "present" moment, or even anticipate a "future" event. But if we are not present to experience it, neither past, present, nor future, is present either!

Nonetheless, Einstein initiated the debate on possible time travel when he showed that even space-time is "curved" by gravity. Which opened the debate on possible FTL (faster-than-light) interstellar space travel across an Einstein-Rosen גשר Gesher "Bridge'" (i.e. through a cosmos traversing "wormhole").

Einstein's contemporary, the genius mathematician Kurt Gödel, proved that time travel is possible in a rotating universe. But we later discovered that our universe does not "spin".

Decades later, physicist Frank Tippler (the possible initiator of the widely circulated "Perplexed Physicist" prose that appears on page 8 of Forest Primeval) mathematically "proved" that time travel is possible by rotating a massive cylinder at near light speed, and flying at great speed around it, thus moving backwards in time.

But problematically the cylinder would need to be infinitely large, and if you tried to make it smaller, a singularity would be created, thus precluding its usefulness as a time travelling machine!

More recently, physicist Kip Thorne has predicted the existence of high mass, quantum sized worm holes that warp the space and time around them. And Physicist Richard Gott also predicts the existence of microscopic "wormhole strings" that are remnants of high density vacuum energy left over from the Big Bang. (Although none have yet been found by the large Hadron Collider, which brings their actual existence into question.)

Ironically however, we can now create extremely small, but ephemeral, worm holes with the extreme power of the improved LHC, which nearly duplicates the power of the Big Bang!

And based upon the latest research and revelations in astronomy and physics, astrophysicist Michio Kaku, also predicts the existence of wormholes in the center of rotating black holes. Rather than compressing down into a "super crushed" one dimensional mass, a "super dense" ring of rapidly spinning neutrons forms at the bottom of a wormhole that leads into another (parallel) universe. Moreover, it also seems likely that in a very large wormhole, like that at the center of our own Milky Way Galaxy, will continue to repeat this process, perhaps indefinitely. Thus providing a possible space-time doorway that leads into our entire Superverse!

So we can readily see that when space is curved sufficiently by extreme gravity, time is compressed relative to space, and time travel becomes possible. But our current biological vehicles (bodies) simply cannot survive the trip!

Yet, if the laws of quantum gravity allow it, by taking advantage of quantum "nonlocality" (being in more than one place at a time on the quantum scale), moving a data stream bi-directionally through time by utilizing (currently "super-cold") *qubit* computing technology, it may well become possible in our not-too-distant future to time-travel. In fact, this may even provide a means to travel between parallel dimensions! (I.e. See "Pale Suns and Bright Moons", dialectic on page 109).

However, it is more likely that what will really be happening is the streaming of data into and out of an already potentially existing (relative to our own time line) parallel universe. In this way the physical laws of our universe are not violated and conundra like the "Grandfather Paradox" are not created (i.e. if you went back in time and killed your own grandfather, your father would never be born, and neither would you!).

Most likely, this technology will allow data gathering of past and future events, and possibly even virtual "Avatar" like travel into the past and future, and maybe even limited interaction with the events, but actual bodily travel by this method remains unlikely.

But the apparent "retro-causality" of this physics does reveal that future events are most likely also quantumly entangled with past events, which is also a modern extension of the venerable Japanese "Rinzai" Zen concept of *jijimuge*.

Which can then be defined as the interrelationship of all events transpiring in apparent space-time with the consciousness by which they are known to occur!

Also (and not to be discounted by any means!), the virtual reality constructions of future computers will no doubt allow the reanimation of past events in such a compelling manner that they will be indistinguishable from our present perceptions of "reality".

Indeed, it is a rather sobering thought to realize that even if this were currently going on, we would have no way of determining that what we are presently experiencing is not even "real"!

TORAH: The Torah is a wonderful written collection of myth, history, law, "civilization building", and teaching tales. But it is also the repository of an ancient wisdom about the deeper, occult nature of our consciousness and its deeper connections with our universe.

In this context the secret "key" to unlocking its mysteries lies in the metaphysical wisdom of the Kabbalah. The Jewish religion, which grew over time from the primeval root of its intuitional revelation, is a masterpiece of the immanent Divine!

The Torah serves the important function of making these teachings accessible to almost everyone, and so assists in the evolution of our universe through the practical application of its occult higher truths in our everyday life (see "Tikkun Olam", dialectic on pg. 261).

By following its *Mitzvoth* "Commandments" the bonds that tie our essential Awareness to a strong identification with our sensorially apparent mind, body and world, are eventually loosened.

Thus, over time's passing, the sincere devotee to this method begins to awaken to the Divine nature of his or her own consciousness, its hidden connections with all of life, and the entire Manifestation. Thus it is not easily dismissed, nor should it be taken lightly!

To gain a better understanding of the midrashic Torah exposition on pages 67 through 69, I encourage you to read the collected works of Wei Wu Wei, who's "Negative Way" premises are applied in this *Midrash* "Truth drawn from Torah", (see "Midrash", on pgs. 551, through 553).

TOTALITY: It is becoming an accepted fact in quantum physics and quantum cosmology that in order for anything in the universe to "exist", it must first be experienced.

Without a doubt there is most certainly something "out there", outside of our mind. But apparently it only has a potential existence until it is exposed to our consciousness and comes to life as an event, as it is "thought into existence" within our mind.

And as if to underscore the fundamental essentialness of consciousness in the multi-Universal Manifestation of our Superverse, fantastically, the very quantum wavicles of which the universe is comprised display certain qualities of consciousness! (Google® "Young's Interference Experiment" and Michelson-Morley's "Double-Slit Experiment" for even more information on quantum intelligence indicators.)

UNIVERSAL-CONSCIOUSNESS: Because all of space is present, all of time is as well! Therefore the future is already present in potential, but it is not yet "fully present" until it is experienced. The same also applies to all past and percipiently present events (i.e. events cannot be experienced in a "real" present because our cognition of them takes time).

Thus, when taken altogether, our universe is always equal to the perfect monism of the primeval singularity that preceded its dispersal into apparent diversity within space-time.

Moreover, prior to the universe's experience of itself in our consciousness', it is still simultaneously present everywhere within *it*-Self as Universal-Consciousness on the quantum level. But with the universe's apparently individual experience of itself as "us", it also becomes diversified in space and time (within its diverse experiencers).

This is a scientific explanation of Universal-Consciousness, and curiously, it only differs from a metaphysical one as a matter of perspective!

VAN GOUGH: Ironically, as it turns out, Van Gough was probably not really "insane", at least not fully so! More likely the prescient artist was simply able to see more directly that whatever we see is "alive" with the Force of Life that brings the "living" objects of consciousness into our subjective cognition of them. Because when we are dead, the objects of our consciousness are too!

WAR: Other than common greed and the venal lust for power, war is most often the unfortunate outcome of persistently unresolved and significantly differing perspectives. Moreover, in a universe whose fundamental physics are often also incorrectly assumed to be governed by a strict dualistic polarity, war is perhaps as unavoidable as the fundamental laws of physics and chemistry are!

But there is a third more inclusive "Cosmic" perspective and a "Fourth Way" of living, which invite peace rather than discord through education and the continued (and soon to be enhanced) evolution of our species (currently rather limited) hominid mind.

Both of which are regularly represented in the various teaching tales and mnemonics that altogether comprise the gist of Forest Primeval's essential meaning.

Nature is relentlessly establishing a "higher" ground in the cognizance of Divine Awareness, which (in spite of sometimes contrary appearances) is increasingly emerging in humankind.

But to understand this somewhat subtle verity, just as when we try to understand the phenomenon of so-called "Global Warming", we need to look at the long term trend, rather than just the immediate "weather" or the current events that (when taken altogether) comprise it.

The resulting inner peace should be allowed to flourish in the world (and with great resolve!). Because, as history also quite clearly shows, even the fruits of a "just", but nonetheless violent victory are (sadly) almost always fleeting!

And failing a deeper understanding of our essential unity with each other and the wondrous universe that we all share, sadly war will soon follow, as surely as the setting of the sun.

But within Forest Primeval I respectfully submit, for your serious consideration, some compelling evidence that we will, and no-doubt within the course of just a few more human generations, soon be doing much better.

Thus, I feel compelled to ask; "When so many brave and selfless men and women have sacrificed so much to ensure a better future for humanity, how could we in good conscience do any less than our very best!"

"And when they fell before the thunder it was as when
the lordly redwood green with boughs goes down with
a great shout upon the hills and leaves a lonesome
place against the sky."

"Lincoln Man of the People"
by Edwin Markham, 1852-1940.

WAVICLE: A *wavicle* (i.e. energy wave/particle) is the particle physicists' nod to the empirical evidence, which strongly suggests that, on the subatomic level, the universe is comprised of a substance that has the qualities of both.

On this level Universal-Consciousness "tips its hand", revealing its presence in several verifiable ways. Eastern metaphysics references quantum "vrittis" as similar constituents of our inner experiential universe, which is comprised of thought energy.

In this work vrittis are also considered to be the principal Universal-Consciousness/Energy presence of which cosmic strings are composed. And by following a metric of heuristic "most likely" expression(s), it thus determines the rate of vibration by which they assume a subatomic expression.

Both are considered interchangeable, depending upon your point of view, and wavicles of course display the same characteristics.

WU: Our individual consciousness is expressed in three aspects, waking experiences, active dreams, and dreamless "deep-sleep" states. We witness these states because our divided subject-object mind subliminally splits the "projected" light of Universal-Consciousness like a holographic projector's prism lens does.

And when our independent Awareness witnesses them, it illuminates them in its ("our") human arrangement with sentience. *Wu* "Nothing" is what remains when the universe finally resolves back into the essential nothingness of its primeval Self-Awareness. Prior to all dualism, in our "Essence", we too are simply Wu!

YAHWEH: *Yahweh* יהוה, (*Allah* الله, *Deüs* G-d, etc.) is the "Supreme" from which the creative force of Universal Consciousness arises, unbidden, like a wave spontaneously rising up from within an endless space-time quantum sea, giving birth to a vast multitude of universes.

When the momentum of the inflationary "wave" of manifestation (that began with the "Big Bang") eventually subsides back into the Supreme, all resumes its true and natural state of *Wu* "Nirvana", which is also the astrophysicist's curious "through the looking glass" *Elsewhere*!

ZEN: Zen has been aptly described as; "A special teaching without scriptures, beyond words and letters, pointing to the mind-essence of man, seeing directly into one's nature, attaining enlightenment." – *"Zen Flesh Zen Bones"* by Paul Reps, published by Charles E. Tuttle Co., Rutland, Vermont, first printing 1957.

ZENO'S PARADOX: If the distance an arrow travels to hit its target is divided, the time for it to traverse that space is also divided. Since the remaining space and intervening time that the arrow must cross to reach its target can be continually divided, the arrow will never reach its target!

But our common experience is that a launched projectile reaches its target in a time that is commensurate with its rate of travel. Yet, if we continually divide the space-time of the arrows flight, the quantum level is eventually reached, and a different set of space-time rules applies.

For example, an unstable particle, representing the quantum-condensed flight of our arrow, will change (decay) sometime during a given interval. But if the particle is continually observed to see if it decays, it will not decay! This phenomenon is accepted as a literal "Zeno's Paradox" in quantum theory.

A mathematical solution has been proposed, which is basically a structure theorem of "semi-groups". But, while this hypothesis may satisfy the rules of mathematics, it ignores the critical, if not absolute, role of consciousness in the (non-infinite) regression of this phenomenon.

Its role in Zeno's Paradox is implied in the Zen tale of two monks seated outside the Zendo "meditation hall" debating the motion of a flag. One monk stated, "Clearly it is the wind that is moving the flag!"

To which the other responded, "No, it is the flag that is moving, and not the wind." But when their Roshi overheard them, he vociferously declared, "It is neither the wind, nor the flag, but your minds that are flapping!"

Tao Te Ching

Chapters 31 and 41

夫唯兵者，不祥之器，物或惡之，故有道者不處。
君子居則貴左，用兵則貴右。兵者不祥之器，非君子之器，
不得已而用之，恬淡為上。勝而不美，而美之者，是樂殺人。
夫樂殺人者，則不可得志于天下矣。
吉事尚左，凶事尚右。偏將軍居左，上將軍居右，言以喪禮處之。
殺人之眾，以悲哀泣之，戰勝以喪禮處之。

上士聞道，勤而行之；中士聞道，若存若亡；
下士聞道，大笑之。不笑不足以為道。故建言有之：
"明道若昧；進道若退；夷道若纇；
上德若谷；
廣德若不足；　建德若偷；
質眞若渝；　大白若辱；
大方無隅；　大器晚成；
大音希聲；　大象無形；　道隱無名。
夫唯道，善貸且成。

Memorial Day

Memorial day, on page 93, is written to honor all of the brave Americans who died in the undeclared Vietnam War, especially my fellow Stepps, who gave their all for home and country. And because the theft of valor is especially egregious, I can only add that each of us fought the war in our own way:

"I returned, and saw under the sun, that the race [is] not to the swift, nor the battle to the strong, neither bread to the wise, nor riches to men of understanding, nor yet favor to men of skill; but time and chance happen to them all. For man knows not his time: and as fishes taken in an evil net, and birds caught in a snare; so [are] the sons of men snared in an evil time, when it falleth suddenly upon them" – Kohelet "Ecclesiastes" 9:11.

"And in that time when men decide and feel safe to call the war insane, take one moment to embrace those gentle heroes you left behind".
Major Michael O'Donnell – KIA, Feb. 7, 1978.

Pia memoriae: Charles Harold Stepp; Donald Eugene Stepp; Dow E. Stepp, Eugene Henry Stepp; Joel Richard Stepp; John Paul Stepp; Paul Robert Stepp; William D. Stepp; William Howard Stepp.

ז'ל

Mountain Men

"Mountain Men" and "Still Water" are written in happy memory of "Merry Prankster" Ken Kesey, who blithely entered the enchanting Santa Cruz Mountains with widely opened eyes, and later emerged with "gift bestowing" hands. (I'll never forget that "magic" little cabin on La Honda Creek!)

Are We Alone?

Where our true nature is concerned, in reality, no questions can be asked, so no questions remain to be answered. Therefore the only real question is one of "choice". Hence, we evidently have no choice (where such arcane things are concerned) but to pursue answerless questions! Or we can learn to eschew all asking, and simply accept that the root of reality lies in the aware Essence of all understanding. Which is evidently also our own universally true Nature.

But as in all effective processes of understanding, skepticism plays an essential role, because without it, something less than the superlative truth might be prematureally accepted. Therefore, we somehow first need to discover cogent reasons to believe that we are indeed on the right track! (Then follow the clues that lead to understanding how it is that we understand the contents of our own asking.)

When we examine our experiences of the universe, which is all that we can ever really "know" of that which only appears to be outside of our consciousness, is anything present that might indicate that we are not merely the insubstantial, yet statistically verifiable, isolates of a transitory (and therefore essentially meaningless), accidental psychosomatic phenomenon?

In other words, are there signs that we may not really be "all alone" in our experience(s) of the universe? To which we can now answer; "Since we have recently discovered hundreds (and soon no-doubt, thousands) of earth-like planets with the right elements and liquid water, orbiting stars much like our own sun, at the right 'Goldilocks' distance for life to thrive, we are most probably not alone, even in our own (Milky Way galaxy) backyard!"

VOLUNTEERS OF THE 101ST SQUAD

Stan Andrews; Eddie Cohen; Leon Frankel; Coleman Goldstein; Lou Lenart; George Lichter; Gideon Lichtman; Harold Livingston; Milton Rubenfeld; Al Shwimmer; Smokey Simon; and Bob Vickman.

TWELVE DAYS BEFORE THE START OF THE Israeli War of Independence, a group of World War II fighter pilots volunteered to help the vastly outnumbered Israelis. With their help the Israelis prevailed.

But what few in America know today, is that the IAF (Israeli Air Force) actually began with only two unarmed, prop driven "Piper Cub" airplanes!

The courage and selfless sacrifice of these brave men was fueled by the horrors of the *Shoa* "Holocaust", and the unwise threat from the advancing foe that they would do far worse to the Jewish people than even Hitler had ever imagined!

These brave and selfless *mitnadev* "volunteers" will never be forgotten by the Jewish people, or by the State of Israel.

להות עם חופשי בארץנו

...lihyōt ahm chōfshi beyartzeynu...

..."to be a free people in our own land"...

(From the National Anthem of Israel, *Ha-Tikva* "The Hope").

History teaches us an important lesson. No matter how much that we might hope that it is so, a personal god quite clearly does not exist! Which is not to say that an impersonal and Supreme G-d (*Yahweh* יהוה, Allah الله, Deüs, etc.) does not. Hence we have with an indelible finality learned that, "G-d helps those, who help their own truer Selves". And in like vein, if we want a 'Shangri-La', neither religion nor academia offer any real solutions. Science and courage offer our best chance of success, but without compassion our "success" could well become our prison!

"Our response to life's unfairness with sympathy and with righteous indignation, G-d's compassion and G-d's anger working through us, may be the surest proof of G-d's reality." – *When Bad Things Happen to Good People,* by Harold S. Kushner.

The Home of the Brave

ON ONE RATHER MEMORABLE "MEMORIAL DAY", when my cousin Jimmy and I were still young boys, we were talking about what it means to be brave. My late uncle, champion bull rider and US Navy Seal, James Stepp, heard us talking and carefully pulled an old creased and blood stained paper from his wallet that he said another "brother" Seal had once given to him.

He was the strong silent type of man who spoke simply and to the point, and when he did, everyone generally listened:

"Live your life that the fear of death can never enter your heart. Trouble no one about their religion. Respect others and their views, and demand that they respect yours. Love your life, perfect your life. Seek to make your life long and be of service to your people.

When your time comes to die be not like those whose hearts are filled with the fear of death, so that when their time comes they weep and pray for a little more time to live their lives over in a different way. Sing your death song, and die like a hero going home."

Tecumseh

Trees Are Sanctuaries

OUR MODERN WORLD is without a doubt a stressful, if not at times *distressing* place to be! Indeed, with all of the violence, bluff and bluster of our "modern" world, it even seems quite difficult at times to understand!

That is, until we begin to ignore our human societies inevitable existential noise and clatter, and then more thoughtfully apply the measured scientific rubric of geology to nature's less hurried method of evolution!

Indeed, by geologies slow measure only a few moments ago our ancestors were still happily swinging around, like tree trimmers still do, in the branches of green trees! So it should come as no great surprise, when our somas and our psyches are revealed to be better suited for a less stressful life spent amongst the trees.

Thus, although we sadly rarely have the time today to earnestly consider it, the welfare of our body, heart and mind still depend upon the quietly supporting branches of healthy green trees!

"Trees are sanctuaries. Whoever knows how to speak to them, whoever knows how to... listen to them, can learn the truth. They do not preach learning and precepts, they preach undeterred by particulars, the ancient law of life". – An excerpt from *Wanderung: Aufzeichnungen*, by Herman Hess, first published in 1920.

Bibliography

Alex Shigo, *A New Tree Biology* (Durham, NH: Shigo and Trees, Associates, 1990).

A. Shigo, *Modern Arboriculture* (Durham, NH: Shigo and Trees, Associates, 1991).

Abraham Harold Maslow, *The Farthest Reaches of Human Nature* (New York: Arkana, 1993).

Adin Steinsaltz, *The Talmud* (New York: Random House, 1999).

Alan B. Wallace, *Contemplative Science* (New York: Columbia University Press, 2007).

Alan Jacobs, *The Wisdom of Balsekar* (London: Watkins Publishing, 2004).

Alan J. Singer, *Teaching To Learn, Learning To Teach* (New York: Routledge, 2014).

Alan Watts, *The Watercourse Way* (New York: Random House, 1975).

Albert Einstein, *Out Of My Later Years* (New York: Philosophical Library, Inc. 1950).

Antonio Damasio, *The Feeling of What Happens* (New York: Harcort Brace and Co., 1999).

Arianna Huffington, *Pigs at the Trough* (New York: Crown Publishers, 2003).

Betty Bethards, *The Dream Book* (Mass: Element Books Inc, 1995).

Brian Greene, *The Fabric of the Cosmos* (New York: Alfred A. Knopf Publishers, 2004).

Brink Lindsey, *The Age of Abundance* (New York: Harper Collins Publishers, 2007).

Burton Watson, *The Complete Works of Chuang Tzu* (New York: The Columbia University Press, 1968).

Carl G. Jung, *Man and his Symbols* (New York: Doubleday, 1964).

Dan Neely and Gary Watson, *The Landscape Below Ground II* (Champaign, IL: International Society of Arboriculture, 1998).

Daniel M. Perrine, *The Chemistry of Mind Altering Drugs* (Washington, DC: American Chemical Society Books, 1996).

Douglas Hofstadter, *I Am a Strange Loop* (New York: Basic Books, 2007).

Ed Rosenthal, *Why Marijuana Should Be Legal* (New York: Thunder's Mouth Press, 1996).

Edward Horowitz, *How the Hebrew Language Grew* (Brooklyn: KTAV Publishing House Inc. 1960).

Ekhart Tolle, *A New Earth* (New York: Dutton, Penguin Group Inc. 2005).

E. Tolle, *The Power of Now* (California: New World Library, 1999).

Francesca Fremantle, *Luminous Emptiness* (Boston: Shambala Publications, Inc. 2001).

Francisco J. Ayala, *Darwin's Gift to Science and Religion* (Washington, DC: Joseph Henry Press, 2007).

Frank M. Stanger, *Sawmills in the Redwoods: Logging on the San Francisco Peninsula, 1849 – 1967* (CA: San Mateo County Historical Association, 1967).

Fred Hoyle, *The Intelligent Universe*, (New York: Holt, Rinehart and Winston, 1984).

G.D. Coughlan and J.E. Dodd, *The Ideas of Particle Physics an Introduction for Scientists* (New York: Cambridge University Press, 1991).

Gary W. Hickman, Ed Perry, *Ten Common Decay Fungi on California Landscape Trees* (Sacramento, CA: Western Chapter International Society of Arboriculture, 1997).

Geoff Simons, *Robots – the Quest for Living Machines* (London: Cassell, Villers House, 1992).

George Thompson, *The Bhagavad Gita* (New York: North Point Press, 2008).

Gunther W. Plaut, *The Torah* (New York: Jewish Publication Society, 1974).

Harold S, Kushner, *When Bad Things Happen to Good People* (New York: Avon Books, 1981).

Helena Cobban, *The Moral Architecture of World Peace* (Charlottesville: University Press of Virginia, 2000).

Igor Aleksander, Piers Burnett, *Reinventing Man* (New York: Holt, Rinehart and Winston, 1983).

J. Allan Hobson, *Consciousness* (New York: Scientific American Library, W.H. Freeman and Co. 1998).

J. H. Conway and N. J. A. Sloane, *Sphere Packing, Lattices and Groups* (New York: Springer – Verlag, 1988).

James H. Austin, *Zen and the Brain* (Cambridge, MA: The MIT Press, 1999).

J. H. Austin, *Zen-Brain Reflections* (Cambridge, MA: The MIT Press, 2006).

James Mooney, *The Swimmer Manuscript Cherokee Sacred Formulas and Medical Prescriptions,* – Smithsonian Institution, Bureau of American Ethnology, Bulletin 99 (Oklahoma City, OK: Reprinted by Noski Press, 2005).

Jane Goodall, *The Chimpanzees of Gombe* (Cambridge, Harvard University Press, 1986).

Jean Dunn, *Consciousness and the Absolute* (Durham, NC: The Acorn Press, 1994).

J. Dunn, *Seeds of Consciousness* (Durham, NC: The Acorn Press, 1982).

Jeffrey M. Schwartz, Sharon Begley, *The Mind, Brain-Neuroplasticity, and the Power of Mental Force* (New York: Harper Collins Publishers Inc. 2002).

John D. Barrow, *The Origin of the Universe* (New York: Basic Books, Harper Collins Publishers, 1994).

John Demos, *The Enemy Within* (New York: Viking, Penguin Group, 2008).

John Hunter Thomas, *Flora of the Santa Cruz Mountains of California* (Stanford, CA: Stanford University Press, 1961).

Joseph Campbell, Pathways *to Bliss – Mythology and Personal Transformation* (Novato, CA: New World Library, 2004).

Karl A. Menninger, *The Human Mind* (New York: Alfred A. Knopf, Inc. 1961).

Robert Powell (Editor), *Talks with Ramana Maharshi* (Carlsbad, CA: Inner Directions Publishing, 2001).

M.R. Franks, *The Universe and Multiple Reality*, (New York: iUniverse, Inc. 2003).

Marilyn Tower Oliver, *Drugs* (Springfield, NJ: Enslow Publishers, Inc. 1996).

Mario Livio, *The Accelerating Universe* (New York: John Wiley and Sons Inc. 2000).

Martin Rees, *Just Six Numbers- the deep forces that shape the universe* (New York: Basic Books – Perseus Book Group, 2000).

Master Sheng-yen, Dan Stevenson, *Hoofprint of the Ox* (New York: Oxford University Press, 2001).

Mathew Greenblatt, *The Wisdom Teachings of Nisargadatta Maharaj* (Carlsbad, CA: Inner Directions Publishing, 2003).

Maureen Caudill, *In Our Own Image – Building an Artificial Person* (New York: Oxford University Press, 1992).

Maurice Frydman, *I Am That* (North Carolina: Durham Press, 1999).

Maurice Simon, Paul P. Levertoff, *The Zohar* – Series (New York: The Sonchino Press, 1984).

McMinn, Maino, *Pacific Coast Trees* (Berkeley, CA: University of California Press, LTD. 1963).

Michael Berenbaum, *The World Must Know* (New York: Little, Brown and Co. 1983).

Michael Talbot, *Beyond the Quantum* (New York: Macmillan Publishing Company, 1986).

N. Matheny, J. Clarke, *Evaluation of Hazard Trees in Urban Areas* (Savoy, Il: International Society of Arboriculture, 1994).

Nyogen Senzaki, Paul Reps, *Zen Flesh Zen Bones* (Rutland VT: Charles E. Tuttle Co. 1980).

Orson K. Miller Jr., *Mushrooms of North America* (New York: E.P. Dutton, 1985).

Paul Davies, *About Time Einstein's Unfinished Revolution* (New York: Simon and Shuster, 1995).

Paul J. Kramer, Theodore T. Kozlowski, *Physiology of Woody Plants* (Orlando, FL: Academic Press Inc. 1979).

Ram Dass, *Be Here Now* (New York: Crown Publishing, 1971).

Ramesh S. Balsekar, *Pointers From Nisargadatta Maharaj* (Durham, NC: Acorn Press, 1982).

R. S. Balsekar, *A duet of One* (Redondo Beach, CA: Advaita Press, 1989).

R. S. Balsekar, *Consciousness Speaks* (Redondo Beach, CA: Advaita Press, 1992).

Randall S. Stamen, *California Arboriculture Law* (Riverside CA: The Law Offices of R. S. Stamen: 909-787-9788).

Richard Fortey, *Life* (New York: Alfred A. Knopf, Inc. 1977).

Richard J. Preston Jr. *North American Trees* (Massachusetts: MIT Press, 1966).

Richard Leakey, Roger Lewin, *Origins Reconsidered* (New York: Doubleday, 1992).

Richard Morris, *Cosmic Questions* (New York: John Wiley and Sons, 1993).

Rita Carter, *Exploring Consciousness* (Berkley: University of California Press, 2002).

Robert Aitken, *Original Dwelling Place* (Washington, DC: Counterpoint, 1996).

Robert Powell, *The Experience of Nothingness* (San Diego, CA: Blue Dove Press, 1996).

R. Powell, *The Nectar of Immortality* (San Diego, CA: Blue Dove Press, 2001).

R. Powell, *The Ultimate Medicine* (Delhi: Motalil Banarsidass Publishers Pvt. Ltd. 2004).

Roy Porter, *Madness – A Brief History* (Oxford: Oxford University Press, 2002), see: "Christian Madness", p.p. 17-25.

Sandra and Matthew Blakeslee, *The Body Has a Mind of its Own* (New York: Random House, 2007).

Shmuley Boteach, *Judaism for Everyone* (New York: Basic Books, 2002).

Sir Herbert Reed et al, *The Collected Works of C. G. Jung* (New York: Pantheon Books, 1954).

Sri Aurobindo, *The Life Divine* (Twin Lakes, WI: Lotus Press, 2000).

S. Aurobindo, *The Synthesis of Yoga* (Silver Lake, WI: Lotus Light Publications, 1973).

Stanly L. Greenspan, *The Growth of the Mind* (Menlo Park CA: Addison-Wesley Publishing Co. 1997).

Stephen Whitney, *Western Forests* (New York: Alfred Knopf, 1985).

Steven Rose, *The Future of the Brain* (New York: Oxford University Press, Inc. 2005).

Swami Nikhilananda, *The Gospel of Sri Ramakrishna* (New York: Ramakrishna-Vivekananda Center, 1977).

The Dali Lama, *The Middle Way* (Somerville, MA: Wisdom Publications, 2009).

Tom Griffith, *Plato, The Republic* (New York: Cambridge University Press, 2001).

Uriel Weinreich, *College Yiddish* (New York: YIVO Institute for Jewish Research Inc. 1984).

Victor D. Merullo, Michael J. Valentine, *Arboriculture and the Law* (Savoy IL: International Society of Arboriculture, 1992).

Victor J. Stenger, *Timeless Reality* (Amherst, NY: Prometheus Books, 2000).

Virginia Satir, *The New Peoplemaking* (Mountain View, CA: Science and Behavior Books, Inc. 1988).

W.E.H. Tanner, *Traditional Aboriginal Society* (Melbourne: Macmillan, 1987).

Wade Rowland, *Greed Inc.* (New York Arcade Publishing, 2005).

Wallace, King, Sanders, *Biosphere* (Glenview, IL: Scott Foresman and Co. 1984).

Wayne Liquorman, *Never Mind* (Redondo Beach, CA: Advaita Press, 2004).

Wei Wu Wei, *All Else is Bondage* (Colorado: Sentient Publications LLC, 2004).

W. W. Wei, *Ask the Awakened* (Colorado: Sentient Publications LLC, 2002).

W. W. Wei, *Fingers Pointing Towards the Moon* (Colorado: Sentient Publications LLC, 2003).

W. W. Wei, *Open Secret* (Boulder, CO: Sentient Publications LLC, 2004).

W. W. Wei, *Posthumous Pieces* (Colorado: Sentient Publications LLC, 2004).

W. W. Wei, *The Tenth Man* (Boulder, CO: Sentient Publications LLC, 2003).

W. W. Wei, *Unworldly Wise* (Colorado: Sentient Publications LLC, 2004).

W. W. Wei, *Why Lazarus Laughed* (Colorado: Sentient Publications LLC, 2003).

William A. Tiller, *Science and Human Transformation* (Walnut Creek, CA: Pavior Publishing, 1997).

William H. Calvin, *How Brains Think* (New York: Basic Books, 1996).

William Karush, *The Crescent Dictionary of Mathematics* (Palo Alto, CA: Dale Seymour Publications, 1962).

Z'ev ben Shimon Halevi, *A Kabbalistic Universe* (York Beach, ME: Samuel Weiser, Inc. 1977).

Z. ben Shimon Halevi, *Adam and the Kabbalistic Tree* (York Beach, ME: Samuel Weiser, Inc. 1990).

Z. ben Shimon Halevi, *Psychology and Kabbalah* (York Beach, ME: Samuel Weiser, Inc. 1991).

Z. ben Shimon Halevi, *The Way of Kabbalah* (York Beach, ME: Samuel Weiser, Inc. 1991).

For a deeper understanding of the science discussed in Forest Primeval I also highly recommend the visionary writings of the following authors:

Stephen Hawking
Michio Kaku
Neil deGrasse Tyson

Index

A

B

C

D

E

F

G

J

K

M

N

O

P

U

X, Y

Z

Afterword

The primeval singularity is by definition an unbreakable "Supersymmetry". And with all due respect to the Higg's Field (see pg. 543) and Boson, by designation, and in fact, it must be of Consciousness-Energy comprised!

We could even not incorrectly say Consciousness *and* Energy, but since both exist within an unbreakable Supersymmetry, again by definition, they are in essence always One!

Firstly, because it is "perfect", its supersymmetry is unbreakable. And secondly, because any subsequent universe could only be known to exist within a sentient consciousness, it must in some fashion either contain, or be permeated by, the consciousness by which it is known!

Thus, when we factor consciousness into the equation of the cosmology of our universe, we are left with an inescapable chain of rather curious facts about its existence that await further scientific substantiation.

This is because, as our modern science continues to penetrate ever deeper into the mysterious quantum realm of our universe, we are rapidly discovering that the infinite potential of the Primeval Singularity can only be expressed virtually. The "reality" of which exists much like a dream, that nonetheless appears quite real to its dream character (*us*)!

Nonetheless, we do experience the universe as if it were real. Which in fact it is, within the conscious expression that is generated by our personal experience of its infinite quantum potential. (For further clarification, see the dialectics for "Supersymmetry Breaking" on pg. 399.

But this much we do know with some certainty, incredible as it may seem, nothing actually exists (quantumly speaking) *until we experience it!* Hence, we are evidently the purveyors of our own reality.

Therefore the most likely answer to the mystery of our universe's existence, is that it simultaneously exists and does not exist! Moreover, when we closely examine the evidence, it also seems most likely that the

Awareness of its existence is not only transcendent, but even *prior* to the Consciousness-Energy by which it is known to exist through us.

In other words, there is a confluence of contingency between the conscious experience of our person and universe, and the quantum existence of the experience itself.

This is, in fact, all that there could ever in "pure potential" be! And although this at first seems quite radical, it is not really even new knowledge.

Yet it is being steadily substantiated by today's unfettered science as we continue to develop ever better instrumentation, and ever more accurate theories, to explain the apparent anomalies and quandaries that are being (almost "daily" now it seems) revealed!

Both ancient esoteric traditions and cutting-edge science are now in near agreement about many of the more arcane aspects of our mysterious universe's emergence from a state of evident "nothingness".

Yet the full impact of what this actually means for science, technology and the individual within the currently unfolding "Singularity of Consciousness" has not yet arrived with full force.

However, its impact is being daily revealed by the advance of consciousness into the everyday technologies that are rapidly transforming our world. Indeed, when we get right down to it, upon closer investigation, physics has recently become quite "metaphysical"!

"We are the universe looking back at itself."
Carl Sagan

ᎾᏍᎩᎾᎢ ᎤᏂᏥ
"For Mom"

Made in the USA
Columbia, SC
27 November 2017